Land Use Simulation for Europe

The GeoJournal Library

Volume 63

The titles published in this series are listed at the end of this volume.

Land Use Simulation for Europe

edited by

JOHN STILLWELL

School of Geography,
University of Leeds,
UK

and

HENK SCHOLTEN

Department of Regional Economics,
Vrije Universiteit Amsterdam,
The Netherlands

SPRINGER-SCIENCE+BUSINESS MEDIA, B.V.

A C.I.P. Catalogue record for this book is available from the Library of Congress

Additional material to this book can be downloaded from http://extras.springer.com.
ISBN 978-94-010-3895-9 ISBN 978-94-010-0466-4 (eBook)
DOI 10.1007/978-94-010-0466-4

Cover illustration:
EuroScanner System structure. See also fig. 16.1.
This illustration has been provided courtesy of the Graphics Unit, University of Leeds.

Printed on acid-free paper

Originally published by Kluwer Academic Publishers in 2001
Softcover reprint of the hardcover 1st edition 2001

TABLE OF CONTENTS

PART II: MODELLING ENVIRONMENTAL AND HUMAN SYSTEMS

PART III: EVALUATION AND VISUALISATION METHODS

PART IV: THE DEVELOPMENT OF A EUROPEAN LAND USE SIMULATION SYSTEM

LIST OF FIGURES

LIST OF TABLES

PREFACE

The processes of development impact on the natural and human landscape, resulting in a spatial mosaic of land cover, use and ownership which geographers and economists have studied and which planners have modified over many years. Identifying and explaining spatial phenomena is the very essence of the discipline of geography and understanding how the complex environments of the modern world have come about and how they change over time is of paramount importance if we want to ensure a sustainable future for our regions and localities across Europe.

The underlying focus of the book is the exciting and challenging new research opportunities that are involved when we attempt to simulate land use change across the urban and rural areas of Europe. The availability of remotely sensed data of improving quality, together with increasingly standardised information about the attributes of different regions means that data resources are no longer as constraining as they used to be. Moreover, much progress has been made on the development of new simulation tools and decision support systems that exploit the benefits of increasingly powerful desktop computing resources and unlimited storage space.

The challenge of developing methods to simulate Europe's future land use is both difficult and ambitious because of the huge complexity of the real world, but it is clear that policy-makers require new systems to help them predict the outcomes of decisions and evaluate alternative scenarios in times when land use change is being driven by physical forces and human processes operating at the global level as well as within countries and regions.

We are grateful to those authors who have contributed to this book, which originates from a project known as SIMILOR (http://www.econ.vu.nl/re/similor/) and from a seminar held in Soesterberg in October 1999. We also acknowledge the support of the Netherlands Remote Sensing Board (BCRS) and those who have worked extremely hard to bring this book to fruition, especially Mark Newcombe and Alison Manson in the Graphics Unit in the School of Geography at the University of Leeds.

John Stillwell and Henk Scholten
June 2001

1 A FRAMEWORK FOR EUROPEAN LAND USE SIMULATION

John Stillwell and Henk Scholten

1.1 Introduction

Land is a finite natural resource and a factor of production whose use by mankind has been determined historically by a wide range of influences. The processes of development impact on the natural and human landscape, resulting in spatial mosaics of land cover, ownership and use that have been studied for many years. Whilst some environmental scientists have argued that land use is determined by physical or climatic factors (Brouwer *et al.*, 1991), there is a long tradition of attempts to explain patterns of land use according to economic factors. This is a tradition that includes the work of the classical scholars of geographical theory and modelling; in particular the contributions made by von Thünen (1826) on agricultural land use, by Weber (1909) on industrial location, by Christaller (1933) and Lösch (1940) on central places and settlement patterns and by Burgess (1927), Hoyt (1939), and Harris and Ullman (1945) on residential land use. There is no doubt that environmental and economic explanations continue to exert major influences on land use. The role of economic factors such as the availability of subsidies, the fixing of quotas on food production, the setting aside of land in return for monetary compensation and schemes to encourage farms to diversify and become less dependent on one activity, all combine to create a complex and often dynamic pattern of agricultural land use. Similarly, the operation of the land market combined with the imposition of planning controls results in very complex urban systems whose components show strong interdependencies (Lowry, 1964). The analysis of these structures and relationships requires the application of appropriate methods to study complex systems (Wilson, 2000). Moreover, as Mather (1986) has argued, land use is the result of human decisions and patterns of land use reflect the decision-making processes that are made by those who control land resources.

Given the complexities of understanding and explaining past and current patterns of land use, the task of predicting what will happen to land use in the future is an even greater challenge. This challenge provides the focus for this book and is a challenge that confronts all those who endeavour to construct visions of what the future spatial distribution of activities and land uses will look like. Currently, increasing attention is being paid to closer and more systematic cooperation through territorial planning across Europe and it is this emphasis that gives the book its European dimension. The first Europe 2000 report (European Commission, 1991) underlined the need for common action on spatial planning in order that the Community might obtain the maximum benefit from the Single Market. More recently, the publication of Europe 2000+ (European Commission, 1994) strengthened the view that sustained and balanced economic development can be achieved through a common framework for territorial planning, a framework which appeared formally as the European Spatial Development Perspective (ESDP) in 1999 (European Commission, 1999).

More specifically, our objective in this volume is to demonstrate that, despite the difficulties involved, it has become possible to provide both physical and strategic planners with the

J. Stillwell and H. Scholten (eds.), Land Use Simulation for Europe, 1–11.

tools that are required to envisage the outcome of particular trends and assess the implications of alternative decisions or planning strategies at different spatial scales. What effects would major new infrastructure investment programmes have on the land uses in the areas that surround them? What would be the impact of a radical reform of the Common Agricultural Policy (CAP) or a major alteration in domestic agricultural policy in particular regions? What are likely to be the consequences of the continued decentralization of people from Europe's major cities? We must strive to answer these types of questions if we want to address the key issues of sustainability and balance in a serious way. This book attempts to show that with the development of new modelling methods to simulate land use change, linked to the enormous power of computer technology and state of the art geographical information systems with which to store and manipulate more extensive and more detailed sets of data (Stillwell *et al.*, 1999), it has become feasible to develop decision support systems to evaluate alternative scenarios of future land use change. In Section 1.2, we outline the conceptual framework that underpins these ideas and the structure of the book, whilst in Section 1.3, we introduce the content of individual chapters.

1.2 The conceptual framework

The environmental and economic factors identified previously are amongst the suite of 'driving forces' that determine land use change in the real world as indicated in the conceptual framework illustrated in Figure 1.1. In some cases the relationship between each factor and land use is simple and direct such as the location of natural resources or the climatic or soils conditions that pertain at a particular location. On the other hand, the precise impact of some driving forces on land use change may be more difficult to distinguish such as the influence of changing technology on agricultural practices or the impact of a tax liberalisation policy. In other cases, some uncertainty may surround the identification of the factors that determine change. Trends in the use of land for recreation, for example, are likely to be a function of a combination of changes in social and economic variables. Demographic drivers are a very important set of forces that include changes in the number of people due to natural change or migration as well as changes in the composition of the population such as ageing, for example. These changes may have labour force or service provision implications as well as implications for housing. The demand for residential land is a key issue in many parts of 'more crowded' Europe and is driven not by increases in population alone but also by changes in social behaviour that influence the translation of population numbers into households requiring dwelling units. Regulatory planning factors such as development control systems or land use zoning may operate to slow down the rate of change in some cases, whereas area designation may stimulate new developments, particularly when supported by financial subsidies or inducements.

In order to measure and monitor land use change, it is necessary to capture data about the real world. This usually comes in the form of data on land cover and requires careful interpretation to provide reliable information about land use. Direct observations in the field using previously surveyed base maps, and aided by aerial photographs, have traditionally provided most land use information. However, the introduction of improved remote sensing techniques using satellites such as the American Landsat series or the French SPOT, and most recently the IKONOS sensor, has led to more detailed and accurate land cover information becoming available. Mixes of mosaics of different land covers can now be mapped accurately, changes in land cover can be monitored on a fine scale, and better judgements can be made or inferred about the use and capability of the land in question. The availability of satellite-based land use maps provides a starting point for many applications in different domains of planning (Donnay, 1999). Satellite images are usually stored in raster geographical information systems (GIS) or, after vectorisation, in vector GIS, where the data may be manipulated or combined with other geographic data and displayed in map form. Geotechnology or geographical information technology is the term used to subsume the whole range of information systems from CAD to GIS that are used to store, manipulate, analyse and display data. Geertman (1999) argues that the use made by these systems in the field of physical planning has been severely limited to so-called lower-order or operation-orientated planning activities and most applications have involved little more than the production of high quality automated maps. Relatively little has been made so far of the potential of geotechnology for higher-order or strategy-orientated planning applications in which more sophisticated spatial analyses and modelling may be undertaken. The technology also remains under-exploited as a framework in which the results of alternative strategies can be compared using various evaluation methodologies, or in which the results can be displayed using innovative forms of visualization. Thus, whilst it is clear that the potential to construct robust and intelligent decision support systems exists, it is also clear that proprietary GIS lack the spatial analysis, modelling and evaluation facilities that are frequently required for particular applications.

Figure 1.1: Conceptual framework

Driving forces	⇐	Projections and scenarios
⇓		⇑
Land use change in the real world	⇒	Theoretical explanation and modelling
⇓		⇑
Data on land use change	⇒	Geotechnology

In order to predict future land use patterns, it is therefore necessary to construct systems that are linked or coupled either directly or indirectly to simulation models. The upward arrow in Figure 1.1 between geotechnology and modelling represents this link. As with the models of the classical scholars, simulation models attempt to explain and replicate what is happening in the real world. They require calibration, testing and verification before being used in a predictive context. There is now an increasing number of computational approaches that are being developed and adopted by different groups of researchers in Europe and elsewhere for applications in the context of land use change. Land use simulation modeling is not new. Many readers will be familiar with *SIMCITY*, a software package for the PC that is an excellent example of an urban simulation system. The *SPARTACUS* (System for Planning and Research in Towns and Cities for Urban Sustainability) project (Commission of the European Communities, 1998) has developed a methodology for assessing the sustainability implications of urban land use and transport policies based on long term simulations in each of a selection of cities. The system is based on a land use-transport interaction model (*MEPLAN*) that provides values for environmental, social and economic indicators of urban sustainability. Cellular automata (CA) models are another type of simulation approach where, in very simple terms, the state of each cell in an n-dimensional array depends on its previous state and on the states of the cells within its neighbourhood, according to a set of stated transition rules. The links between GIS and CA modeling have been developed by White and Engelen (1993), Batty and Xie (1994) and Clarke *et al.* (1997), for example, and the Research Institute for Knowledge Systems in the Netherlands has developed *Geonamica* (Engelen *et al.*, 1999), a simulation package geared to build and run CA models.

An alternative simulation approach developed by a consortium in the Netherlands is the *Land Use Planner*, a grid-based system in which the suitability of land use of a particular type in a grid cell changes during the simulation by adding or subtracting scores that represent policy intervention (Scholten *et al.*, 1999). The model is based on determining the probabilities that each grid cell will be suitable for a particular type of land use (Hilferink and Rietveld, 1999). It is formulated as a spatial interaction model and may incorporate constraints to ensure that groups of cells do not exceed certain land use totals. The simulation of future land use change using this approach relies on projections and scenarios that reflect changes in the forces that drive what happens in the real world and these links complete the cyclical framework shown in Figure 1.1.

This conceptual framework underlies the SIMILOR (SIMulation and Integration of Land-cOveR dynamics) project that provides the context and focus for this book. The aim of SIMILOR has been to develop a prototype spatial information system that simulates future land use over a relatively large area, using the spatial interaction modelling approach introduced previously. The emphasis of the project has been on the interaction between land use change in urban and rural areas in different parts of Europe. The primary aim of the project in practical terms has been to develop a geographic information system, that has been called *EuroScanner*, that would be used to help policy-makers to identify the spatial implications of different sectoral and macro policies under different economic scenarios, rather than to predict the future *per se*. Whilst the structure of the model-based GIS is explained and exemplified in the last part of the book, the SIMILOR project has also tried to develop a broader perspective and to consider some of the other empirical, theoretical and methodological dimensions implied in the conceptual framework in more detail.

The success of a model depends on the quality of the data as well as on the insights and understanding that we have of the driving forces behind land use change and their effects on the environment. Consequently the first part of the book, which is in four parts altogether, contains a variety of contributions that discuss various drivers for change (including the role of spatial policy) and that consider various data availability, quality and consistency issues. The second part of the book focuses on recent methodological research and draws together some examples from both environmental and human systems of how the impacts of particular factors on land use change can be modelled. In the third part, the contributions focus on methods and examples of evaluating and visualising alternative scenarios. The three initial parts of the book provide a series of state-of-the-art reviews or reports of research that, although by no means comprehensive, illustrate and exemplify the contexts relevant to the project. The fourth part of the book then contains an explanation of the modelling approach that underpins EuroScanner and the presentation of two case study applications of the model-based system. In the remaining section of this introductory chapter, we seek to explain the structure of the book in more detail by outlining the focus and content of each of the chapters.

1.3 Structure and content

1.3.1 *Part 1: Policy perspectives, driving forces and data considerations*

The chapters in Part 1 are designed to demonstrate the practical and strategic importance of the SIMILOR project to planners and policy-makers, to develop a better understanding of some of the underlying determinants or 'drivers for change' that impact on land use, and to review types of data and data infrastructures relating to land use in Europe.

The first chapter in Part 1 by *Peter Ungar* reviews how the European Spatial Development Perspective (ESDP) has come into existence over the last decade as the main vehicle for the European Commission's thinking about the future of EU territory. This process began with the preparation for the Single Market in the late 1980s and has resulted in the ministerial agreement in October 1999 over a programme of actions to be implemented on a voluntary basis in the next century. The chapter emphasises the importance of co-operation in bringing about a consensus on what the EDSP should contain. The author draws on the economic theory of the calculus of consent to demonstrate how the costs involved in the decision-making process can be divided into preference and resource costs that both vary with the number of participants involved. The relationship between the two cost components provides a mechanism for identifying a point of equilibrium at which the number of participants is optimised and total costs are minimised. Thus in theory, participation in the decision-making process, a key Commission objective, should increase if resource costs are reduced as a result of initiatives such as the development of information and modelling systems and the generation of simulated predictions using the most advanced computer technology and the sets of remotely sensed data that are now becoming available. In Chapter 3, *Hans ten Velden* and *John Stillwell* argue that, in order to exploit opportunities, spatial planners must consider policy from a transnational or European perspective, putting aside national or regional interests. Once professional assessments of transnational spatial planning for the future have been delivered, the next challenge is to communicate the messages necessary for politicians to take a less nationalistic or regional position. The ESDP emphasises the necessity to take a functional rather than a national approach and advocates a polycentric development model. The chapter identifies some of the issues and questions that spatial planning at a European level raises for national planners in the Netherlands and regional planners in Yorkshire and Humberside in the UK.

In contrast to the spatial policy focus of Chapters 2 and 3, the next three chapters deal with trends and mechanisms closely associated with land use determinants in three different contexts: changes taking place in agricultural and rural areas; long term demographic and household projections; and the operation of the land market. In Chapter 4, *Arnold Bregt* and *Kees de Zeeuw* establish some of the major trends that have taken place in European agriculture, forestry and nature areas in recent years. These trends include the reduction of overproduction, diversification, intensification, urbanisation and the formation of ecological corridors. The authors indicate that, in order to build effective simulation models, it is necessary to ask three questions: What land use is where? What land use changes are occurring where? What land uses will be where in future? They emphasise that knowledge of the status of agriculture, forest and nature land use requires good data collection in the first instance.

Key issues for planners in the United Kingdom over the next decade are the amount of land that will be used to build houses to accommodate the projected growth in households and the location of the sites that are used for this purpose. These issues are also of paramount importance in all member states. Although the population of Europe is likely to remain fairly stable overall, cohort effects, migration trends and social changes will create new demands for residential development and new pressures for land use change both on greenfield and brownfield sites. In Chapter 5, *John Stillwell* and *James Debenham* illustrate the 1995-based population projections for EU NUTS-2 regions to show how the all-age projections conceal very diverse patterns of change by age group that will have to be taken into consideration by those concerned with housing provision and the allocation of land for residential development. Whilst these demographic changes will underpin household formation over the next 30 years, that will in turn impact on the demand for housing and residential land, there are no regional projections across the EU of these other variables. Consequently, the authors review the latest rounds of national population and household projections in the UK and of sub-national population and household projections in England. The technical and political processes for identifying regional claims on land use for housing from projections of population and households are discussed with reference to the region of Yorkshire and Humberside in the north of England.

As well as demographic and household projections, land markets are also the subjects of political debate in many countries. *Joost Buurman, Piet Rietveld* and *Henk Scholten* give an overview of the way in which the market for land works from a spatial economic perspective in Chapter 6. Using the economic theory of demand and supply, a single price for different uses of land in a homogeneous market is derived. However, in reality, different land prices can be observed from one parcel of land to another. These are the result of externalities and government intervention. Governments can use taxation and supply restrictions to interfere with the land market. When adding a spatial perspective to the analysis, variations in land use prices and price structures can be explained. Classic theories of von Thünen, Mills, Wingo and Alonso explain higher land prices closer to city centres and arrive at homogeneous zones of land use. These models are mainly developed to simulate urban land use and transportation interaction, although models for more types of land use have also been developed. The message that sophisticated models can and should be combined with decision support tools for policy-makers is critically important.

Chapter 7, by *Henk Kramer* and *Sander Mücher*, is the first of two chapters that consider data types and availability. It provides a review of the remote sensing images currently available at different levels of resolution and the three main land cover databases that have been constructed: CORINE, PELCOM and DISCover. The difficult problem of using satellite images to extract information on land uses from data on land cover is considered and a methodology for deriving a basic land use classification is presented. The chapter demonstrates how land use change maps can be generated. In the final chapter of Part 1, *Joana Abreu* and *Henk Scholten* explain progress that has been made towards the development of an Internet-based system providing access to metadata and data in such a way as to facilitate spatial decision-making. This chapter specifies some of the requirements of a metadata infrastructure and explains some of the problems that are associated with gaining access to geographic information across Europe. The authors review some of the geographic information infrastructures that exist in Europe and elsewhere before describing, in more detail, the progress achieved through the European Spatial Metadata Infrastructure (ESMI) project which allows direct access to data using Web mapping technologies.

1.3.2 *Part 2: Modelling environmental and human systems*

Part 2 focuses on explaining and exemplifying new modelling methods for environmental and physical systems on the one hand, and for human systems on the other. Clearly, understanding environmental change processes and modelling impacts on land use in rural areas are critically important but it is also necessary to acknowledge, understand and model the impacts of demographic and economic forecasts and the relationships between environmental and human systems.

Chapter 9, by *Jean-Paul Hettelingh, Maximilian Posch* and *Peter de Smet,* is about land cover information for European environmental modelling. Data on land cover, soil chemicals and plant physiological characteristics have been successfully used to quantify thresholds in Europe expressing acceptable deposition and ambient concentration thresholds for atmospheric pollutants. These so-called 'critical' loads and levels were used in an integrated assessment model (RAINS), enabling the assessment of environmental impacts of the reduction of emissions of compounds which are the precursors to acidification, eutrophication and the formation of tropospheric ozone. Results were used to support European environmental policy developments within the UN/Commission of Europe and the European Commission of the EU. The inclusion of national environmental characteristics into European environmental modelling involves close scientific and technical collaboration in a network of currently 24 National Focal Centres, enabling the assessment of critical loads reflecting more than 1.3 million ecosystems covering more than 5.5 million square kilometres of Europe. This chapter describes the science behind this work, addressing the use and potential of land cover information when appropriate. A structured approach to environmental modelling (DPSIR) is recommended to further explore relationships between environmental modelling, land cover and land use.

In Chapter 10, *Stan Openshaw* and *Andy Turner* describe the development of a prototype Synoptic Prediction System (SPS) for the EU funded Mediterranean desertification and land use project (MEDALUS III). The prototype SPS was designed to forecast the possible impacts of global climatic change on agricultural land use patterns across the Mediterranean region of the EU. Designing such a system is a challenging task because many of the theoretically desirable data sets are either unavailable or do not exist, whilst significant uncertainties are apparent in the data that are available. Additionally, process knowledge is woefully deficient, as virtually all the principal mechanisms for linking the dynamics of the climate and physical environment with the associated socio-economic systems are poorly understood. In order to make the best predictions of the geographical impacts of climate change for around 75 years hence, an integrated, synoptic, GIS, scenario-based modelling approach was considered to be the best option. This chapter describes the development of a prototype SPS that employs a mix of GIS, neurocomputing and fuzzy logic technologies that attempt the almost impossible, yet nevertheless important, task of forecasting agricultural land degradation risk under various climatic scenarios. The chapter focuses on methods developed to enrich the available data, the quantitative approach to modelling and forecasting land use using neural networks and the fuzzy logic based translation of the modelling results into land degradation terms.

The remaining two chapters on this part of the book report the results of two projects that involve human systems modelling. Chapter 11, by *Leo van Wissen* and *Corina Huisman*, describes the structure of a mathematical model used to simulate regional demographic and economic dynamics and linkages with urban land use in north west Europe. Each of the components of the model is described in detail and various indicative results are presented. Chapter 12, by *Beatrice Eiselt, Nicla Giglioli* and *Bob Peckham* initially describes some recent attempts to model interregional migration in European countries using socio-economic data available from the REGIO database of the European Statistical Office, EUROSTAT. This work gives an indication that interregional migration can be modelled in some countries using GDP and unemployment as explanatory variables, but it also points to the need for better data. A GIS-based tool for visualization of the actual modelled and predicted migration flows has also been developed in this work. The chapter raises some ideas on how this type of migration modelling might be used as an additional input to the simulation of urban expansion, hence linking economic driving forces, through migration, to land use changes.

1.3.3 *Part 3: Evaluation and visualisation methods*

Part 3 contains two chapters that provide some indications of the types of methods that are required to evaluate alternative scenarios and one chapter that provides a state-of-the-art review of visualisation methods, including multi-media and virtual reality systems. In recent years a wide variety of decision support methods and expert systems have been developed to comply with the policy need for sustainable spatial development. Cities have increasingly gained an important position in such sustainability policies. Chapter 13 by *Adele Finco* and *Peter Nijkamp* aims to investigate whether multicriteria decision support methods are able to offer a clear analytical contribution to the identification of promising sustainable city initiatives in a given area. The chapter offers an overview of various key issues in the current discussion on urban sustainability policy. It also offers a concise survey of recently developed decision support methods for sustainable urban land use management, in particular multicriteria methods. The usefulness and feasibility of multicriteria methods for the evaluation of sustainability initiatives at the urban level is illustrated by an extensive empirical application to sustainable city planning of the Italian city of Cremona, seen from the perspective of contrasting choices on urban development options.

Chapter 14, by *Euro Beinat,* is concerned with the spatial evaluation and conflict management in transport policies. The evaluation of a transport policy, plan or infrastructure requires a systematic analysis of the multiple consequences of the development, and of their acceptability for policy-makers and stakeholders. In most cases, this analysis has a spatial dimension due to the geographical distribution of the policy effects, and the possibility of drawing different benefit-cost balances at different spatial scales. The mix of multiple policy goals, multiple actors, spatial distribution of policy effects and uncertainty about the policy effects, makes this process problematic. A typical consequence is a poor articulation of policy objectives at different spatial levels. For instance, supranational objectives are not taken into account at the national level, and local objectives are not accounted for at a higher policy level. Because of this, it is often difficult to understand and predict conflicts between different policy levels (e.g. nations, local authorities, citizens), which complicates the decision process and may interrupt or stop policy development. This chapter presents a methodology that can be used to articulate policy objectives at different spatial levels, to assess and compare policy alternatives and to support the analysis of spatial conflicts in a transparent way. The methodology is based on multicriteria analysis and spatial analyses. The chapter illustrates the methodology, describes spatial conflicts, identifies the factors that originate conflicts and provides a general scheme for their analysis. Four large projects in Europe are used as examples to show the spatial distribution of the decision actors, their different concerns and objectives and the way this drives the evaluation of the project.

The visualisation of Europe's land uses in the early twenty-first century may be based on virtual representations. Virtual reality technology and its state-of-the-art applications in the visualisation of planning exercises are reviewed in Chapter 15 where *Antõnio Câmara* demonstrates the advantages of such a solution. Suggestions for professional solutions based on desktop computers and high-end visualisations relying on large screens are also discussed.

1.3.4 *Part 4: The development of a European land use simulation system*

The final part of the book contains a series of three chapters that explain, and then demonstrate through case studies, the prototype system called *EuroScanner* that has been developed for simulating land use change across Europe. The structure of the mathematical model that underpins the simulation process is outlined in Chapter 16 by *Piet Rietveld, Henk Scholten* and *John Stillwell*. The model is based on spatial interaction theory concepts and involves predicting the expected amount of land in an individual grid cell that will be used for a particular type of land use, taking into account the supply side constraints; i.e. a known total amount of land in that category across all the grid cells in the system. Chapter 17, by *Kees Schotten* and *Camiel Heunks*, reports on a national planning application of *EuroScanner* in the Netherlands. Spatial or physical planning can be seen as strategic policy aimed at the optimisation of space and society and which comprises a set of government practices for the development and implementation of spatial strategies, plans, policies and projects by regulating the location, timing and form of development. Land use modelling can be used to understand the effects of autonomous or planned developments and sets of indicators can be defined and implemented to show the impact of land use change on different aspects of space and the environment in order to facilitate the (political) discussions that are also an essential part of strategic planning. This chapter focuses on the application of a GIS-based simulation model in the framework of the Fifth National Physical Planning Report in the Netherlands. The simulation model generates future land use in the Netherlands given several growth scenarios and a spatial strategy that comprises both foreseen strategic and autonomous developments. The simulation results, for the year 2020, are used to assess the effects of land use changes on a wide range of environmental indicators.

Finally, Chapter 18 by *Rui Julião, Alfred Wagtendonk* and *Kees Schotten*, reports on the application of the land use simulation system in the Lisbon and Tagus Valley region. This area was one of those chosen for a pilot project on the use of *EuroScanner* in regional planning. The study area is introduced, some of the physical and socio-economic characteristics of the area are identified and some of the key driving forces for land use change are identified, including future spatial planning intentions. The chapter reports in detail on the correlation and regression analyses required to identify the variables that determine urban expansion and 'densification' in the Lisbon test area. The modelling system is validated by simulating both these types of land use change between 1984 and 1998, and comparing the results with observed data. The results are encouraging and consequently the authors use the system to carry out a series of future simulations under different scenarios. Initially, a reference scenario that extrapolates historic trends into the future (2012) is used to generate simulated land use changes to produce a base against which simulations based on alternative scenarios can be compared. Secondly, E*uroScanner* is used to simulate the effects the development of a new airport to the north-east of Lisbon and the implementation of restriction zones on residential development. Finally, the authors report the results of an experimental simulation that involves applying a set of policy-based transition rules derived from urban development elsewhere (the Randstad in this case) to assess the effects of a 'compact city' approach to urban development in the Lisbon and Tagus Valley region.

In conclusion, readers will be aware that this book draws on the knowledge and experience of leading experts in the fields of data management, scenario planning, forecasting, GIS, simulation modelling, evaluation and visualisation, but it also contains perspectives from those policy-makers at regional, national and European levels of administration who are responsible for thinking about the future and making strategic decisions. The utility of *EuroScanner* as a decision support tool for policy-makers is very important and therefore it is the policy context that provides our starting point in the next chapter.

References

Batty, M. and Xie, Y. (1994) From cells to cities, *Environment and Planning B*, 21: 31-48.

Brouwer, F.M., Thomas, A.J. and Chadwick, M.J. (1991) *Land Use Changes in Europe Processes of Change, Environmental Transformations and Future Patterns*, Kluwer Academic Publishers, Dordrecht.

Burgess, E.W. (1927) The determinants of gradients in the growth of a city, *Publications, American Sociological Society*, 21: 178-84.

Christaller, W. (1933) *Die Centralen Orte in Suddeutschland*, Gustav Fisher, Jena; translated by Baskin, C.W., *Central Places in Southern Germany*, Prentice Hall, Englewood Cliffs, NJ.

Clarke, K.C., Hoppen, S. and Gaydos, I. (1997) A self-modifying cellular automata model of historical urbanization in the San Francisco Bay area, *Environment and Planning B*, 24: 247-261.

Commission of the European Communities (1998) *Final Report on SPARTACUS System for Planning and Research in Towns and Cities for Urban Sustainability*, EC Environment and Climate Research Programme: Human Dimensions of Environmental Change, ECDGXII.

Donnay, J-P. (1999) The use of remote sensing information in planning, Chapter 13 in Stillwell, J., Geertman, S. and Openshaw, S. (eds.) *Geographical Information and Planning*, Springer, Berlin, pp. 242-260.

Engelen, G., Geertman, S., Smits, P. and Wessels, C. (1999) Dynamic GIS and strategic physical planning support: a practical application, Chapter 5 in Stillwell, J., Geertman, S. and Openshaw, S. (eds.) *Geographical Information and Planning*, Springer, Berlin, pp. 87-111.

European Commission (1991) *Europe 2000: Outlook for the Development of the Community's Territory*, Office for Official Publications of the European Communities, Luxembourg.

European Commission (1994) *Europe 2000+ Cooperation for European Territorial Development*, Office for Official Publications of the European Communities, Luxembourg.

European Commission (1999) *ESDP European Spatial Development Perspective Towards Balanced and Sustainable Development of the territory of the European Union*, prepared by the Committee on Spatial Development, European Commission, Luxembourg.

Geertman, S. (1999) Geographical information technology and strategic physical planning, Chapter 4 in Stillwell, J., Geertman, S. and Openshaw, S. (eds.) *Geographical Information and Planning*, Springer, Berlin, pp. 69-86.

Harris, C.D. and Ullman, E.L. (1945) The nature of cities, *Annals, American Academy of Political and Social Sciences*, 242: 7-17.

Hilferink, M. and Rietveld, P. (1999) Land Use Scanner: an integrated model for long term projections of land use in urban and rural areas, *Journal of Geographical Information Systems*, 1: 155-177.

Hoyt, H. (1939) *The Structure and Growth of Residential Neighbourhoods in American Cities*, Federal Housing Administration, Washington DC.

Lösch, A. (1940) *Die Raumliche Ordnung der Wirtschaft*, Gustav Fisher, Jena; translated by Woglam, W.H. and Stopler, W.F. (1954) *The Economics of Location*, Yale University Press, New Haven, CT.

Lowry, I.S. (1964) A Model of Metropolis, RM-4035-RC, The Rand Corporation, Santa Monica.

Mather, A.S. (1986) *Land Use*, Longman Scientific and Technical, Harlow.

Scholten, H.J., Velde, R. van de, Rietveld, P. and Hilferink, M. (1999) Spatial information infrastructure for scenario planning: the development of a Land Use Planner for Holland, Chapter 6 in Stillwell, J., Geertman, S. and Openshaw, S. (eds.) *Geographical Information and Planning*, Springer, Berlin, pp. 112-134.

Stillwell, J.C.H., Geertman, S. and Openshaw, S. (eds.) *Geographical Information and Planning*, Springer, Berlin.

Thünen, J.H. von (1826) *Der Isolierte Staat in Beziehung auf Landwirtschaft und Nationalökonomie*, Gustav Fisher, Stuttgart; translation by Wartenburg, C.M. (1966) *The Isolated State*, Oxford University Press, Oxford.

Weber, A. (1909) *Uber den Standort der Industrien*, Tubingen; translated by Friedrich, C.J. (1929) *Theory of the Location of Industries*, University of Chicago Press, Chicago.

White, R. and Engelen, G. (1993) Cellular dynamics and GIS: modelling spatial complexity, *Geographical Systems*, 1: 237-253.

Wilson, A.G. (2000) *Complex Spatial Systems. The Modelling Foundations of Urban and Regional Analysis*, Prentice Hall, Harlow.

John Stillwell
School of Geography
University of Leeds
Leeds LS2 9JT
United Kingdom

Henk Scholten
Department of Spatial Economics
Vrije Universiteit Amsterdam
De Boelelaan 1105
1081 HV Amsterdam
The Netherlands

PART I:
POLICY PERSPECTIVES, DRIVING FORCES AND DATA CONSIDERATIONS

2 THE EUROPEAN SPATIAL DEVELOPMENT PERSPECTIVE: PROCESS, POLICIES AND CONSENSUS*

Peter Ungar

2.1 Introduction

European integration and globalisation are accelerating; national boundaries are progressively losing their significance; news about cross-border co-operation, mergers and acquisitions increasingly fill the columns of daily newspapers; and more and more European companies consider the entire territory of the European Union (EU) as their market. Spatial development and planning policies have to keep pace with and attempt to provide some control over these developments.

The *European Spatial Development Perspective* (ESDP) is a response at the European scale to the challenges resulting from integration. It takes a spatial development and planning perspective and it identifies and makes operational the European dimensions of territorial development. It is based on non-binding co-operation and it does not change the current distribution of competencies in spatial and land use planning. The ESDP is about participation, coordination and consent.

A description of the political process leading to the ESDP in its final published version is presented initially in Section 2.2. The need for EU-wide co-operation in spatial development is outlined in Section 2.3 and action programmes required for its application are reviewed in Section 2.4. In Section 2.5, the results of the process are measured as a 'calculus of consent', a theoretical approach developed by Buchanan and Tullock in the early sixties. Conclusions are drawn in the final section.

2.2 The evolution of the European Spatial Development Perspective

EU regional and cohesion policies became significantly more important in the context of preparations for the Single Market in 1992. In quantitative terms, the Structural Fund budget doubled and, as a consequence, spatial impacts of Community structural policies became much more focused than previously. The 1988 reform of the Structural Funds was based on an integrated and comprehensive approach for dealing with cohesion in a Europe of accelerated integration. Consequently, the Council regulation 4254/88 on the European Regional Development Fund (ERDF) included a Community view on spatial development. In fact, Article 10(a) of the regulation indicated the need for studies on the spatial impacts and effects of the policies of member states that occur beyond national boundaries, on the specific measures for border regions, and on *"the elements necessary to establish a prospective outline of the utilisation of Community territory"* (Council regulation 4254/88, art. 10).

* In this chapter the author expresses his personal opinions rather than the views of the European Commission.

J. Stillwell and H. Scholten (eds.), Land Use Simulation for Europe, 15–23.

Ministers responsible for regional policy and spatial planning met for the first time in Nantes in November 1989, and called for a discussion on spatial planning at European level. The European Parliament confirmed this need by its decision in October 1990, to ask for a concept of spatial planning at Community level to be developed. In *Europe 2000* (European Commission, 1991), the European Commission (EC) spelled out the basic perspectives for future spatial planning in the Community. From a situation of relatively loose co-operation on spatial development issues in the Community, a more structured discussion emerged. The establishment of the Committee on Spatial Development (CSD), agreed at the Ministerial Conference in The Hague in 1991, helped to encourage the development of a perspective on spatial planning. In 1993 in Liège, the ministers responsible decided to launch the elaboration of a *European Spatial Development Perspective* (ESDP). Meanwhile, infrastructure measures were introduced for the first time on a clear trans-European basis through the development of TransEuropean Networks (TENs). The TENs implied Community-wide planning activities, but they took place as a sectoral planning exercise without an integrated and comprehensive spatial development perspective across Europe as a whole. At the Lisbon Ministerial Conference in May 1992, the development of TENs for spatial planning was discussed intensively.

By 1994, the EC had assembled progress made so far on issues of spatial planning in a report entitled *Europe 2000+* (European Commission, 1994) which set out political priorities for the EC and provided quite comprehensive and in-depth analysis of findings and guidelines for co-operation on territorial planning. Under the heading 'Spatial planning at European level and co-operation between member states', *Europe 2000+* states that "*there is a growing acceptance of the need for spatial planning at a European level among Member States*" (European Commission, 1994, p.13). In the same year, the Leipzig Ministerial Conference defined the political principles and main objectives for spatial development in the EU. Based on these principles and objectives, a first official draft of the ESDP was presented at the Ministerial Conference in Noordwijk in June 1997.

The decision by the Commission to organise together with the member states eight transnational seminars on main issues of the ESDP in 1998 gave a substantial boost to the European dimension in the first draft of the final ESDP. There was intense and wide-ranging co-operation in its development. The European Parliament, the Committee of the Regions, the Economic and Social Committee, the member states, the regional and local authorities, non-governmental organisations and social partners participated. This was demonstrated also at the final Forum on the EDSP, held by the EC in Brussels in February 1999, where more than 600 participants represented this broad spectrum. The Commission assembled the results gained from the transnational seminars and the forum (European Commission, 1999a).

The ESDP was finalised (European Commission, 1999b) under the German presidency and was adopted at the Potsdam Ministerial Conference in May 1999 (Presidency Conclusions, 1999a). It spells out a shared vision for the future European spatial development, based on three fundamental objectives: "*economic and social cohesion; conservation of natural resources and cultural heritage; and more balanced competitiveness of the European territory*" (European Commission, 1999b, p.10). Figure 2.1 indicates schematically that this involves reconciling the economic and social claims for sustainable spatial development with the environmental and cultural claims.

Figure 2.1: The triangle of ESDP objectives

Source: European Commission (1999b)

All three objectives must be met simultaneously if spatially balanced development is to be achieved. The ESDP is therefore centred around three policy guidelines:

- *"the development of a balanced and polycentric urban system and a new urban-rural relationship;*
- *securing parity of access to infrastructure and knowledge; and*
- *sustainable development, prudent management and protection of nature and cultural heritage"*

For each of the three guidelines and their sub-themes, concrete policy options are given and proposals for the application of the ESDP are made. The EDSP is non-binding, but its concrete and comprehensive policy options make it a tool for decision-makers at all levels of spatial planning in the EU. At the Ministerial Conference in Tampere in October 1999, a programme for the application of the ESDP and a paper on the future co-operation in spatial development matters was adopted (Presidency conclusions, 1999b).

Three conclusions can be identified from this brief historical resumé. Firstly, it took six years to reach consent on the ESDP document, but less than half a year to agree on the practical steps for its application. Thus, the political process obviously accelerated substantially and has gathered a momentum. Secondly, the process offers some proof that it is possible to achieve a consensus about a rather complicated and comprehensive issue such as spatial development on the basis of inter-governmental co-operation. Thirdly, the emergence of an EDSP illustrates that a new spatial development 'culture' is emerging, combining top-down and bottom-up elements in a coherent and integrated manner.

2.3 The need for co-operation on spatial development at the EU scale

The EU Treaty does not require a Community 'competence' for spatial or land use planning and such a responsibility was never on the agenda of the EC. However, policies at the EU level have impacts on spatial development, particularly when it comes to the activities supported by the Structural Funds. Therefore, a common strategic vision of spatial development in the EU is required. It goes without saying that such a common perspective has to be non-binding and based on a broad range consensus between the parties able and willing to participate.

The challenges from integration, globalisation and technological change for spatial development are profound and complex. The need to develop more integrated structural policies, based on a shared vision with member states, local and regional entities and other partners, has been recognised. The speed of world-wide structural change is accelerating and spatial planning has to keep pace in its objectives and in the policies and methods that can be applied. This requires improving Europe's position in the world economy, reducing regional disparities and attempting to achieve balanced and sustainable development. Appropriate visions could not be elaborated by summing up member states' individual policy responses and reconciling them. It became clear they need a strategic consensus in a true European sense. The Commission has tried to play a leading role in their formulation.

The perspective that has been elaborated reconciles the needs of competitiveness (globalisation), a polycentric and decentralised development model and the sustainable integration of economic and spatial development. The broad based discussion and support in the preparation of the ESDP allowed for the accommodation of an integrated and consistent spatial planning policy approach at all levels (from the Community to local entities) and the selection of appropriate priorities.

2.4 The ESDP action programme

The conclusions of the informal meeting of EU ministers responsible for spatial planning and urban/regional policy held in Tampere in October 1999 enumerate twelve voluntary actions required to put the application of the ESDP on track (Presidency conclusions, 1999b, p.3). These are as follows:

- the orientation of ESDP policies in Structural Funds mainstream programmes;
- INTERREG III and ESDP demonstration projects;
- the orientation of ESDP policy in national spatial planning;
- the spatial impacts of Community policies;
- territorial impact assessment;
- urban policy application and co-operation;
- the establishment of the ESPON co-operation;
- the ESDP as a basis for geography books for secondary schools;

- the 'Future Regions of Europe' award;
- guidelines on integrated strategies for coastal regions;
- a pan-European framework for spatial development; and
- the spatial impacts of enlargement on EU member states and non-member states.

The inclusion of ESDP aims in the Structural Funds guidelines for the next programming period (2000-2006) is a first step towards a more integrated approach. Good practice in this respect might improve the linkages between regional and spatial development planning, when at least some member states or sector policies apply the ESDP guidelines when drafting their programmes. Also, during the next programme period, the transnational co-operation under INTERREG will be the main instrument for applying the ESDP and spatial planning in a European perspective, thereby promoting a more harmonious territorial integration. Operational spatial development strategies to be elaborated on a transnational scale should cover the strategies proposed in the ESDP for balanced polycentric and sustainable development.

When it comes to national spatial planning some member states have started to improve their systems by applying the ESDP, taking steps towards more integrated approaches (particularly the better inclusion of sector policies) and taking into account the European instead of the national perspective where appropriate. This exercise will necessarily be performed with different time horizons and intensities, because national systems are comparatively heterogeneous.

The hostility of ministries responsible for Structural Fund programmes to apply the ESDP in certain countries could be persistent. Structural Fund programmes are often developed in a ministry other than the one responsible for national spatial planning schemes. An improvement of exchange of information and of co-operation between ministries seems being a precondition for a successful application of the ESDP.

In some member states, a huge part of the responsibility for spatial planning is allocated at the level of regions. Depending on the degree of their political autonomy, the vertical co-operation in spatial planning matters might face substantial impediments, particularly because regions might fear that their competencies could be exposed to erosion when vertically integrated approaches such as the ESDP will be applied. This kind of reluctance is already very marked in the national context in some cases, but is even stronger when it comes to the Community level.

The conclusions of Tampere defined member states' views on the role of the EC in a concrete manner in this context and this could help to avoid misunderstandings. They called for a continued co-operation between member states and the EC on spatial development and planning. They devoted a leading role to the EC for the impact assessment of Community policies and EU enlargement, in co-operation with member states. The EC, in close co-operation with member states, "*should monitor territorial changes on a European scale; it should participate in developing strategic issues in European spatial development; and it should enhance the co-ordination of relevant Community Policies in respect of their spatial impacts and impacts on the territorial cohesion*" (Presidency conclusions, 1999b, p.4).

2.5 Co-operation in European spatial development: a matter of consent

The progress made in creating a European perspective on spatial development might be explained by increasing economic, social and environmental pressures, such as intensified competition through globalisation, polarisation of economic activities and settlement structures or global climate change. These pressures are clearly fundamental driving forces, but we must also recognise that progress may also be a function of the ongoing integration in the EU and at a global level. Another influential factor on the progress in developing an ESDP may also be the process of co-operation and consensus. In this section, a theoretical approach is elaborated which may shed some light on this process from what we know about public choice.

Buchanan and Tullock (1962) developed the basic concept of the *'calculus of consent'* that lies at the heart of co-operation. This concept explains the overall cost of co-operation by introducing two different components. The first component is the *preference cost*, comprising the frustration of being overruled and either being excluded or being insufficiently included in the decision-making process. The second component is the *resource cost*, which includes the traditional cost of public measures (such as capital formation or transfers) but also the costs of information, negotiation and decision-making. Figure 2.2 depicts the combination of preference and resource costs in relation to the number of individuals participating in the co-operation on a specific issue, e.g. the planned construction of an airport.

The resource cost of the construction itself is not shown in the figure, in order to isolate the co-operation cost. The curve for the preference cost declines as the number of participants increases. On the left-hand side of the graph, costs are very high when only one or a few individuals are involved in the decision-making. Costs then decline comparatively rapidly as those individuals that are most affected are included. Then there is a slow-down in decline, when individuals participate for which the decision is of lesser importance. Preference cost will not touch the baseline because, even under ideal conditions for co-operation, a certain number of participants will consider the choice made as being incompatible with their preferences.

The resource costs (C_1) develop in the opposite direction. They are lowest when one or a small group of individuals decide. They are typically concerned with the problem on a professional basis; thus their cost for obtaining the required information will be comparatively low and cost for bargaining and deciding will be very limited. The cost increase remains fairly moderate for the next participants, those that are familiar with the problem and feel directly concerned. They may have collected the necessary information anyway and could try to influence the decision by various means without being explicitly included in the co-operation process.

As the number of participants comes closer to the total number of people concerned, so the resource costs for information and participation accelerate. Someone without any basic knowledge or preference about the discussed measure could suffer enormous resource costs in getting the necessary information and in participating in the co-operation process. Stressing again the example chosen, somebody who never thought or informed himself about the pros and cons of an airport, who has technical difficulties in gathering relevant information, who does not know at all how to participate in the planning process through hearings of citizens or other public participation schemes will face comparatively high resource cost in case he wants to participate.

Figure 2.2: The calculus of consent: an issue for co-operation in spatial development

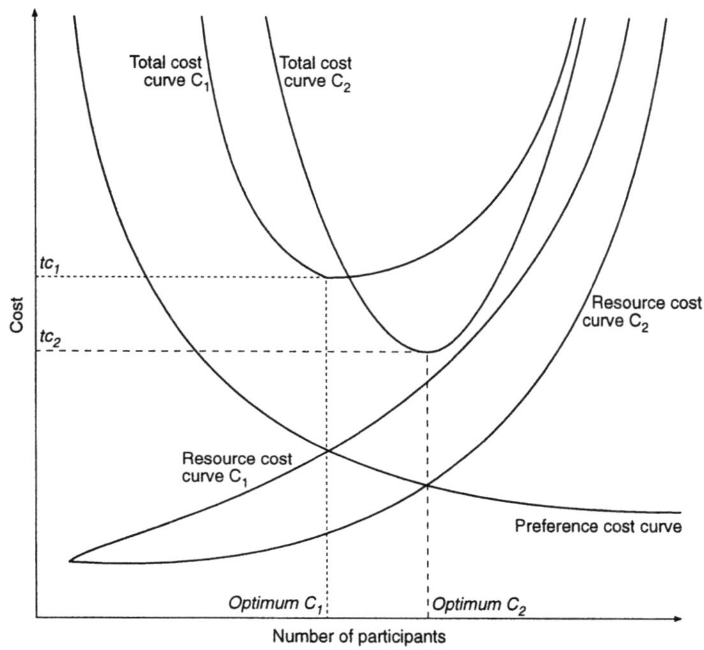

Source: Based on Buchanan and Tullock (1962)

The result of aggregating the preference and the resource costs identified by curve C_1 is a u-shaped total cost curve and there is an optimal number of participants in the co-operation (Optimum C_1) which operate under the minimum total cost of tc_1 (preference and resource cost are equal at the point the two curves cross each other; taken together they reveal the total cost). Below the C_1 curve, another resource cost curve C_2 is depicted, indicating resource cost gains in a world with improved conditions for co-operation. One outstanding reason behind the lower resource cost is the development of information technology that affects information cost in a dramatic way. *Ceteris paribus,* the higher the density of Internet and IT-access, the lower the cost for the necessary electronic equipment, the lower the tariffs for being online, the more people learn how to use IT, the lower resource cost will be. Extended and intensified co-operation is possible on previous cost levels. The gains will be higher, as co-operation gets broader and more intensive, both in a geographical sense and with respect to the number of participants. Political integration with the setting of common rules, standards, institutions also helps reduce resource cost, particularly bargaining and decision making cost. However, in most cases it may also reshape the curve for preference cost, which then makes assessing the calculus of consent more difficult.

European co-operation on spatial development clearly plays its role in lowering resource cost component. The ESDP and the related action programme give good examples of this process. The solution found and described above follows this type of calculus of consent.

The shared vision of spatial development in the Union provided by the ESDP and its political options could reduce resource cost in a variety of cases. In this respect, its role is a suitable information tool for all levels of governance, planners and other actors involved in the now enlarged co-operation perspective. By encompassing a series of political options, it helps improve planning activity. By strengthening the consensus on overall targets and orientations, it could avoid mismatch between actors and thus possibly reduce bargaining and decision making cost. This list is not exhaustive. Last but not least, improved co-operation on spatial planning should lead to resource cost gains in the classical sense that are to be added. Therefore, the conclusions of Potsdam and Tampere open promising perspectives to be exploited in the near future.

New techniques to simulate Europe's land use based on geographical information systems and spatial modelling, together with a variety of other IT-based tools currently developed, could help with co-operation on spatial development and land use planning issues through providing information at reasonable cost. The envisaged creation of a European Observatory (ESPON) for spatial development at a European scale follows the same logic. Modifying the calculus of consent in spatial development co-operation towards reduced resource cost allows for extended co-operation at reduced total cost.

2.6 Conclusion

New structures for better co-operation regarding spatial development could strengthen economic and social cohesion, not only in the field of proven classical measures, but also contribute overall to sustainable and well balanced economic and social development. The precondition is avoiding a substantial increase in preference cost, so that the calculus of consent will be mainly determined by gains in resource cost.

Measured against this theoretical background, the process of co-operation on spatial development in the EU obtained so far suggests a promising future. That applies both to the implementation of the ESDP by the action programme and the future co-operation between the Commission and the member states, being beneficial for the further strengthening of internal integration and competitiveness in the worldwide markets.

References

Buchanan, J. and Tullock, G. (1962) The Calculus of Consent, Ann Arbor.

Council Regulation (EEC) 4254/88 of 19 December 1988, laying down provisions for implementing Regulation (EEC) No 2052/88 as regards the European Regional Development Fund, Official Journal L 374, 31/12/88, pp. 15-20.

European Commission (1991) Europe 2000 – Outlook for the Development of the Community's Territory, Luxembourg: Office for Official Publications of the European Communities.

European Commission (1994) Europe 2000+ – Co-operation for European Territorial Development, Luxembourg: Office for Official Publications of the European Communities.

European Commission (1999a) European Spatial Development Perspective, A Strategy for Balanced and Sustainable Development in Europe, European Issues in the Debate, Luxembourg: Office for Official Publications of the European Communities.

European Commission (1999b) European Spatial Development Perspective – Towards Balanced and Sustainable Development of the Territory of the European Union, European Commission, Luxembourg: Office for Official Publications of the European Communities.

Presidency Conclusions (1999a) Final Conclusions issued by the German Presidency at the close of the Informal Council of EU Ministers responsible for Spatial Planning held in Potsdam on 10-11 May, Internet version.

Presidency Conclusions (1999b) at the informal meeting of EU ministers responsible for spatial planning and urban/regional policy held in Tampere on 4-5 October 1999, Internet version.

Peter Ungar
Commision Europeéenne
Direction Générale XVI
Avenue de Tervuren 41
B-1040 Bruxelles
Belgium

3 EUROPEAN SPATIAL PLANNING: NATIONAL AND REGIONAL PERSPECTIVES

Hans ten Velden and John Stillwell

3.1 Introduction

Individual member states continuously define, redefine and defend their national interests within the European Union (EU) policy arena. On many occasions, national interests and EU policies involve a spatial dimension with implications for physical land use planning across one or more sectors. Spatial planners are relatively inexperienced in how to operate in this new context of European and national power relations, finding themselves in a unique position because the EU has no formal competence in spatial planning. However, it is acknowledged that sectoral policies exert a significant impact on the way the European territory is used. The creation of a European Spatial Development Perspective (ESDP) means that there is now the beginnings of a spatial policy for the entire territory of the EU.

Such an agreed policy perspective has consequences for national and regional spatial policy making in each of the constituent member states. In the Netherlands, this has been recognised by national planners as being a continuous process of defining and redefining the role of the national territory and the domestic planning system within the wider European territory and in relation to both transnational (between two or more regions within groupings of regions) and cross-border (between neighbouring areas) policy making. National interests have to be reconsidered in a new European territorial context but the strength of national opinion depends on the statutory requirements, organisational structures and characteristics of the planning system in each nation. In the Netherlands, there is a longstanding tradition of national spatial planning, for example, whereas in the UK, spatial planning at the national level has never been seriously embraced. However, regional planners in the UK have begun to consider how the wider issues associated with the ESDP have relevance for their own territorial units and their own more localised spatial planning activities.

The substance of European spatial policy is introduced in this chapter in Section 3.2 with a sketch from professional and political perspectives. The essentials of the first ESDP are highlighted in Section 3.3 whilst in Section 3.4, the imperative to collect better information and to undertake more effective analysis is emphasised. In Section 3.5, we assess how the European dimension of spatial policy making impacts on national spatial planning in the Netherlands from the perspective of the Dutch National Spatial Planning Agency and we also discuss the implications of the ESDP on regional planning in the UK. Conclusions are presented the final section.

3.2 The need for European spatial planning

It is important to recall that spatial policy making at a transnational or European scale is not really a new activity. The creation of the EU itself can be seen as a major act of spatial

J. Stillwell and H. Scholten (eds.), Land Use Simulation for Europe, 25–36.

policy formulation, having enormous impacts on the behaviour of large numbers of people and organisations across the whole of Europe. More specific examples of spatial policy making and planning in operation include the design and implementation of Trans European Networks (TENs) and the evolution of the Common Agricultural Policy (CAP). However, some of the policies emerging from these frameworks have often appeared in the form of relatively crude guidelines and measures, paying only limited respect to the varying qualities and functions of the European territory as a whole and its regions.

Spatial planners claim that all kinds of sectoral policies do have a major impact on the quality and the functionality of the European territory and the way it is used. Community competition policy, involving a range of measures designed to encourage the integration of national markets, has distinct effects on the geographical distribution of economic activities and patterns of trade. The Structural Funds, and the European Regional development Fund (ERDF) especially, have had very significant geographical impacts as they have endeavoured to reduce economic and social disparities outlined in successive periodic reports (European Commission, 1999a). The environment is another sector where a variety of Community activities have direct and indirect effects on land use and spatial development. The EC Nitrate Directive, for example, is aimed at reducing nitrate pollution from agricultural land. A significant part of the EU's research and technology development policy is currently focused on the Fifth Framework Programme designed to support the development and implementation of various Community policies. Here again, the territorial or spatial dimension is frequently clearly defined.

The claim to be justified is that consideration of the territory of Europe as a whole offers good opportunities for more integrated economic, social and environmental development through recognition of the existing internal geographical diversity. There seems little doubt that, in order to exploit spatial opportunities, it is frequently necessary to forget national or regional boundaries and put national or regional interests to one side. This is often very difficult, of course, especially for national politicians. However, with the ESDP, national planners are in a process of redefining national interests *vis à vis* European interests in relation to the future quality of the European territory. In the Netherlands, some national interests appear to be consistent with European scenarios or at least with certain transnational development issues or proposals. These include, for example, the treatment of water systems, threats stemming from a rising sea level, and the development of seaport areas in connection with ever higher volumes of hinterland transport. It is likely that these types of issues are more readily articulated in countries with small and densely populated regions (like the Benelux countries). There is general acknowledgement that in the larger member states, European monetary union will play a key role in removing the obstacles that national borders currently present. This will consequently require urgent reconsideration of the relationship between national and European interests.

Some of the highlights in the history of European spatial policy making and of the formulation of the current ESDP that were discussed in the previous chapter by Peter Ungar, are summarised in Figure 3.1 which is part of a schematic framework prepared for the 1999 Yearbook of the Dutch National Physical Planning Agency. As a result of this rather long drawn out process (Williams, 1999), in which both the Netherlands and the UK have played their respective parts, there is now more widespread professional awareness ('understanding') as well as the beginnings of a political awareness ('intervening') that many spatial policy issues can no longer be effectively treated at the national level only. Political awareness means that it is acknowledged that all the separate planning decisions of member states, and of the Council of Ministers at the EU level, do have unexpected and often undesired impacts on the way the territory is used. In many cases these impacts require concerted policy action. Central ideas feeding the political process are 'competitiveness', 'sustainability' and 'cohesion' which reflect the main goals of EU policies in general. In terms of the ESDP, the goals are identified as more balanced competitiveness between different countries and regions, conservation of natural resources and cultural heritage, and economic and social cohesion.

Figure 3.1: The dimensions and key events in the development of the ESDP

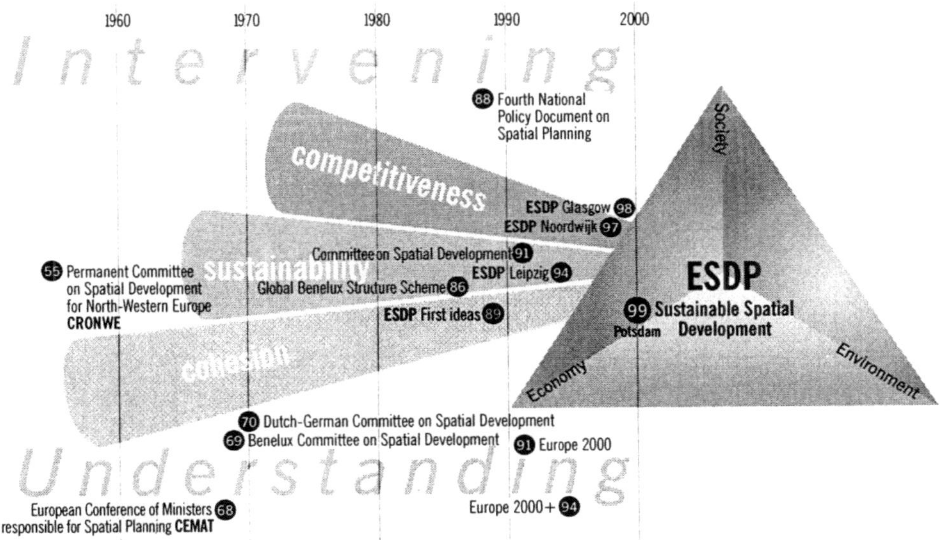

Source: National Spatial Planning Agency (1999)

The challenge is to link professional assessments of spatial developments to political assessments. One way of doing this is to highlight the relationships between spatial phenomena of an economic, environmental and social nature that, in many cases, transcend national borders. The added value of a spatial planning approach is that these three basic themes are actively related whenever the focus is on specific geographical areas or types of area. Such an area-specific approach reveals the necessary transnational actions required. Take the management of the rivers in the low countries as an example. It is obvious that solutions to reduce the risk of flooding in the Netherlands are partly found upstream in the catchment areas in neighbouring countries where spatial planning policies are needed to increase the water retention capacity of the rivers. Where a transnational or European spatial perspective has been absent, opportunities have been missed to link such a policy to territory. Another example of this is the lack of transnational or European policy instruments to guide investments in seaport and airport infrastructure and hinterland facilities. The scenario of ever increasing flows of goods through densely populated territory, and the need for efficient use of land and existing infrastructure, turns national economic interests into transnational issues. The trends of intensification, concentration and specialization in agriculture under the regime of the CAP and of market globalisation lead to further decay of territorial quality if the desired standards of areas are not predefined. The need to prepare political decisions to that end is a major demand on the basic professional skills of spatial planners.

For planning agencies like the Dutch National Spatial Planning Agency and its Minister, it is a real challenge to communicate messages of this kind to politicians who are responsible for transport, agriculture or economic affairs, for example. Despite the fact that the unification of Europe is now rapidly turning into a reality, it remains difficult for national politicians to admit that opportunities and threats regarding their territories in many cases can only be tackled properly in partnership with other countries and at the level of the EU. To gain experience in cooperation in the field of transnational planning, the European Commission launched the INTERREG IIC programme in 1996 which enables regions within the different member states to develop spatial planning projects together. In contrast to INTERREG IIA, which involves cross-border collaborations, transnational cooperation under INTERREG IIC covers much larger areas such as the North Sea Region and South-Western Europe. There are several programmes for transnational cooperation for spatial development as well as for flood mitigation and drought prevention.

3.3 The European Spatial Development Perspective

The ESDP document (European Commission, 1999b) and its basic ideas were agreed in Potsdam in May 1999. It is important to bear in mind that the EU does not have any formal competence for spatial planning whatsoever. On the other hand, the Treaty of Rome talks about a *"harmonious development of the European territory"* as a goal to strive towards. Given the absence of formal powers, it is all the more remarkable that the ministers of the member states responsible for spatial planning have now reached an agreement on a common document on spatial development, a document approved by the 15 member states and the European Commissioner for Regional Policy as a reference for the development of European and transnational spatial policy development. The document respects the principle of subsidiarity but, on the other hand, places an emphasis on the need to cooperate.

An important starting point in the ESDP process was the acknowledgement that the whole of the EU territory was now the playground for people and companies, as it has always been for natural elements. This leads to the three planning objectives set out in the EDSP:

- a more balanced pattern of spatial development and new urban and rural relationships;
- more equal access to physical and new communications networks and improvements to foster innovation; and
- the sustainable management of Europe's resources and natural, built and cultural heritage.

The EDSP gives some new assessments of issues that require international cooperation in order to avoid unnecessary environmental damage or financial risks, and issues regarding the exploitation of the comparative advantages of regions which should keep or put them on the global economic map. A basic premise of the ESDP is that opportunities and threats regarding the three goals of competitiveness, sustainability and socio-economic cohesion of Europe in a global context can be captured much better if functional areas or area-specific systems are taken as a starting point for policy development instead of national boundaries. From an economic point of view, monetary union in Europe turns the concept of competition between regions and states upside down. With monetary union, it is critically important that the less developed regions should be put in a position such that they can develop themselves and become as competitive as the economic core regions of the EU. The concept introduced in the ESDP to achieve this balance is of a 'polycentric development model' at the European scale. From a political point of view, this may be a 'long shot' because it radically changes the way of looking at the European territory. From an environmental perspective, it focuses attention on issues such as the carrying capacity of the soil, continental water systems and the European natural heritage. It is interesting to observe the increasing recognition of the economic significance of these environmental qualities and functions at the present time.

The problems of adverse social conditions found in the larger cities and of social marginalization in specific rural areas are identified in the ESDP. These problems clearly have an economic impact through mechanisms of state financing and investment, but these topics need further research to give better insights into the relationships between urban structures and social exclusion, for example, and to establish what combination of economic, environmental and social factors is required to preserve characteristic landscapes across Europe. The ESDP summarizes a whole range of these issues in the three basic policy aims from which a series of policy options is derived.

3.4 Information and analysis imperative

During the process of formulating the ESDP it became very clear that there is an important lack of systematic, comparative data and knowledge of various aspects of spatial development in Europe. There is fairly good documentation of variables directly related to EU funds like regional gross domestic product (GDP) per capita and unemployment. On the other hand, comparable indicators of the quality and performance of cities, rural areas or other types of area, are very poorly documented. Other indicators, such as those developed to measure accessibility, are questioned because they can easily be manipulated. The quality, comparability and reliability of statistical data for many variables or indicators are not good. One solution is to use satellite images to provide land cover data. However, there are limitations associated with this type of data when related to physical phenomena.

The 'knowledge problem' for spatial planning has led to the implementation of a pilot research programme on spatial planning issues at the European scale. This so-called European Spatial Planning Observatory Network (ESPON), involving 15 institutions from the member states, in cooperation with EEA and EUROSTAT, has finalized its test phase in the middle of 2000. The EPSON network covers three streams of research. The first of these is directed at a set of spatially relevant criteria to be translated using indicators such as geographical position, economic strength, social integration, spatial integration, land use pressure, natural assets and cultural assets. It is expected that some combination of these criteria would be helpful in evaluating the opportunities and threats to specific areas more fully *vis à vis* the traditional, mainly economic assessments underlying European regional policies. A second stream of research is directed at the broad theme of urban-rural relationships. This theme covers all the types of issues that relate to urban-rural interactions at different spatial scales. Subthemes that have been investigated include the main trends shaping the European territory, typologies of cities, issues such as the territorial impacts of agricultural change, settlement structures and accessibility, conservation and development strategies for natural heritage. Lastly, attention is paid to cartography and infographics because, especially in a political environment, maps have proven to be an emotional subject.

3.5 Impacts of the ESDP at national and regional levels

Experience to date with policy development and research at the European level has raised a number of basic questions about how to move forward, including:

- how to study spatial phenomena with only relatively little harmonised and consistent data for NUTS regions across Europe;
- how to develop effective two way communication between professional researchers and politicians when analysis and politics have always tended to be distanced from one another at the European level;
- how to establish permanent knowledge development for spatial planning at a European scale; and
- how to deal with professional and political emotions emerging from mapping the alternative scenarios of Europe's future.

Figure 3.2 highlights how the main policy goals of the ESDP relate to current national policy development in the Netherlands. Firstly, the (re)introduction of 'layers' as a basis for analysis and the expression of policies will lead to more explicit spatial policies. The 'ground layer' draws political attention to water (both quality and flooding), soil and nature. The 'occupation layer' claims to treat seriously policies regarding more balanced 'red/green' patterns of human occupation. And that will be a hell of a job especially in the western part of the country. A recent inventory of sectoral demands for space concluded that towards 2030 the Netherlands may need an additional 170,000 hectares of urbanization, of which 50% could be necessary in the western part. But also green and blue functions have 'claimed' substantial additional space. From the networks layer, more explicit policies will be derived to transform the Randstad Holland into a 'deltametropolis' with shorter internal travel times. The decision making process will be focused both on issues stemming directly from the separate layers but mainly to issues stemming from a confrontation between the layers.

Due to the borderless character of the layers, transnational policies are very explicitly dealt with in the Fifth National Policy Report that has been approved in December 2000. Implicitly, the layers reflect the structure in the ESDP. The ESDP recognizes water and natural elements, infrastructure networks and the occupation of the territory through urbanization as different layers for spatial policy development. It very explicitly takes a number of spatial scales into account. So does the new national spatial policy in the Netherlands. It is recognized that the determination of scales related to policy issues is a major political act. Moreover, spatial scales will reflect the international political ambitions of spatial policy. Some items may be scaled upwards to the transnational or European level; others may be scaled downwards to the regional level. This is both a national and a transnational or even European political process and it will leave the national level(s) with new roles ranging from 'intermediary' to 'interventionist'. This process than reflects the struggle with rethinking national interests in a wider European territorial context.

Figure 3.2: ESDP policy goals related to current Dutch policy development

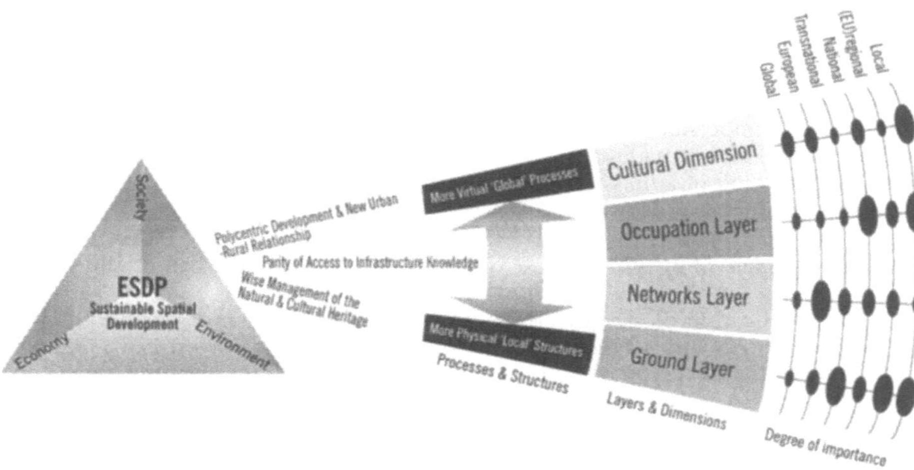

Source: National Spatial Planning Agency (1999)

Perspectives on the ESDP at either a national or regional level will vary according to the particular planning structures, processes, cultures and issues pertaining to an individual nation or region a particular point in time. Yorkshire and Humberside is one of the ten regions in England and Wales whose regional institutions have only been established since the election of the Blair Government and whose regional planning strategy documents are currently in draft form. Responsibility for spatial development in Yorkshire and Humberside rests with a number of key organisations at the regional level, including the voluntary *Regional Chamber* which is a public private partnership that attempts to focus on policy coordination across the region by bringing together all the stakeholders, the *Regional Assembly* (RAYH) representing all the constituent local authorities in the region, the *Government Office* (GOYH) representing central government and *Yorkshire Forward*, the Regional Development Agency (RDA). There are, of course, a host of other regional and sub-regional agencies and stakeholders within Yorkshire and Humberside that constitute what is a very complex institutional environment.

As in every other region in England, there are now two main regional planning documents available in Yorkshire and Humberside. The first of these is Regional Planning Guidance (RPG), currently in draft form, which is essentially a physical planning strategy, concentrating on sustainable development, employment, land, transport and cultural, natural and environmental resources (RAYH, 1999). RPG is a 'regional spatial strategy' for the period 1998-2016 as requested by the Department of the Environment, Transport and the Regions (DETR) and provides the context for development plans produced by local authorities as well as the spatial context for the economic development activities of the RDA and the plans of infrastructure and service providers, industry and commerce in the region. RPG in Yorkshire and Humberside is intended to provide a clear sense of the direction in which the region should be moving and of the processes required for managing and shaping change.

The second of the planning documents is the Regional Economic Strategy (RES) prepared as one of the first tasks required by central government of each of the RDAs established in every English region in April 1999. The RESs were prepared in consultation with other regional partners and were envisaged as influencing existing programmes and budgets as well as having implementation plans (Action Plans) elaborating their own policies. In Yorkshire and Humberside, the RES built on a previous study of the state of the region (ERBEDU, 1999) and provides a 'road map' indicating how the region will achieve its vision. It is the result of a massive consultation exercise involving thousands of people and businesses (Yorkshire Forward, 1999) and it consists of:

- an assessment of the present state of the economy and a statement of aspirations;
- the specification of six objectives: grow the region's business; achieve higher business birth and survival rates; attract and retain more investment; radically improve the development and application of education, learning and skills; implement targeted community-based regeneration programmes; and get the best out of the region's physical and cultural assets and conserve and enhance its environmental assets;
- the identification of five themes that cross-cut the objectives: sustainability, partnership, geographic adaptation, social inclusion and creativity, innovation and technology;
- an outline of the priority actions for the first two years of the period covered by the strategy;
- proposals for monitoring and review; and
- headline and supporting indicators.

There are various other strategy documents in various stages of preparation in Yorkshire and Humberside including the Regional Transport Strategy, which is an integral part of RPG, the Regional Skills Strategy, and CoMPRIS, the region's Information Society Strategy which deals with the implications of information and communications technologies (ICTs) for teleworking and business development, learning and skills, knowledge, health and public services. In addition, there are the EU programmes associated with the funding of Objective 1 and 2 areas in the region as well as a host of other more localised planning and policy documents such as the Unitary Development Plans (UDPs), planning briefs, design guides and local transport plans.

A similar structure of organisations and strategic planning documents described in the brief synopsis outlined above is replicated in each of the other English regions, although Yorkshire and Humberside is unique in the extent to which its draft RPG and RES have been prepared and launched in parallel. In fact, both the RPG and RES have been subjected to the same sustainability appraisals in Yorkshire and Humberside and both have been prepared though partnership arrangements which were singled out for special comment in the Government's response document published by the Department of the Environment, Transport and the Regions (DETR, 2000). A comparison of the aims and objectives of the RES in Yorkshire and Humberside (Figure 3.3) that focus on business, people and the environment, correlate directly with the three objectives in the ESDP of economy, society and the environment (Figure 3.2).

Figure 3.3: Objectives of the RES for Yorkshire and Humberside

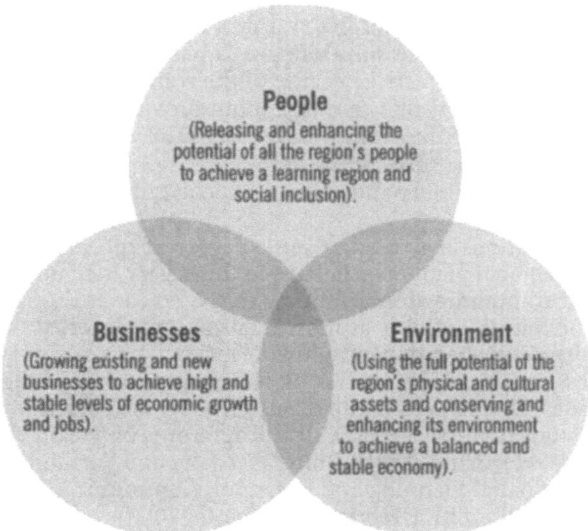

Source: Adapted from Yorkshire Forward (1999)

So it is apparent that spatial planning in the region is already in line with the basic principles of the ESDP but what are the implications of the ESDP for spatial planning in the region of Yorkshire and Humberside? The integration and linkages between strategies promoted in the ESDP are seen as being particularly important in gaining consensus. However, this is not easy to achieve because of the conflicts of interest between different stakeholders (such as local authorities and property developers) and the inconsistent objectives or approaches that are suggested in different strategies or programmes. In fact consistency in terms of whether or not strategies or policies meet sustainability criteria is a major issue raised by the sustainability assessments undertaken in Yorkshire and Humberside. It is necessary for regional planners to ensure that consistency is achieved downwards between regional strategies and sub-regional or local plans, but it is also necessary to achieve consistency upwards between regional objectives and programmes submitted to draw down funds from the EU. In the latter case, there is evidence to suggest that objectives associated with bids for Structural Funds associated with Objective 1 and 2 areas in South Yorkshire are at odds with those set out in RPG and RES. This occurs because EU programmes have often been designed with the purpose of maximising revenue at all costs, thereby squeezing as much money out of EU coffers as possible. The answer here is to attempt to appraise bids for EU funds using the same sustainability criteria as used to appraise RPG and RES.

This extends the notion of improved linkage or 'joined-up' thinking that is favoured so much by the current UK Government and that is implicit in the ESDP.

One of the major impacts of the ESDP on planning at the regional level is that it forces the planners and the communities as a whole to think very carefully about the range of issues that are important for the future and to attempt to prioritize these issues. In many cases, the issues refer to major decisions about dealing with problems. Do we develop greenfield or brownfield sites or what percentage of new housing should be on each? How do we deal with the problems of traffic congestion or social exclusion? How do we achieve better integrated transport systems or better access to industrial sites? For many of the questions, it is possible to define solutions on paper but it is extremely difficult to implement these solutions in practice because of the complexities of the real world. The solution to the congestion problem is not just about providing more roads, for example; it is about understanding the forces that are continually driving businesses and the public to use road vehicles the way they do. Central to the exercise are debates about the definition and measurement of sustainability and about what constitutes a better quality of life.

In other cases, the questions may require some degree of speculation about the future. Thus, for example, regional planners will have to think very hard about the implications of information and communication technologies (ICTs) for enabling change. How will ICTs impact on population mobility as individuals become less tied to workplaces? Will ICTs create new business clusters or new industrial location patterns? How will they change lifestyles, cultures, productivity, modes of learning and knowledge acquisition or dissemination? Issues relating to ICTs and the use of space are currently being investigated in a transnational study in the INTERREG IIC programme by a team of regional partners from North Holland, Flevoland, North Rhine Westfalia and Yorkshire and Humberside. The project is called Spatial Planning and Emerging Communication Technology in the Regions of Europe or SPECTRE (Dabinett and Booth, 2000).

There are other issues, however, that provoke a fundamental reconsideration of existing systems of planning and policy making rather than speculating about future change. In the British case, there is real concern over the credibility of the existing planning and development control system which is seen as being far too rigid, inflexible, bureaucratic and slow moving. It is apparent that the world is changing very rapidly; globalisation means that footloose businesses locate and relocate according to regional or local economic and resource circumstances; clusters of high tech companies may emerge at short notice in response to major investment decisions; call centres and out of town shopping centres have proliferated. The speed of change has left the British planning system way behind and needing to catch up. Some changes are accentuated by the way in which ICTs are collapsing or redefining time and space. The result of having a byzantine land use control system is disenchantment and frustration on behalf of developers and inward investors as well as planners themselves; opportunities for regeneration are missed. All this points to the need for a more flexible and dynamic system; a system with more criteria-based policies; a system which is about planning for change.

In Yorkshire and Humberside, the ESDP has raised other concerns. Although the focus on integration, coherence, partnership and linkage has a number of important benefits, there are also some downsides. One of these is the perceived volume of bureaucracy associated with the planning process, together with the enormous amounts of time that are spent in consultation to achieve consensus. Encouraging participation is an admirable principle but in practice, it means that the whole process is further slowed down as every represented group has its say and as the route to consensus is extended. Another related concern is the declining transparency that occurs in the decision making process as participation increases.

Finally, the ESDP provides the stimulus for regional benchmarking and for searching for examples of best practice elsewhere in Europe that might be adopted within the region. More importantly, the policy programmes that operate under the ESDP initiative provide substantial funding to support transnational research and development studies that enable visionary ideas to be thought through and frameworks created that enable regional partners to shape future development. In some cases the studies involve a suite of projects such as the North European Trade Axis or NETA, based on the concept of a braided corridor from Ireland, across northern England (TransPennine) through the Netherlands and on to Germany and involving of projects on eco-cultural assets, political/professional networks, trade and transport, logistics, inter-model chain management and sustainable industrial areas, for example. It is encouraging to know that funds allocated to INTERREG III over 2000-2006 will be three times those supported under INTERREG IIC.

3.6 Conclusion

The ESDP is a broad-based approach that encourages national and regional planners to think transnationally about the implications of their own internal plans and policies. In member states such as the Netherlands and France, with longstanding traditions of national physical or economic planning, it represents a logical progression from thinking spatially about the development of national territories or economies. In countries such as the UK, where national planning is almost anathema, it may prove more difficult for national planners to relate to with conviction. However, in the context of the resurgent regionalism now occurring in the UK, and the priority accorded to the notion of a 'Europe of Regions', the importance of spatial planning is becoming increasingly significant as strategic planners consider how to maximise the assets and comparative advantages of the regions for which they have responsibility. The ESDP has an important role to play in focusing on issues, developing ideas and getting regions to shape their futures. Use of the term 'spatial' emphasises the importance of place and draws attention to the need to reconcile place competitiveness with place cooperation, be it at town, city or regional scale. The term 'development' highlights the need to embrace the notions of economic competitiveness, social cohesion and environmental sustainability in a more integrated, coordinated and comprehensive manner.

The ESDP is not a statutory planning document and was never intended to be so. It is a conceptual framework for providing indicative guidance about the implications of major economic and social change. It encourages planning to be more visionary and for planners to think beyond the confines of their administrative boundaries. It emphasises both the necessity to understand the driving forces behind change in land use across Europe as well as to contemplate the possible scenarios that might be used to simulate land use change over the next 20 years.

References

Dabinett, G. and Booth, C. (2000) Creating new spaces and places: future spatial planning in Yorkshire and Humberside for the information age, *The Regional Review*, 10(2): 5-7.

Department of the Environment, Transport and The Regions (2000) *Regional Development Agencies' Strategies – All Systems Go*, DETR, London.

European Commission (1999a) *Sixth Periodic Report on the Social nad Economic Situation and development of the Regions of the European Community*, Office for Official Publications of the European Community, Luxembourg.

European Commission (1999b) *ESDP European Spatial Development Perspective Towards Balanced and Sustainable Development of the Territory of the European Union*, prepared by the Committee on Spatial Development, European Commission, Luxembourg, pp. 82.

National Spatial Planning Agency (1999) *Spatial Perspectives in Europe*, Ministry of Housing, Spatial Planning and the Environment, The Hague.

Regional Assembly for Yorkshire and Humberside (1999) *Advancing Together Towards a Spatial Strategy*, Draft Regional Planning Guidance, YHRA, Wakefield.

ERBEDU (1999) *The State of the Region*, Report prepared for the Yorkshire and Humber Development Agency, Leeds Metropolitan University, Leeds.

Williams, R.H. (1999) Constructing the European Spatial Data Perspective: consensus without a competence, *Regional Studies*, 33(8): 793-797.

Yorkshire Forward (1999) *Advancing Together: Towards a World Class Economy*, Draft Regional Economic Strategy for Yorkshire and the Humber, Yorkshire Forward, Leeds.

Hans ten Velden
National Spatial Planning Agency
Postbus 30940
2500 GX Den Haag
The Netherlands

John Stillwell
School of Geography
University of Leeds
Leeds LS2 9JT
United Kingdom

4 AGRICULTURE, FORESTRY AND NATURE: TRENDS AND DEVELOPMENTS ACROSS EUROPE

Arnold Bregt and Kees de Zeeuw

4.1 Introduction

Land use across Europe is influenced by a large number of factors operating at different spatial levels. Some of the land use changes are the result of autonomous economic, social and cultural developments, while others are a result of policies or spatial plans. At local level, land use may be affected by the construction of a new road or the creation of a new built-up area, for instance. At the European level, changes in the Common Agricultural Policy (CAP) have pronounced effects on agricultural land use. In this chapter, the main trends in agriculture, forestry and natural environment management are discussed, together with their effects on land use patterns in Europe. The requirements in terms of tools for describing the current situation, monitoring developments and creating scenarios for future land use are indicated. The trends and effects presented in the chapter are based on literature research and consultation with experts in the field.

4.2 Land use in Europe

Changes in land use have occurred frequently in Europe. Expansion of the agricultural area was for many years the only way to supply sufficient food for the increasing population. At the beginning of the Middle Ages, more than 80 per cent of the population was working in agriculture. Around 1400, Europe was confronted with the Black Death. Some 30 per cent of the population died and it was for that reason that the cultivated area contracted. After this period of contraction, the expansion continued until the end of 17th Century when a new period of contraction of the cultivated area began as a result of the increase of productivity per hectare. After some 100 years of contraction, the agricultural area expanded again as the European population increased more than the productivity per hectare. In the mid of the 20th Century, a sharp increase in productivity per hectare occurred, and the agricultural area started to contract again (after Rabbinge and van Diepen, 2000). In European history, the size of the agricultural area was in balance with the population and productivity per hectare. For instance, a higher productivity per hectare with the same population leads to an overall reduction in agricultural land use.

In Table 4.1, some data on the change of land use in Europe (EU15) in the last forty years is presented. A steady reduction in agricultural area in Europe can be observed of about 5 million hectares every 10 years. At the same time, a slight increase in forest and woodland can be found to have occurred.

4.3 Trends in agriculture, forestry and nature

Agriculture in Europe is strongly influenced by the CAP of the European Union which was implemented in 1960 to guarantee enough food for European citizens and to provide the

J. Stillwell and H. Scholten (eds.), Land Use Simulation for Europe, 37–44.

farmers in Europe with an acceptable level of income (Baldock *et al.*, 1996). Price support for agricultural products has been the main measure of the CAP for years. This policy was deemed appropriate since farmers' incomes were guaranteed and the price support for agricultural products favoured efficient farming, resulting in higher average yield per hectare as well as the cultivation of marginal lands. However, the policy resulted in an overproduction of agricultural products in Europe in the 1980s, which needed to be stored or exported at high cost. In 1992, the Mac Sharry reform of the CAP changed to a lower level of price support for agricultural products, direct farm income support and set-aside regulations. The change in CAP has had a number of effects on agriculture and agricultural land use. Some farmers have tried to compensate the loss of income by increasing farming efficiency, by enlargement of the farm and creation of higher yield per hectare. Other farmers have tried to compensate the loss of income by diversifying and undertaking additional activities outside agriculture. This leads to a group of so-called 'part-time' farmers, whose holdings are relatively small and whose lands are farmed more extensively.

Table 4.1: Land use in Europe

Land use type	Area in year (1,000,000 hectares)				
	1961	1970	1980	1990	1998
Agriculture area	164	159	153	147	142
Forest and woodland	101	107	111	113	no data

Source: FAOstat

In less favourable areas, abandonment of agricultural land has occurred (Figure 4.1). This is not only a result of reduced subsidies for agricultural products, but also a result of demographic developments in certain regions of Europe. Overall, we see a reduction in the number of farmers in Europe; between 1965 and 1985 the number of farmers in the Europe of twelve countries dropped from 15 million to 8 million, a reduction from 17 per cent to 6 per cent from the workforce (Rijksplanologische Dienst, 1997). Scenario studies of the change of land use in Europe for the time horizon 2000-2020 (Kamminga *et al.*, 1993; WRR, 1992; FAO, 1993; Latenstein, 1999) all indicate a reduction of land use for agriculture. The Netherlands Scientific Council for Government Policy (WRR, 1992), consider a reduction of 80 per cent of agricultural land use in Europe as one possible scenario.

The FAO has carried out, on a more or less regular basis, prospective studies of the production of and the demand for timber in Europe. The first study was published in 1953. In 1986, the prospects for 2000 and beyond suggested a small percentage increase in the consumption of wood products between 1980 and 2000 (FAO, 1986). From 2000 to 2025, a more stable rate of consumption is forecasted. Assuming that the present forest production system is in equilibrium, no big changes in forest area are expected from a timber supply point of view. Forested areas are not only considered as a source for wood products, but also as a place for recreation and a source of biodiversity. The last two functions of the forest will result in a more diverse character of the forested areas. One possible increase in forested area in Europe might occur as a result of the Kyoto protocol (1997). In order to create national carbon sinks, the planting of trees is considered as an option. It is expected therefore that, due to international pressure to meet the national carbon reductions, more areas of Europe will be forested.

Figure 4.1: Evidence of the abandonment of marginal land

For centuries the nature areas in Europe were those areas not very useful for other types of land use. In recent decades, however, there is increasing public and political attention paid to nature conservation. It is increasingly realized that natural areas are essential for our society and environment, for maintaining biodiversity, recreation and environmental control. Recently, policies were made to strengthen nature areas in Europe by creating ecological corridors. Despite these measures to increase the nature areas, the quality of nature areas is still declining due to urbanization, new infrastructure and environmental pressure.

In 1997, the first official draft of the European Spatial Development Perspective (ESDP, 1997) was produced. This perspective serves as a framework for spatial integration, which takes account of the various levels of political and spatial organization in Europe. Three main driving forces, which play a major role in shaping future trends in spatial development, are identified:

- demographic and urbanization trends;
- characteristics and trends in the European economy;
- longer-term environmental trends.

According to the ESDP, these forces led to the following changes in spatial development. First, whilst agriculture has been the predominant activity in the rural areas for many years, these areas are now becoming increasingly dependent on the development of urban settlements. Shifts in the economic base of the rural area will, however, vary from region to region, reflecting contrasting processes of agricultural intensification, diversification, extensification and marginalisation (ESDP, 1997). Second, the pattern, form and use of infrastructure will undergo continual and rapid changes in the years to come. Advances in telecommunications technology will change the way infrastructure is used. The location of production and sale (e-commerce) is becoming less relevant leading to a more intense use of the transport infrastructure in Europe. Third, the fragmentation of habitats and protected zones continues to be a major threat of the nature areas in Europe.

In the next section, we will discuss the effects of the reported trends on the spatial environment and structure.

4.4 Influence of trends on the spatial environment

Some of these major trends in agriculture, nature and forestry do have an influence on the spatial patterns of land use in Europe while others do not. In Table 4.2, the possible spatial impacts of these trends are summarized. Some of the trends are more manifest in certain parts of Europe, e.g. the abandonment of marginal areas will be more pronounced in hilly areas where there are no other sources of economic activity, whereas the influence of urbanization will be more pronounced in the densely populated parts of Europe.

Table 4.2: Main trends and spatial impacts

Trend	Spatial impact
Agriculture-political	
• Reduction of overproduction (e.g. Mac Sharry)	Reduction in agricultural area
• Diversification of activities in rural areas	Change in land use/landscape
Agriculture-autonomous	
• Intensification on potentially productive grounds	More production per hectare and abandonment of marginal areas
• Indoor agricultural production	Formation of agricultural production landscapes
• Farm enlargement and hobby farmers	Larger parcels and dispersed urbanization
Forestry	
• From production forest to multi-purpose forest (recreation, nature, water conservation, erosion control)	More differentiation in forest
• Formation of ecological corridors	Change in landscape patterns
• 'Red pays for green'	Scattered urbanization in forest More forest close to urban areas
• Continuation of wood production	-
• 'Kyoto protocol'	Increase in forested area
Nature	
• Urbanization and development of infrastructure	Loss of biodiversity and fragmentation of nature areas
• Environmental pressure	Loss of biodiversity
• Formation of ecological corridors	Change in landscape patterns

Overall, the following main effects in Europe can be distinguished:

- trends in agriculture will lead to a reduction in agricultural area and a simplification of the rural landscape;

- trends in forestry will lead to a more differentiated and slightly larger forested area. Patterns in the landscape will be more scattered; and
- trends in nature areas will lead to loss of biodiversity and spatial fragmentation of nature areas.

4.5 From trends to data and tools

The spatial organization of Europe and its component parts are of constant concern to policy and decision makers and planners at local, regional, national and European level, as well as to the public. Generally, they require three different types of information. Initially, they are interested in the status of the land. Where are the forested areas in Europe? Where are the soils that have highest and lowest productivity? What is the population density in different regions of Europe? These are the *'what is where'* questions answered by one-off surveys. In Europe, a number of these surveys have been undertaken or are currently being executed, such as the survey on soils (Bullock *et al.*, 1999) and on land cover/land use (EC, 1999; Mücher *et al.*, 1999).

Secondly, planners are interested in the change of a particular variable over time. What is the change to the forested area in Europe? How many hectares of agricultural land have been abandoned since 1990? What has changed in the nature area over the last ten years? These are the *'what is changing where'* questions that can be answered by multiple time observations of land use and related themes (monitoring). This can be done by recording, periodically, certain parameters at selected spots. For environmental quality information, this is often the practice. However, if one is interested in area changes, a full survey at regular time intervals is often necessary.

Thirdly, planners are interested in future projections. What will happen with the area of agricultural land due to the Mac Sharry regulations? What will be the effect of urbanization on nature development in densely populated areas? What scenarios are there for the marginal areas in Europe? These are the *'what will be where'* questions, which are the hardest to answer of all the types of questions. Besides knowing the initial state of land use (the 'what is where' question), an understanding of the processes of change and the ways different factors interact is needed. This is extremely difficult, especially when biophysical and social-economical processes are considered together. In practice we see that it impossible to answer the question 'what will be where' with any degree of precision. Thus, instead of a crisp prediction, we see the development of scenarios that represent possible lines of development.

It is possible to identify the requirements in terms of data for the different land use types (agriculture, forestry and nature) associated with each type of question (Table 4.3).

General requirements concerning tools can also be identified. For the *'what is where'* question, storage of spatial data in a geographical information system (GIS) is necessary. The commercially available GIS systems support the storage and analysis requirements for this question fully. The main problems are more in the domain of data access and data accuracy. The European initiative (ESMI) to develop a national clearinghouse for geoinformation might solve the problem of data access. In most cases, however, data availability appeared to be more of a political than a technical problem. The application of remote sensing techniques is a possible tool to obtain a full coverage of land use in Europe relatively quickly.

Table 4.3: Data requirements for different questions and land use types

Land use type	Questions What is where? (status)	 What is changed where? (monitoring)	 What will be where? (scenarios)
Agriculture	Area	Change in area	Area Political agenda Transformation rules Process models
Forestry	Area	Change in area	Area Transformation rules
Nature	Area, quality	Change in area	Area Transformation rules

For the '*what is changed where*' questions, multiple observations of land use are required to observe change. Remote sensing techniques seem to be attractive but due to uncertainty and imprecision in classification of at least 5 per cent, they are unsuitable for recording small land cover changes per time period. Only major shifts in land use can be detected by remote sensing. The analysis techniques in GIS for analysing spatial time series are also limited. More methodology needs to be developed to survey, manage and analyze spatial data in time.

For the '*what will be where*' questions, a combination of data, representing the initial status, and some rules or models describing the change of land use over time, are required. These rules range from relatively simple expert tables describing land use change in discrete intervals over time to complex dynamic simulation models describing change for continuous time intervals. Complex dynamic simulation models are more frequently used when the processes under study can be described in physical or chemical laws, such as the transport of water or the acidification of the soil. However, when we are dealing with socio-economic processes, it is difficult to model them in a deterministic, quantitative way due to the complexity of the system and lack of fundamental knowledge of the underlying behavioural processes. We tend to model change in the form of expert rules or regression equations. Change of land use is a complex process due to the large number of factors influencing it. Also, the direct influence of the individual factors on land use is quite often unclear. Therefore, in practice, the transformation rules for land use have to be formulated in the form of expert rules or regression equations, describing change without a fundamental understanding of the underlying processes.

For exploring the '*what will be where*' questions, data and transformation rules are combined in what is often called a spatial decision support system (SDSS). These systems generally consist of a standard GIS with some additional software components to facilitate decision support. As shown by Bregt and Bulens (1998), standard GIS was, for a long time, not very well equipped to support analysis over time. Due to the component-based development of DSS in recent years, combining software components of different origin, we are better able to develop such systems. The key issue in the actual implementation and use of these systems at the European scale is in the formulation of the transformation rules for land use change.

4.6 Conclusions

Land use in Europe is constantly changing due to social, economic, bio-physical and political factors. Focussing on the agricultural, forestry and nature land use types, it is expected that agricultural land use will reduce, especially in the marginal regions. The magnitude of the reduction is difficult to predict, but we expect a reduction of at least 10 per cent in the next 25 years. The remaining agricultural area shows an increase in parcel size and production per hectare. The forest land use type in Europe will slightly increase in area and the forest itself will be of a more differentiated character. The natural land use type will increase in area in future. Due to environmental pressure and development of infrastructure, the natural areas will be more fragmented.

Sound information on the speed, location and character of change is important for policy-makers at all political levels. For monitoring these land use changes and evaluation of future scenarios for spatial development, GIS-based systems are useful tools. In order to guarantee relevant information, more attention should be given to the monitoring methodology. The chance for monitoring noise in the data instead of real world changes is manifest (see Zeeuw *et al.*, 1999, for example). For 'scanning the future' of land use in Europe, a spatial decision support system is a useful tool. Again the quality of the data and the land use transformation rules determine the difference between a nice toy and a useful instrument.

References

Baldock, D. Beaufoy, G., Brouwer F. and Godeschalk, F. (1996) *Farming at the Margins, Abandonment or Redeployment of Agricultural Land in Europe*, IEEP and LEI-DLO, London, The Hague.

Bregt, A.K. and Bulens, J.D. (1998) Integrating GIS and process models for land resource planning, in: Heineke, H.J. Eckelmann, W, Thomasson, A.J., Jones, R.J.A., Montanarella, L. and Buckley, B. (eds) *Land Information Systems: Developments for Planning the Sustainable Use of Land Resources*, European Soil Bureau Research Report No.4, pp. 293-304.

Bullock, P., Jones R.J.A. and Montanarella, L. (eds.) (1999) *Soil Resources of Europe*, EUR 18991 EN, EC, Luxembourg.

EC (1999) *Land Cover and Land Use Information Systems in European Union Policy Needs*, EC, Luxembourg.

ESDP (1997) *European Spatial Development Perspective*, The European Commission.

FAO (1986) *European Timber Trends and Prospects to the Year 2000 and Beyond, Volume I and II*, UN/FAO, New York.

FAO (1993) *Agriculture: Towards 2010*, 27th Session, 6-25 November 1993, Rome.

Latenstein, C. H. van (1999) *Land Use in Europe. A Methodology for Policy-oriented Future Studies*, Sdu, The Hague.

Kamminga, J.T., Jenkins, G.J., Slangen, L.H.G., Bischoff, N.T. and van Hoorn, A.S. (1993) Toekomst verkenningen ruraal grondgebruik, *NRLO rapport 93/20*, Den Haag (in Dutch).

Mücher, C.A., Steinnocher, K. Champeaux, J.L. Griguolo, S. Wester K. and Loudjani, P. (1999) Land cover characterization for environmental monitoring of pan-Europe, in Proc. 18th EARSEL Symp. on Operational Remote Sensing for Sustainable Development, ITC, Enschede, 11-13 May 1998, pp. 107-113.

Rabbinge, R. and C.A. van Diepen (2000) Changes in agriculture and land use in Europe, *European Journal of Agronomy*, 13: 85-99.

Rijksplanologische Dienst (1997) *Landelijke gebieden in Europa*, Ministerie VROM/RPD, Den Haag, (in Dutch).

WRR (Netherlands Scientific Council for Government Policy) (1992) Ground for choices. Four perspectives for the rural areas in the European Community, *Report nr 42, SDU uitgeverij*, The Hague.

Zeeuw, C.J.de, Bregt, A.K., Sonneveld, M.P.W. and van den Brink, J.A.T. (1999) Geo-information for monitoring land use from map overlay to object-structured noise reduction, in Cammelli, A., Tjoa, A.M. and Wagner, R.R. (eds.) Tenth international workshop on database and expert systems applications, September 1-3, 1999, Florence, Italy. Los Alamitos CA (USA), IEEE Computer Soc., 1999, pp. 419-425.

Arnold Bregt
Wageningen UR,
Centrum voor Geo-informatie
Postbus 47
6700 AA Wageningen
The Netherlands

Kees de Zeeuw
Wageningen UR
Centrum voor Geo-informatie
Postbus 47
6700 AA Wageningen
The Netherlands

5 THE PROJECTION OF POPULATION, HOUSEHOLDS, HOUSING AND RESIDENTIAL LAND USE

John Stillwell and James Debenham

5.1 Introduction

Understanding the changing demographic composition and spatial distribution of the population throughout Europe presents a major challenge to the European Commission and, in particular, to those concerned with formulating policies that strive to meet European Union (EU) objectives of reducing inequality across member states and regions. In general, these policies aim to redistribute revenues towards the less developed states and towards the poorer regions, many of which are peripheral in terms of geographical location with respect to the core of the EU. Demographic statistics play an important role in the procedures and equations through which funding is allocated and population volume counts are used as measures for standardizing familiar indicators of development such as gross national product and unemployment.

In the context of planning, the projection of future populations is critically important for two main reasons. Firstly, there is a social imperative that we all endeavour to plan for pension schemes and health care provision in the face of an increasingly ageing population and likely geographical variations in future demographic structures. Secondly, population projections based on key assumptions of how fertility, mortality and migration will change over time, provide the main inputs for the projection of other variables such as households and labour supply. Thus, on the demand side, knowledge of population projections by age and sex is required to inform the production of household projections i.e. projections of the future numbers of different types of household (such as married couples and single parents). However, household projections are also influenced by trends in social behaviour. Bonvalet and Lelievre (1997), for example, have documented the transformation of household structures in Britain and France due to an increasing number of non-married cohabiting couples, increased divorce and separation rates, later marriages, more single parents and the propensity for more elderly people to remain in their own homes, often on their own. Thus, the future demand for dwelling accommodation of different types will be determined partly by the number of people requiring homes and partly by the demographic structure of that population, but also by the social trends that influence the way that people will live as individuals or together as groups, often as couples or families. Spatial variation in projected households of different size and complexion will imply spatial variation in the demand for different types of housing.

On the supply side, it is the local planning authorities, housing associations and private developers whose role it is to ensure an appropriate provision of dwellings to meet the future changes in demand. Physical or land use planners have a responsibility to ensure that housing of decent quality can be made available to all households who require dwelling accommodation in the future. This means that they have to support the contribution made by the existing housing stock as well as to specify the distribution of new housing and, consequently, the land required for additional housing. The geographical location of

J. Stillwell and H. Scholten (eds.), Land Use Simulation for Europe, 45–64.

new housing development is a difficult and frequently a politically contentious issue in most countries as different groups debate the necessity to protect the green belt. Currently, in the UK, the Government is seeking to find houses for an additional 3.8 million projected households over the next 20 years (DETR, 1999a) and debates about the regional distribution of new households and the location of new developments on brownfield or greenfield sites are ongoing, as local authorities plan the release of land to housing development. In some countries national policy plays an important role. The British Government's Urban Task Force, for example, aims to raise the proportion of new homes built on previously developed land or that make use of existing buildings to 60 per cent over the next ten years (Urban Task Force, 1999).

Population change is therefore one of the key drivers of land use change. Reliable population projections are a pre-requisite for land use simulation modelling in areas where residential development is a dynamic component of the urban system. This chapter aims to present an overview of population projections with particular emphasis on the European region scale in Section 5.2 and the changes that are projected to occur over the next 25 years. Corresponding projections of households, housing and land for residential use are not generated for European regions, but left for national and sub-national agencies to produce. The chapter draws on the example of the UK to illustrate the hierarchy of projections that are produced from national down to small area scales in Section 5.3. Section 5.4 shows the type of 'official' national and sub-national projections of households in the UK are produced by Central Government, whilst Section 5.5 indicates the way in which sub-national projections of new housing and of land required for residential use are typically generated by the regional planning process in Britain. Some conclusions are presented in Section 5.6. We should emphasise that this chapter does not attempt to review the detailed modelling approaches that underlie the projections of populations and households constructed by different agencies.

5.2 Population projections for EU regions

Demographic developments across the EU have changed considerably in the past decades as declining fertility levels and expected life expectancies have altered regional age structures and as the EU as a whole has become the destination for international immigrants, including large numbers of asylum seekers and visitor switchers. Overall, population growth of the 15 countries that comprise the EU has been slow, rising from nearly 349 million in 1975 to around 372 million in 1995 (van der Gaag *et al.*, 1999). The total fertility rate dropped from 1.96 children per woman in 1975 to 1.45 in 1995, due largely to the sharp fall in births among young women, and having declined even more dramatically in the early 1970s (Masser *et al.*, 1992). Between 1980 and 1995, life expectancy at birth increased by over 3 years; based on 1995 mortality rates, men are expected, on average, to live 73.9 years, and women, 80.4 years. International migration levels tend to vary significantly from year to year but the annual average for 1990-94 was 1,056 thousand compared with 55 thousand in 1980-84 (van de Gaag *et al.*,1999). Immigration into western Europe has served to offset the population decline resulting from lower fertility, and this effect has been accentuated in several EU states in recent years by the arrival of asylum seekers in large numbers. In Germany, immigration increased particularly sharply, accounting for half of total net migration to the EU 15 in the early 1990s. By the end of the 1990s, the UK was receiving more asylum seekers than anywhere else.

There are, of course, some interesting contrasts between the demographic experiences of different regions as well as countries. Ireland has maintained the highest total fertility rates across Europe whilst fertility rates have fallen most dramatically in Spain. Male life expectancy in 1995 is highest in Greece and Sweden and lowest in Portugal; female life expectancy is lowest in Denmark and highest in France. Internal migration becomes important at the regional scale and experience varies significantly between countries. There is some evidence that migration responds to major disparities in economic development among regions consequent on structural differences, resulting in major streams from southern to northern Italy, from northern to southern France, from eastern to western Germany, for example (Rees *et al.*, 1996). In some countries, such as the UK, internal migration over the past three decades has been dominated by counterurbanisation tendencies – major losses of migrants from all metropolitan areas to neighbouring or more distant non-metropolitan areas – but such processes were only beginning to occur in other countries, such as Spain in the early 1990s, where return migration has been identified as of increasing importance in inter-provincial flows (Stillwell and García Coll, 2000).

The responsibility for commissioning population projections for EU 15 regions falls on the European Commission's Directorate General for Regional Policy and Cohesion (DG XVI) and Eurostat. So far, there have been four rounds of regional projections. The first two were commissioned by DG XVI and produced by the Netherlands Economic Institute (NEI). These projections, based on 1980 and 1985 respectively and with horizons of 2010 (DEMETER2010) and 2015 (DEMETER2015) used an age-time period interval of five years and were produced for the NUTS 2 regions in 12 EU countries (NEI, 1986; Haverkate and van Haselen, 1990). DEMETER refers to DEMographic Evolution through Time in European Regions. The third set, DEMETER2020, produced by NEI and the Netherlands Interdisciplinary Demographic Institute (NIDI), were 1990-based and provided more detailed projections by single year of age for 170 NUTS 2 regions (NEI, 1994). Finally, the 1995-based projections commissioned by DG XVI and Eurostat have been produced by NIDI and Statistics Netherlands (SN) and cover the 18 countries (202 regions) of the European Economic Area (EEA) over the period to 2050. Some countries, such as Luxembourg and Ireland, have no equivalent regional projections. These projections are referred to as EUROPOP1995 and are reported in Eurostat (1996).

In order to develop projections, it is necessary to make certain assumptions about the way in which the components of population change will alter in the future. These assumptions are referred to as scenarios. The most recent 1995-based projections make use of five scenarios: baseline, low, high, young and old. The baseline scenario assumes that current trends will continue, whereas the low scenario describes a demographic future in which current low fertility levels persist, life expectancies increase marginally and total net international migration drops by 50 per cent (van der Gaag *et al.*, 1999). The high scenario assumes a recovery of fertility levels, increasing life expectancy and an increasing net inflow of international migrants. The young and old scenarios refer to assumptions about ageing. The young scenario combines high fertility and high net migration with low life expectancies whilst the old scenario combines low fertility and low net migration with high life expectancies. Each of these scenarios relates to national population characteristics. In addition, three regional scenarios have been developed at NUTS-2 level for the 15 EU countries – baseline, high and low – that correspond with the national scenarios (van der Gaag *et al.*, 2000). In the low scenario, regional population growth is projected to be slow and regional imbalances large, whereas in the high scenario, there is rapid population growth and imbalances are small.

The modelling methodologies used in the respective sets of projections and the detailed assumptions that underpin the various scenarios have been evaluated by Rees *et al.* (1999).

The evaluation suggests that whilst improvements in the successive projection procedures have been significant, the central mortality assumptions have consistently under-estimated the growth of the elderly population. Suggestions are made by Rees *et al.* for the fifth round of projections that will be based on populations in 2000.

In order to illustrate the order of magnitude of population change across Europe and to emphasise the importance of the demographic driver, we have mapped the changes in population projected to occur between 1995 and 2025 in the NUTS 2 regions across EU 15. These figures are set out as graphs in van de Gaag *et al.* (1999). Figure 5.1 indicates the percentage change in total population using the baseline scenarios and shows that a third (67) of the 204 NUTS-2 regions will increase their populations by over 10 per cent, whereas only 14 regions will have growth over 20 per cent. Fifty-two regions, including several in northern Spain and northern Italy, as well as in the former East Germany, will experience population decline.

Figure 5.1: Projected change in total population, NUTS-2 regions, 1995-2025

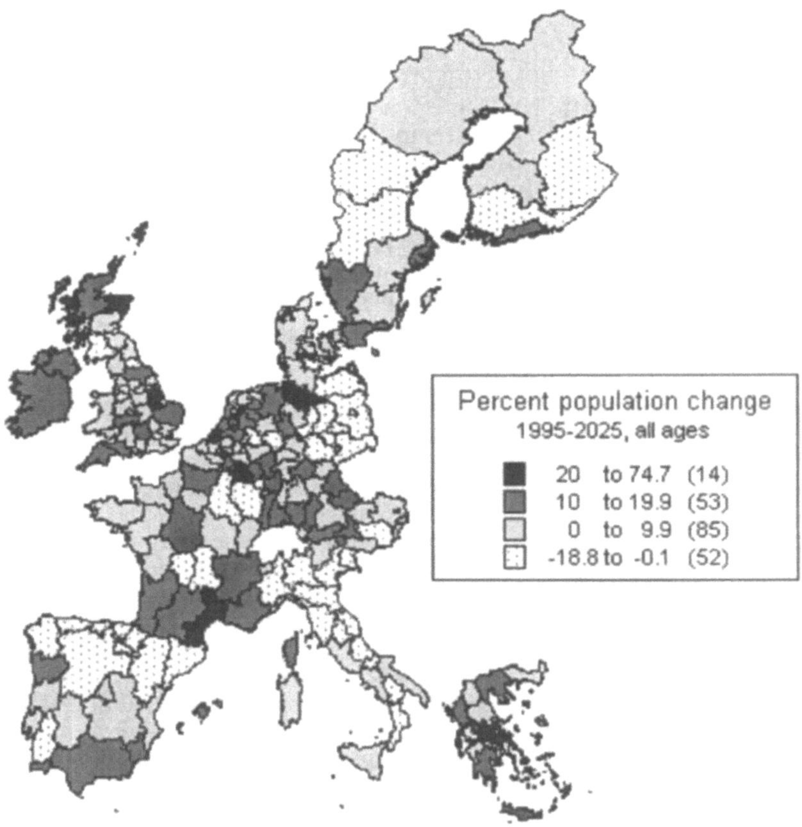

Source: Eurostat (1996)

The spatial pattern of projected aggregate population growth and decline is complex and difficult to interpret because it reflects the net changes taking place at different ages that are themselves the function of a number of influences. Not only will the population changes reflect changes brought about at different ages by cohort replacement that transmit fluctuations in earlier levels of fertility, but they will also reflect trends in inter-regional and international migration and improvements in mortality. In order to demonstrate the change variation between by age, Figures 5.2-5.4 indicate proportional change in three broad age groups, 0-19, 20-59 and 60+, under the baseline scenario assumptions. The levels and patterns of change are radically different in each case. The impact of declining fertility in the last 20 years is observed in the first age group for which only 38 regions are projected to have population growth over the 30-year period. Large areas of northern Spain and eastern Germany, as well as Eire and several regions of France and Italy, have rates of population decline in excess of 20 per cent.

Figure 5.2: Projected change in population aged 0-19, NUTS-2 regions, 1995-2025

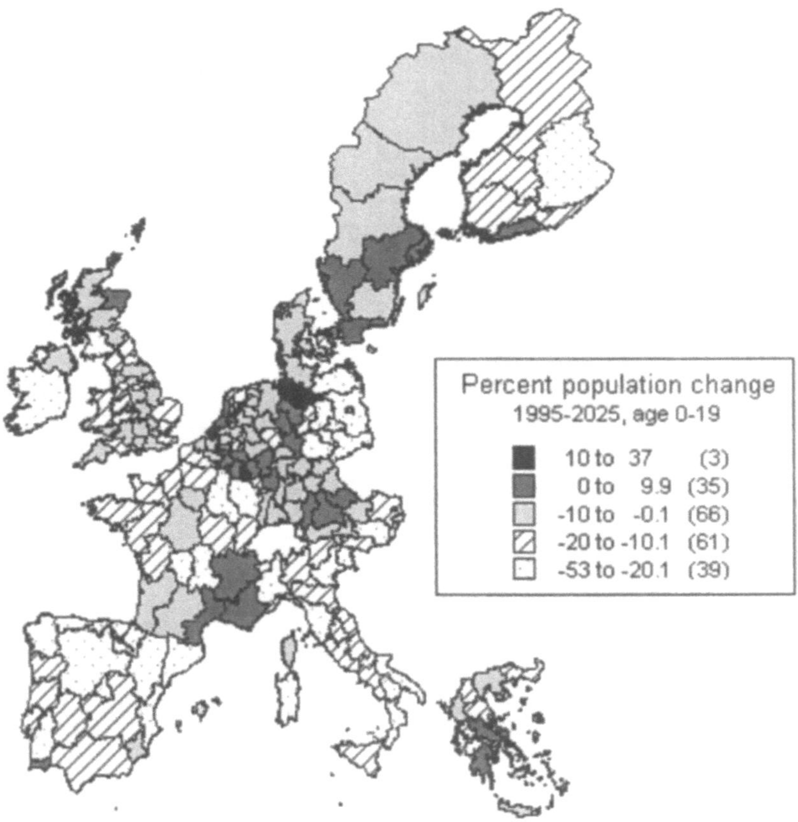

Source: Eurostat (1996)

The pattern of change for the middle ages has a more mixed geographical distribution with population declines throughout much, though by no means all, of the EU core contrasting with population increases in the peripheral regions of Greece, southern Spain, southern France and Eire. The proportion of the EU population in the 20-59 age group is projected to drop from 55 per cent in 1995 to 51 per cent 2025 (van de Gaag *et al,* 1999). Moreover, there will be an ageing of the working population such that the percentage aged 40-59 will peak between 2010 and 2020 as the baby boomers of the 1960s reach retirement age. Whilst the cohort effect will be important in many regions, the projected changes in this age group are also influenced to a certain extent by migration trends since it is in the young working ages where migration propensities are highest, determined frequently by economic developments.

Figure 5.3: Projected change in population aged 20-59, NUTS-2 regions, 1995-2025

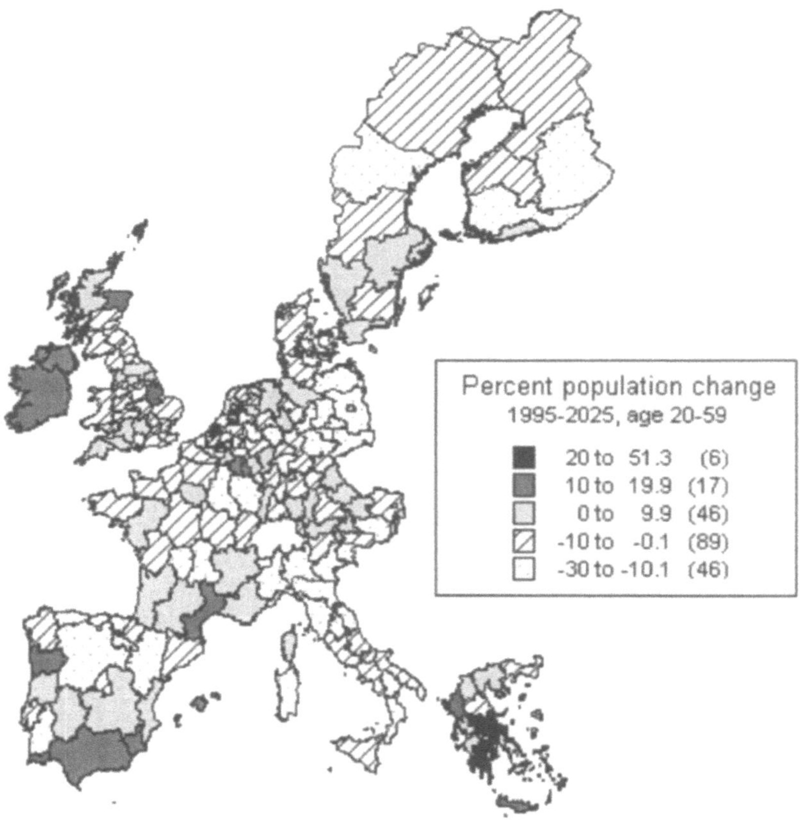

Source: Eurostat (1996)

In contrast to the decreasing number of young people, the projected population change of those aged 60 and over in the EU is dramatic with a projected increase of almost 50 per cent by 2025. The largest relative increases in numbers of elderly are projected to take place in regions that currently have relatively low percentages of old people. These regions are found in parts of Finland, Germany, the Netherlands, Austria, Belgium and France. Almost half the NUTS-2 regions are expected to have more than a 50 per cent increase in this age group, with only two regions projected to have a decline. Moreover, the percentage of the very old (aged 80 and over) is expected to increase during this period although it will be after 2025 that the populations of the large 1960s birth cohorts reach older old age.

Figure 5.4: Projected change in population aged 60+, NUTS-2 regions, 1995-2025

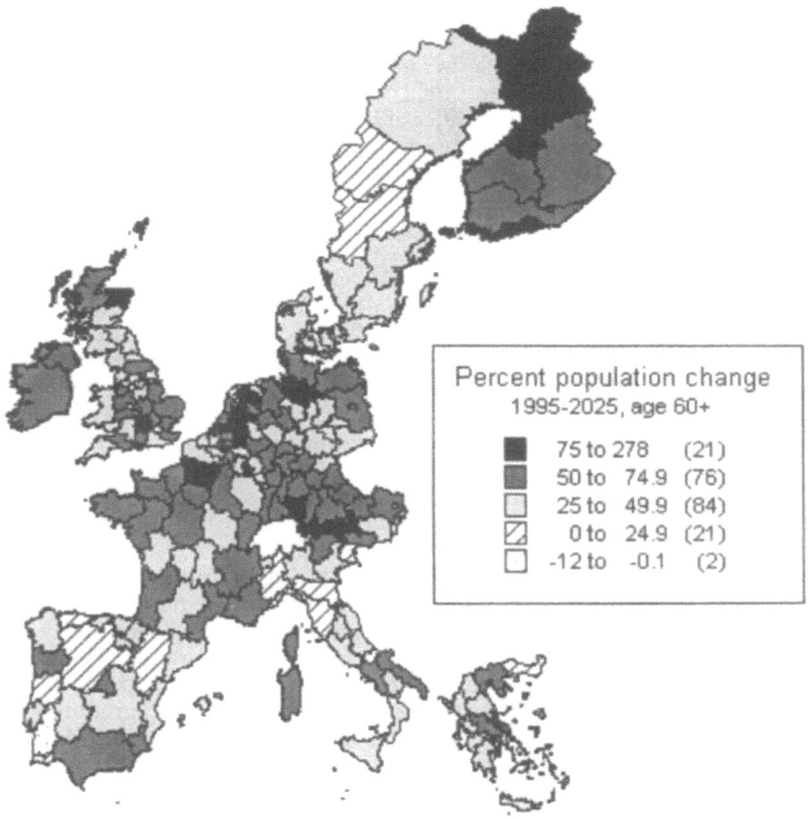

Source: Eurostat (1996)

So what implications do these demographic projections have for household change, housing provision and planning for residential land use? The answers are not clear. In many regions of the EU, physical planners will need to consider the implications of a declining and ageing populations and to reconcile their regional projections with trends towards living in different types of households and requiring different types of housing as household size continues to reduce. In major urban areas, declining population numbers will help to reduce the housing pressures and allow planners to maintain policies of using residential development as a means to regenerate inner city areas rather than continue to bring forward new peri-urban sites. However, if the desire to counterurbanise continues and commuting ties to the urban workplace become even more relaxed, planners will have to confront the issues of increasing demand for housing provision in rural areas. This is likely to be enhanced with higher numbers of older people reaching retirement age and seeking to live away from the big cities.

5.3 National and sub-national population projections in the UK

The baseline scenarios of the EUROPOP1995 projections for each country tend to coincide with the population projections made by the relevant national statistics institutes. In the UK, a cohort survival model is used by the Government's Actuary Department (GAD) to create the official national population projections which are published by the Office of National Statistics (ONS), formerly the Office of Population Censuses and Surveys (OPCS). The most recent are the 1998-based projections (ONS, 2000; Shaw, 2000) that follow earlier projections based on mid-year estimates for 1996 (ONS, 1999a), 1993 (OPCS, 1995), 1991 (OPCS, 1993) and 1989 (OPCS, 1991). The UK national population projections are comprised of individual projections for each of the constituent countries (England, Wales, Scotland and Northern Ireland). They are based on the Registrar General's estimates of the mid-1998 resident population of the UK that, in turn, are derived from the mid-1998 population estimates for England and Wales produced by ONS (1999b), the General Register Office (GROS, 1999) and the Northern Ireland Statistics and Research Agency (NISRA, 1999). Each of these offices has responsibility for national projections for England and Wales, Scotland and Northern Ireland separately and there is no single set of projections for nations or regions of the UK based on the same methodology. The UK therefore has a rather unusual arrangement by which country estimates for England, Wales, Scotland and Northern Ireland are aggregated in 'bottom up' fashion to provide UK estimates, whilst the same national estimates are used as constraints to control sub-national estimates in each country in a 'top down' manner.

The national estimates for mid-1998 (Table 5.1) are based on counts from the last census in 1991, updated for ageing, births, deaths and migration in the intervening period. The national projections in Table 5.1 are made for successive years from one mid-year to the next and based on assumptions about fertility, mortality and migration. The 1998-based projections assume that the total fertility rate will decline from 2.4 to 1.8 children per woman and that mortality will fall by 0.5 per cent. International migration is trended forward using historical data from the International Passenger Survey (IPS), data on flows to and from Eire and inflows of asylum seekers and visitor switchers. The changes projected to 2021 that result in a 7.4 per cent increase in the UK as a whole are not consistent across the four nations. England gains by 8.5 per cent due primarily to immigration. Northern Ireland's population grows by 7.8 per cent because fertility is assumed to remain higher than elsewhere. The population of Wales continues to grow but at a lower level than that of England, whilst Scotland's population declines by 1.2 per cent as in-migration is insufficient to counterbalance natural decline.

Separate sets of population projections for sub-national or local authority areas in each of the countries that constitute the UK are produced by the relevant statistical offices using standard cohort component models (Capron and Corner, 1990). The first sub-national projections were made in 1965 for the regions of England. In 1977, projections were produced for the main planning units, including London boroughs, metropolitan districts and shire counties (Wood *et al.*, 1999). ONS now publishes projections every three to five years that look ahead 25 years into the future and the most recent round of 1996-based projections for England (ONS, 1999c) include projections for the unitary authorities (UAs) and local authority districts (LADs) and health authorities (HAs) introduced in 1996. In addition, short-term projections are published every two years for the Department of Health in order to give advance indications of change.

Table 5.1: National mid-1998 population estimates and projections to 2021

	Population (in thousands)					
	1998	2001	2006	2011	2016	2021
England	49,495	50,187	51,052	51,922	52,831	53,715
Wales	2,933	2,950	2,969	2,993	3,021	3,047
Scotland	5,120	5,109	5,098	5,087	5,078	5,058
N. Ireland	1,689	1,708	1,742	1,771	1,799	1,821
UK	59,237	59,954	60,860	61,773	62,729	63,642

Sources: ONS (1999b; 2000), GROS (1999), NISRA (1999)

The last round of sub-national projections involved an overhaul of the projection system to accommodate advances in computing and forecasting technologies. The new English Sub-national Population (ESP) projection system retains some features of the previous model but now produces projections for a wider range of areas and meets more customer needs (Wood *et al.*, 1999). The modelling system allows for the revision of the basic migration assumptions and incorporates some scenario building to account for change in migration processes. Projections are developed for areas known as 'building bricks' that are subsequently aggregated and reported for local government and health authority areas. A 1996-base population for each 'building brick', disaggregated by sex and age group, is aged one year ahead. Age-specific fertility rates are applied to derive birth numbers and age-specific mortality rates are used to estimate deaths. International and internal migration are then accounted for and the components are constrained to ensure that the populations projected are in line with the projections produced independently by the national model. The internal migration sub-model is seen as being particularly important and makes use of annual data from the National Health Service Register (NHSCR) on patient re-registrations between health areas or Family Health Service Authorities (FHSAs) as they were previously called (Stillwell *et al.*, 1992). The assumptions that underpin the projections are based on recent local demographic trends in each component. This trend-based approach is advantageous in that future demographic behaviour is based on past experience, but at the same time, this highlights the major disadvantage that the projections are not responsive to national, regional and local policy variables. In the context of internal migration, this is one reason why the Department of Environment, Transport and the Regions (DETR) commissioned a study of the determinants of migration within the UK (Champion *et al.*, 1998) and the construction of a policy-sensitive migration modelling system (Champion *et al.*, 2000).

Figure 5.5 illustrates the percentage changes in the aggregate populations of local authorities in England that are projected to occur over the period from 1996 to 2021 (ONS, 1999c). Highest percentage population growth is projected to occur in the rural districts of the south of the country and midlands, particularly in East Anglia, the East Midlands and along the south coast. Population declines are expected in most of the provincial metropolitan districts as well as in some coalfield and industrial areas in the north midlands and in the remainder of the north.

Figure 5.5: Projected population changes by local authority in England, 1996-2021

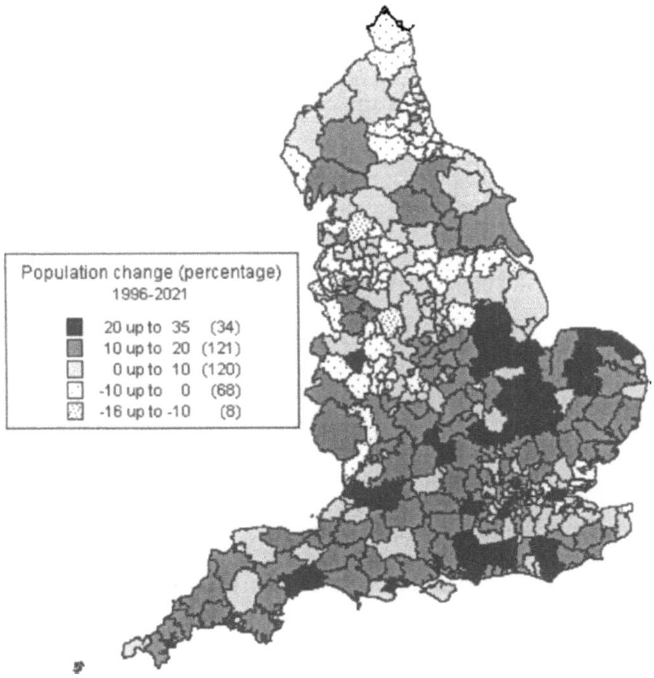

Source: ONS (1999c, Table 8)

Although the sub-national population projections for the countries of the UK are well established and subject to regular review and modification, there are two limitations that are of particular importance. The first of these is that separate projections are made for local areas in each country based on different modelling methods. The internal migration models for Scotland and Northern Ireland are net migration models, for example, whereas ESP is a multi-regional migration model. Work is currently underway on the development of a new sub-national population projection model for the whole of the UK (UKPOP) that will overcome the problem of differences in the style of modelling and produce consistent projections for all the local areas involved (Wilson, forthcoming).

The second major shortcoming is that the current sub-national projections do not extend to small areas (such as wards) and are therefore of only limited use to local authority planners who require more detailed projections of localities within the sub-national units for which projections are available. Consequently, the production of small area projections has been the responsibility of local authorities, along with their own projections for the authority as a whole to compare with ONS projections. In some cases, projections have been produced 'in house' by the authorities themselves; in other cases they have been delivered by commercial or academic agencies. Relatively little is published in the literature on the methods that are used by local authorities for these projections (Joshi and Diamond, 1990). A full account of the 'Chelmer Model', produced by the Population and Housing Research Group at Anglia Polytechnic University and used widely in local government, is not available. There is evidence that a range of different methodologies exist and thus the decision about which of these should be used depends very much on the requirements of the end product. The success of a population projection is closely linked to the success of the plan it supports, whether this plan is a forecast of future school rolls, of housing need or of land for residential use (Simpson, 2000).

Before a small area projection model is run, it is necessary to estimate the base year population because no 'official' central government mid-year estimates are made, as yet, for small areas; ONS are currently considering the issues involved and producing mid-year estimates for small areas after the 2001 Census. The accuracy of small area projections is initially dependent on the accuracy of the base year estimates (Tayman, 1996). It can be argued that projections produced towards the end of an inter-censal period are at a disadvantage because mid-year population estimates or 'nowcasts' for small areas become less reliable as the period of time since the last census lengthens (Wilson and Stillwell, 1998). Different ways of deriving small area population estimates have been appraised by Ghosh and Rao (1994) and methods used in local government such as apportionment, ratio change and cohort survival estimation have been reviewed by Simpson *et al.* (1997).

Some small area projection models have estimation techniques built into them so that they can be run from any base year. One example is the information and modelling system developed for the districts in the former metropolitan county of West Yorkshire (Rees, 1994a). Built along similar lines to its predecessor for the City of Swansea (Rees, 1994b), the West Yorkshire model projects populations by sex for 91 ages for single year iterations for wards across the metropolitan districts of Leeds, Bradford, Calderdale, Kirklees and Wakefield. One innovative feature of this applied model is the way in which internal migration is dealt with using spatial interaction models that are responsive to housing developments; 'turnover' migration between existing housing units is distinguished and modelled separately from migration into new housing and migration away from housing demolitions.

This short review emphasizes that population projection is undertaken either at national level by statistical offices or at local level by local authorities. The regional tier of administration has been almost non-existent until recently and consequently, regional projections are simply constructed as the aggregation of projections produced either centrally or by the local authorities for local government areas. Most regions do not yet have their own regional population projections systems although this might be a very appropriate step forward for the new regional institutions (Regional Assemblies and Regional Development Agencies) to take to support preparation of Regional Planning Guidance. Another important characteristic is that population projection in the UK at the national level involves a demographic model that is not integrated with projections of other variables such as households, housing or employment. In the next section, we illustrate how population projections are used as inputs to determine future household projections. At a local level, there appears to be more integration: in addition to conventional migration-led forecasting, the Chelmer model offers a housing-led model in which the forecasting of population is evaluated in terms of the capacity of an area to accommodate households; the West Yorkshire model provides projections of households as well as individuals (Stillwell and Rees, forthcoming); and Simpson (2000) has outlined a software system (POPGROUP) used by local authorities for projecting populations, that also contains modules for generating household projections and forecasts of the labour force.

5.4 National and sub-national household projections

Projections of households are of particular importance to the DETR and to local authorities because of their involvement with housing provision (Corner, 1989). Even a static population tends to generate growing numbers of households because household formation represents the outcome of complex social behavioural processes. In recent years, the characteristics of households have changed due to increasing cohabitation of unmarried partners, increasing divorce and separation rates, later marriages, increasing single parenthood and elderly people living alone. The net result of these changes has been a significant rise in household numbers, a trend that is expected to continue into the foreseeable future. However, not all the changes identified necessarily lead to more and smaller households. Cohabitation is a substitute for marriage and divorce and may or may not create new households depending on whether or not those involved join existing households. The factors responsible for increasing household numbers are the events that encourage single person household formation. These include the postponement of co-habitation or marriage among younger adults who have an increasing propensity to live alone (Hall *et al.*, 1997), the instability of marital relationships or informal unions and the death of a spouse in older age.

There is a wide range of methods for producing household projections that extends from aggregate models (Corner, 1989) to microsimulation approaches (Hooijmeijer and Oskamp, 1999). In the UK, the use of 'headship rates' is probably the most frequently adopted method. A headship rate is the proportion of members of a population (defined by age, gender and marital status) who act as heads of specific types of households such as married couple, lone parent and single person households. The proportions for all household types for any population group should sum to one. However, a useful summary is the overall headship rate which is calculated by adding together the various household head proportions. The headship rate is applied to population projections disaggregated by age, sex and marital status.

Household projections do not appear to be made across European regions. In England and Wales, "*household projections are compiled by applying projected household membership rates to a projection of the private household populations disaggregated by age, sex and marital/cohabitation status and summing the resulting projections of household representatives*" (Department of the Environment, 1995, p. 63). The headship rate method has been used since the first projections were produced in 1931. The methodology has been constantly improved over time, although it was not until the 1989-based projections that any explicit account of unmarried cohabitations was made. Projections are now produced on a regular basis (every two years) by DETR and normally look forward 25 years. The most recent projections for England (Table 5.2) were based on the 1996-based national and sub-national population projections discussed in the previous section and project household numbers in five household types to 2021. A household is the term used to represent one person living alone or a group of people who share common housekeeping or living space, a definition designed to exclude those living in communal establishments (King *et al.*, 2000).

Table 5.2: 1996-based estimates and projections to 2021 of households for England and Wales

	Households (in thousands)					
Household type	1996	2001	2006	2011	1016	2021
Married couple	10,190	9,829	9,535	9,357	9,251	9,157
Cohabiting couple	1,479	1,896	2,251	2,509	2,660	2,761
One parent	1,168	1,258	1,293	1,299	1,296	1,288
Other multi-person	1,543	1,676	1,836	2,012	2,172	2,286
One person	5,806	6,334	6,819	7,342	7,934	8,509
All households	20,186	20,992	21,773	22,519	23,313	24,000

Source: DETR (1999a)

The official 1996-based projections show some significant changes over this period in which the number of households increases by 3.8 million or nearly 19 per cent by 2021. This is likely to be an underestimate since the 1998-based national population projections indicate an additional 1.2 million people by 2021 over and above those suggested by the 1996-based figures. Whereas the number of married couple households declines by over a million or 10 per cent, the number of one person households increases by 2.7 million or over 46 per cent during the projection period. These projections have been produced by taking the 1996-based resident population projections at national and sub-national level, disaggregated by age and sex, and passing them through the household projection methodology outlined by King *et al.* (2000). Breakdown of the projections into marital status categories and subtraction of projected institutional populations leaves projections of private households by age, sex and marital status. Rates are then computed for 11 types of household using data from previous censuses and from the Labour Force Survey, and these rates are used to give 'uncontrolled' projections of household representatives, where the projections in a given area are not expected to equal the sum of the constituent sub-areas. Thus, there is a final process of 'regional controlling' applied in a top-down manner so that the household projections for all the regions sum to the total for the country as a whole (Capron and Corner, 1990). The Department of Environment published a set of 1992-based sub-national household projections in 1996, the last to be produced by the Building Research Establishment (BRE). Subsequently, the Housing Projection Service (HPS) assumed responsibility and a set of 1996-based projections to 2021 was produced by Anglia Polytechnic University that are available at county level from the DETR web site (DETR, 1999b). Figure 5.6 indicates the spatial variation in the percentage change in household numbers between 1996 and 2021. The highest growth areas, with over 30 per cent increase in households, are located in three groups of counties running diagonally across the southern half of the country, centred on Cambridgeshire, Buckinghamshire/Berkshire and Dorset. The lowest household growth is projected to occur in the provincial metropolitan conurbations. The 1996-based projections have evoked memories of Ebenezer Howard's famous 'three magnets' diagram of the advantages and disadvantages of urban and country living (Howard, 1898) and forced Government and planners to reconsider once again where new housing should be located (Secretary of State for the Environment, 1996). The district projections in Figure 5.6 suggest that new housing should be provided in new and expanded settlements along viable transport corridors in the places where people want to live rather than in existing concentrations of high density. The next set of sub-national household projections is scheduled to be published in 2002, incorporating the results of the 2001 Census.

Figure 5.6: Proportional change in the household projections, 1996-2021

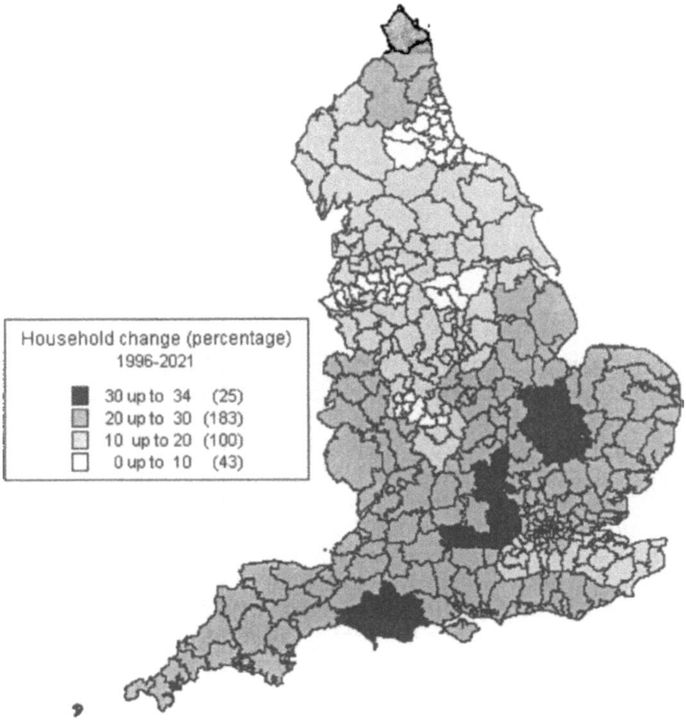

Source: DETR (1999b)

Alternative methods and projections of households in the UK are not commonplace. Some local authorities produce their own projections and others use the Chelmer model, which was originally conceived as an academically independent local demographic/housing forecasting system offering a 'what if' user service to potential users in both the public and private sectors. The model is now available for use on PC compatible computers and has been purchased by over 35 organisations, including local authorities and the Housebuilders Federation. Since the beginning of 1995, it has been able to produce projections at district level for England Wales and Scotland that allow the testing of a variety of assumptions, particularly relating to migration and household formation.

Relatively little progress has been made in the development of systems to build projections of households for small areas (within local authorities). The West Yorkshire model and the POPGROUP software described earlier are exceptional in that they have facilities to do this. Moreover, the development of microsimulation methods for updating small area populations and households (Duley and Rees, 1990) holds considerable promise for future research. It is evident that, whilst the behavioural processes behind household change are complex and that conventional sources of data are somewhat limited, the demand for household projections (for small areas) continues to develop because of the profound implications of current trends. In the next section we consider how household projections are converted to projections of dwellings and residential land use.

5.5 Regional projections of housing and residential land requirements

In the UK, the strategic planning of housing provision is undertaken through a process known as the preparation of Regional Planning Guidance (RPG), which, as its name implies, is a process that involves regional agencies working in conjunction with local local authorities and other organisations. In this section, we use the example of one English region, Yorkshire and Humberside, to illustrate some of the issues involved.

The Labour Government has promoted a comprehensive approach to regional planning in England during the last few years. As well as establishing Regional Development Agencies (RDAs), charged with the responsibility of fostering regional economic development, it has also enhanced the role of RPG (DETR, 1999c) requiring a process of planning that involves greater linkage between economic, social and environmental development, greater concern with sustainability and more attention to monitoring and managing change. RPG is therefore a spatial strategy in each region that provides the context for development planning by local authorities, the economic development activities of the RDA and the plans of infrastructure and service providers, industry and commerce. In Yorkshire and Humberside, the 'draft' RPG entitled *Advancing Together Towards a Spatial Strategy* has been produced by the Regional Assembly of Yorkshire and Humberside (RAYH, 1999) in parallel with the regional economic strategy produced by the RDA (Yorkshire Forward). The future provision of housing and the distribution of the population form a key component of the draft RPG for Yorkshire and Humberside.

The sub-national projections referred to earlier anticipate an additional 330,000 households in the region over the period from 1991 to 2021 and the key question is where the new dwellings to accommodate these new households should be located. This decision-making process is complicated by the fact that regional planners have to take into account recent trends in net migration, have to be keenly aware of issues such accessibility to jobs, shops, transport and services, and must recognize that a large proportion of new households cannot afford to buy or rent housing that meets their need on the open market. Furthermore, the strategic planners are also constrained by government policy. In the light of the uncertainty surrounding the household projections, the Government has moved away from a 'predict and provide' approach towards a 'plan, monitor and manage' approach that requires more sophisticated planning for housing land than in the past. Moreover, the Government has also adopted a national target that at least 60 per cent of land for housing should be 'brownfield' or previously developed land. The definition of brownfield land has been clarified by Alker *et al.* (2000). This criterion is particularly important in Yorkshire and Humberside since the region has one of the highest proportions of brownfield land in the country.

The draft RPG for Yorkshire and Humberside considers the provision of land for household growth until 2016 and emphasises that the distribution of an additional 251,700 dwellings across the region between 1998 and 2016 (RAYH, 1999, Appendix 1) has to reflect an increase in brownfield development as well as a focus on urban renaissance (Urban Task Force, 1999). The proposed targets for new housing in the strategic planning areas of the region are identified in Table 5.3, together with the percentage of housing development that is proposed to take place in brownfield land.

Table 5.3: Projected housing in areas of Yorkshire and the Humber, 1998-2016

Planning area	Housing total	Housing %	Brownfield %
Barnsley	14,600	6.2	49
Doncaster	13,200	5.5	70
Rotherham	14,400	6.1	68
Sheffield	13,900	5.9	80
South Yorkshire sub-regional total	56,100	23.8	67
Bradford	25,000	10.6	57
Calderdale	8,100	3.4	74
Kirklees	20,900	8.9	65
Leeds	25,700	10.9	66
Wakefield	17,100	7.2	61
West Yorkshire sub-regional total	96,800	41.0	63
York/North Yorkshire sub-regional total	42,700	18.1	53
Hull/East Riding	40,300	17.1	50
North/North East Lincs	15,800	6.3	56
Humberside sub-regional total	56,100	22.3	52

Source: RAYH (1999)

The distribution of the total number of houses (235,900) within the region between the strategic planning areas has been debated extensively by the local authorities within the context of the Regional Assembly and has been informed by an independent study of settlement potential and development options commissioned by the Regional Assembly (Baker Associates, 1998). The Baker Study examined a range of policy options and scenarios and the local authorities have subsequently pursued a 'middle of the road' option in RPG. The West Yorkshire sub-region is allocated 41.8 per cent of new houses and South Yorkshire gets 23.8 per cent with 22.3 per cent going to Humberside and the remainder to North Yorkshire and York. The figures in Table 5.3 for brownfield development assume that sites available as a result of clearance of existing housing will be used for new housing. It is clear that if the region's and the Government's aspirations for the use of brownfield land and for urban sustainability are to have any chance of being achieved, then much tighter controls will have to be used by the local authorities in their Development Plans.

RPG therefore proposes a sequential approach to the release of land for housing with the highest priority given to urban brownfield land and conversions of existing buildings to housing use. Thereafter, increasingly lower priority will be given to infill within towns, villages and urban areas, developments on the periphery of main settlements, developments on the periphery of minor settlements, and then greenfield and non-urban brownfield sites. The draft RPG suggests that targets for the provision of social housing in each part of the region should be developed in consultation with housing authorities and other social housing providers. It also indicates that the release of additional greenfield land for private sector housing in each area should take account of the need for different types and sizes of housing to meet the needs of those concerned. As might be expected, the leading volume house builders in Yorkshire and Humberside have joined together to form a consortium to put forward objections at the Examination-in-Public (EIP) to the housing projections contained in draft RPG on the grounds that they seriously underestimate the housing needs of the region. The report of the EIP held during the summer of 2000 is not yet available and the final projections will be published subsequently in the definitive RPG for Yorkshire and Humberside.

5.6 Conclusions

In this chapter, we have attempted to show that the use of land for residential development is determined by a number of demand and supply side factors. Although the overall size of population change in Europe has not been dramatic in the last two decades, the effects of migration on redistributing the existing population, the demographic effects of cohort size variation and of ageing and the effects of changing household behaviour have meant that the demand for housing of different types and therefore for residential land has been very dynamic. On the assumption that past and current trends will continue over the next two decades, planners have had to think very carefully about the supply of land and the spatial patterns of residential development that need to be created in order to meet housing needs. In the UK, the rate of release of new land for housing development has been influenced by Government emphasis of the use of existing land and buildings in built up areas and by the need to consider issues of sustainability and quality of life in the context of housing provision.

It is clear that the forces driving land use for residential development are complex and reliant on series of projections of the numbers of individual persons, the households to which those persons belong and the dwellings in which they will live. It is also clear that these projections are not made in a comprehensive manner for spatial units at any particular scale. We have shown how population projections are produced for the European Commission that are consistent across EU level 2 regions, but there are no parallel projections of households and required dwellings. Using the UK as a case study, we have shown how national and sub-national projections of population and households are derived for use by local authorities for development planning but that projections for smaller areas are left to the local authorities themselves. Finally, we have indicated how, given the projected population and household totals for each region, the projections of housing units emerge from a political process of negotiation and debate between local authorities at a regional level through the preparation of Regional Planning Guidance. The assumptions underlying the official projections are critically important because of the political sensitivity that accompanies the demand for households in different parts of the country to which planners must try to respond. Housing supply strategies are required that will endeavour to be sensitive to demand factors and align with Government policy. Strategic planners are faced with increasing pressures to answer important questions about whether policies to retain the green belt and maintain control over 'city edge' developments are sustainable or whether to encourage the emergence of new 'suburban super-crescents', new housing developments in rural areas for people escaping from the cities in the UK.

Planning is currently facing one of its biggest challenges: how to meet the rising need for more homes and, at the same time, protect the best of our environment and ensure that development is sustainable. For many local authorities, some very hard decisions lie ahead. Yorkshire and Humberside represents a microcosm of the tensions and contradictions between planning policy, market performance, housing investment and housing need that are likely to exist across many parts of western Europe. Inevitably it is impossible to avoid the conclusion that model-based decision support systems that enable planners to simulate the effects of different projected scenarios on land use will be of immense value.

Acknowledgement

The authors are very grateful to Phil Rees for his valuable comments on an earlier version of this chapter.

References

Alker, S., Joy, V., Roberts, P. and Smith, N. (2000) The definition of brownfield, *Journal of Environmental Planning and Management,* 43(1): 49-69.

Baker Associates *et al.* (1998) *Settlement Potential and Development Options Study Final Report,* Report prepared for the Yorkshire and Humberside Regional Assembly, July.

Bonvalet, C. and Lelievre, E. (1997) The transformation of housing structures in France and Great Britain, *International Journal of Population Geography,* 3: 183-201.

Capron, D. and Corner, I. (1990) Subnational population and household projections by central government, in *Occasional Paper No 38,* OPCS, London.

Champion, A., Fotheringham, S., Rees, P., Boyle, P. and Stillwell, J. (1998) *The Determinants of Migration Flows in England: A Review of Existing Data and Evidence,* Report for DETR, University of Newcastle upon Tyne, Newcastle upon Tyne.

Champion, A. *et al.* (2000) *Development of a Migration Model: Final Report,* Report for the DETR under Contract RADS 5/9/22, University of Newcastle upon Tyne, Newcastle upon Tyne.

Corner. I. (1989) Developing centralized household projections for national and sub-national areas, in Congdon, P. and Batey, P. (eds) (1989) *Advances in Regional Demography,* Belhaven Press, London.

Department of the Environment (1995) *1992-based Projections of Households in England to 2016,* HMSO, London.

DETR (1999a) *Projections of households in England to 2021,* DETR, London.

DETR (1999b) *New Projections of Households to 2021,* On-line Press Release 29 March 1999, Department of Environment, Transport and the Regions. Available at http://www.coi.gov.uk/depts/GTE/coi3561f.ok.

DETR (1999c) *The Future of Regional Planning Guidance,* DETR, London.

Duley, C. and Rees, P.H. (1990) Incorporating migration into simulation models, Chapter 13 in Stillwell, J. and Congdon, P. (eds.) *Migration Models Macro and Micro Approaches,* Belhaven Press, London: pp. 228-261

Eurostat (1996) *Demographic Statistics 1996,* Eurostat, Luxembourg.

Ghosh, M. and Rao, J.N.K. (1994) Small area estimation: an appraisal, *Statistical Science,* 9: 55-93.

GROS (1999) *Mid-1998 Population Estimates: Scotland,* General Register Office for Scotland, Edinburgh.

Hall, R., Ogden, P. and Hill, C. (1997) The pattern and structure of one-person households in England and Wales and France, *International Journal of Population Geography,* 3(2): 161-181.

Haverkate, R. and van Haselen, H. (1990) *Demographic Evolution in Time in European Regions (EMETER 2015),* Final report to the European Commission, DG XVI, Regional Policy, NEI, Rotterdam.

Hooimeijer, P. and Oskamp, A. (1999) Advances in the microsimulation of demographic behaviour, Chapter 8 in van Wissen, L. and Dykstra, P. (eds.) *Population Issues: An Interdisciplinary Focus,* Kluwer Academic/ Plenum Publishers, New York, pp. 229-263.

Howard, E. (1898) *Tomorrow: A Peaceful Path to Real Reform*, Swan Sonnenschein, London.

Joshi, H. and Diamond, I. (1990) Demographic projections: who needs them?, in Population Projections: Trends, Methods and Uses, *OPCS Occasional paper No 38*, OPCS, London, pp. 31-42.

King, D., Hayden, J., Jackson, R., Holmans, A., Anderson, D. (2000) Population of households in England to 2021, *Population Trends*, 99: 13-19.

Masser, I., Svidén, O. and Wegener, M. (1992) *The Geography of Europe's Futures*, Belhaven Press, London.

NEI (1986) *Long-term Regional Demographic developments up to the Beginning of the Next Century and Regional Policy*, Commission of the European Communities and NIE, Rotterdam.

NEI (1994) *Regional Population and Labour Force Scenarios for the European Union. Part 1: Two Long-term population Scenarios*, NEI, Department of Regional and Urban Development; Erasmus University, Department of Public Health; NIDI, Rotterdam, June.

NISRA (1999) *Annual Report of the Registrar General for Northern Ireland: 1998*, The Stationery Office, London.

ONS (1999a) National population projections: 1996-based, *ONS Series PP2, No 21*, The Stationery Office, London.

ONS (1999b) Mid-1998 population estimates: England and Wales, *ONS Series PE No 1*, ONS, London.

ONS (1999c) 1996-based subnational population projections England, *ONS Series PP3 No 10*, The Stationery Office, London.

ONS (2000) National population projections: 1998-based, *ONS Series PP2, No 22*, The Stationery Office, London.

OPCS (1991) National population projections: 1989-based, *OPCS Series PP2, No 17*, HMSO, London.

OPCS (1993) National population projections: 1991-based, *OPCS Series PP2, No 18*, HMSO, London.

OPCS (1995) National population projections: 1993-based, *OPCS Series PP2, No 20*, HMSO, London.

Rees, P.H. (1994a) Estimating and projecting populations of urban communities, *Environment and Planning A*, 26: 1671-1697.

Rees, P.H. (1994b) The projection of small area populations: a case study in Swansea, in Hooimeijer, P., van der Knaap, G.A., van Weesep, J. and Woods, R.I. (eds) *Population Dynamics in Europe: Current Issues in Population Geography*, Royal Netherlands Geographical Society, Utrecht.

Rees, P.H., Kupiszewski, M., Eyre, H., Wilson, T. and Durham, H. (1999) The evaluation of regional population projections for the European Union, *Final Report prepared for the European Commission DG XVI and EUROSTAT*, University of Leeds, Leeds.

Rees, P.H., Stillwell, J.C.H., Convey, A. and Kupiszewski, M. (eds.) (1996) *Population Migration in the European Union*, Wiley, London.

Regional Assembly for Yorkshire and Humberside (1999) *Advancing Together Towards a Spatial Strategy*, RAYH, Wakefield.

Secretary of State for the Environment (1996) *Household Growth: Where Shall We Live?*, Cm 3471, HMSO, London.

Shaw, C. (2000) 1998-based national population projections for the United Kingdom and constituent countries, *Population Trends*, 99: 4-12.

Simpson, L. (2000) Population forecasting: old issues, new software, in *Proceedings of the LARIA Conference*, York, March 27-29.

Simpson. L., Middleton, L., Diamond, I. and Lunn, D. (1997) Small area population estimates: a review of methods used in Britain in the 1990s, *International Journal of Population Geography*, 3: 265-280.

Stillwell, J.C.H. and García Coll, A. (2000) Inter-provincial migration of the Spanish workforce in 1988 and 1994, *Regional Studies*, 34(7): 693-711.

Stillwell and Rees, P.H (forthcoming) Applied population projection for regional and local planning, in Clarke, G and Madden, M. (eds) *Regional Science in Business*, Springer, Heidelberg.

Stillwell, J.C.H., Rees, P.H. and Boden, P. (eds.) (1992) *Migration Processes and Patterns Volume 2 Population Redistribution in the United Kingdom*, Belhaven Press, London.

Tayman, J. (1996) The accuracy of small area population forecasts based on spatial interaction land use modeling systems, *Journal of the American Planning Association*, 62(1): 85-98.

Urban Task Force (1999) *Towards an Urban Renaissance*, Final Report of the Urban Task Force, DETR, London.

van der Gaag, N., van Imhoff, E. and van Wissen, L. (2000) Internal migration scenarios and regional population projections from the European Union, *International Journal of Population Geography*, 6(1): 1-19.

van der Gaag, N., van Wissen, L. van Imhoff, E., and Huisman, C. (1999) National and Regional Population Trends in the European Union, *Eurostat Working Paper 3/1999/E/n°8*, European Commission.

Wilson, T. (forthcoming) A new sub-national population projection model for the United Kingdom, Unpublished PhD Thesis, School of Geography, University of Leeds.

Wilson, T. and Stillwell, J.C.H. (1998) Population estimates for Yorkshire and the Humber's new local authority areas, *The Regional Review*, 8(3): 17-19.

Wood, J; Horsfield, G and Vickers, L. (1999) The new sub-national population projections model: methodology and projection scenario, *Population Trends*, 98: 21-28.

Woodhead, K. (1985) Population Projections, Chapter 2.5 in, England, J. *et al.* (eds.) *Information Systems for Policy Planning in Local Government*, Harlow, Longman.

John Stillwell
School of Geography
University of Leeds
Leeds LS2 9JT
United Kingdom

James Debenham
School of Geography
University of Leeds
Leeds LS2 9JT
United Kingdom

6 THE LAND MARKET: A SPATIAL ECONOMIC PERSPECTIVE

Joost Buurman, Piet Rietveld and Henk Scholten

6.1 Introduction

Land markets are the subject of political debate in many countries where discussions focus on themes such as externalities, urban growth and the role of land speculation. Examples of negative externalities can be found in agriculture where the use of pesticides can reduce biodiversity and downgrade the quality of nature areas. Other examples include industries that cause pollution in nearby residential areas or the construction of new residential areas that diminish the amount of open space for existing residents. Open space is becoming increasingly important in many densely populated areas as prosperity increases. As a result, the valuation of open space has started to become a research topic among economists. Spatial patterns of urban development have been another important theme in the political debate about land markets for several decades. A related issue is the relationship between accessibility, transport infrastructure projects and land values. Another area of political debate involves the allocation of ownership titles and the role of speculators, particularly where markets are booming. Market forces play an important role in all these phenomena but, depending on the institutional arrangements in place in any one country, the involvement of governments is also critically important. An economic analysis of both market forces and government intervention in the land market will help improve our understanding of these issues.

Land is a gift of nature that does not have any production costs as such, except for the costs of land improvement. However, in market economies, acquisition of a parcel of land requires that a price is paid and landowners receive a rent for their land. The price of land reflects its scarcity, although it is better to say that location is scarce because at some locations land is abundant. Economists have always studied scarce phenomena and one of the earliest, yet most influential, contributions to the theory of land markets was Ricardo's theory of land rent (1817, in Kruijt *et al.*, 1990). The idea that the value of land is the residual value is still being used to determine the price of land today.

This chapter provides an overview of the land market from a spatial economic perspective. Economic theory will be used to analyse the functioning of the land market and the spatial dimension will receive special attention as many phenomena on the land market have a spatial component. In addition, the consequences of government intervention will be analysed. In Section 6.2, we begin with the somewhat unrealistic assumption of a homogeneous market and introduce various types of land use and government intervention. The spatial dimension of the land market will be analysed in Section 6.3 where we consider models that combine spatial and economic perspectives. Operational models of land use will be discussed in Section 6.4 which are mathematical models that can be used to analyse and simulate land use. In Sections 6.3 and 6.4, attention will be paid to urban areas, as these have been the focus of research in most models. Finally, some conclusions will be drawn in Section 6.5.

J. Stillwell and H. Scholten (eds.), Land Use Simulation for Europe, 65–82.

6.2 The land market

The land market is a special type of market since land has some properties that make it different from other goods (Centraal Planbureau, 1999). Firstly, the amount of land is fixed. It is possible to reclaim some land from the sea or from lakes, but this amount is very small compared to the total available. Secondly, every parcel of land has a fixed location that is a unique property and a certain quality, e.g. its soil fertility or aspect, which is usually difficult to change. Thirdly, the use of a parcel of land also affects the use and value of the surrounding parcels. This is known as the externality of the land use, and is the effect that frequently gives rise to government intervention. We begin the discussion with an explanation of market homogeneity and subsequently introduce more types of land use, actors and government intervention.

6.2.1 *Homogeneous Land Market*
A homogeneous land market means that all parcels of land have the same price per unit area and one market participant or actor (which could also be a homogeneous group) operates on each side of the market. The price of land can be calculated as the present value of the land rent into the future. Land rent is the price for the services yielded by land. (Mills and Hamilton, 1994). All parcels of land are equal in this case. This situation is shown in Figure 6.1.

Figure 6.1: Homogeneous land market

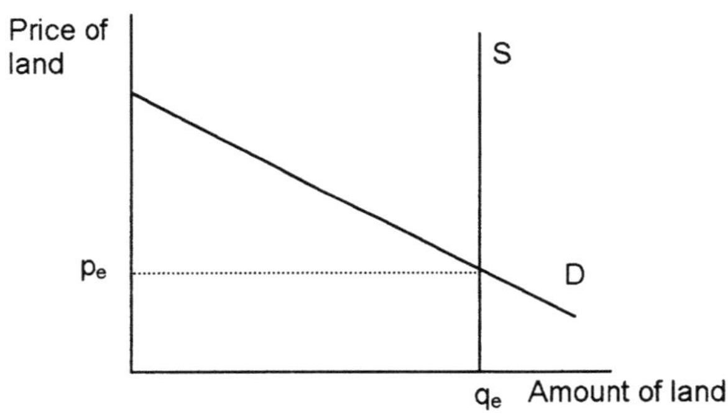

The supply of land, which is depicted by the vertical supply curve S, is fixed at amount q_e. The demand for land, shown by the falling demand curve D, increases when the price of land diminishes. The equilibrium price of land will be p_e in which case all of the supplied land will be used. This situation holds under the assumption of a competitive market, which means that an individual actor cannot influence prices.

In Figure 6.1, the intersection of the demand and the supply curves leads to a positive price. It is possible that the intersection would be in the negative price segment (Figure 6.2). In reality, the price cannot be negative; it would imply a subsidy of the owner of land to the user, which does not make sense in this context. There are situations in which a negative price exists and subsidy of the owner to the user does make sense, e.g. if a parcel of land is polluted. The value of that parcel could be positive without pollution. In this situation, the result of a free market would be that the price of land is zero, and that an amount Q_1 is used, whereas an amount L-Q_1 remains unused. The latter can be interpreted as wasteland. Throughout the history of mankind, there have existed extensive areas of wasteland for which demand has been zero.

Figure 6.2 illustrates a situation where there is no landowner (land is public). The absence of suppliers forcing users of the land to pay an amount of money before they are allowed to use it does not lead to conflicts as long as the demand curve intersects the supply curve at a negative price level. The situation changes when the intersection of the demand and supply curves occurs at a positive price level and there is no owner of the land. In this case, the lack of ownership titles would lead to a rationing problem because at zero price, there would be an excess demand (for example, all members of a community might agree to decreasing their demand for land by a certain percentage to guarantee a scarcity problem). This would be the solution of a public land ownership system: scarcity of land is solved by the public sector agency that owns the land and distributes it according to some principle (for example, distribution according to 'need for land'). Another solution would be to introduce a land market via the distribution of ownership titles leading to the creation of a market for land where some people with a lower willingness to pay will sell the land to people with a higher willingness to pay. Land prices will not be zero in this case.

Figure 6.2: Land market with zero market price

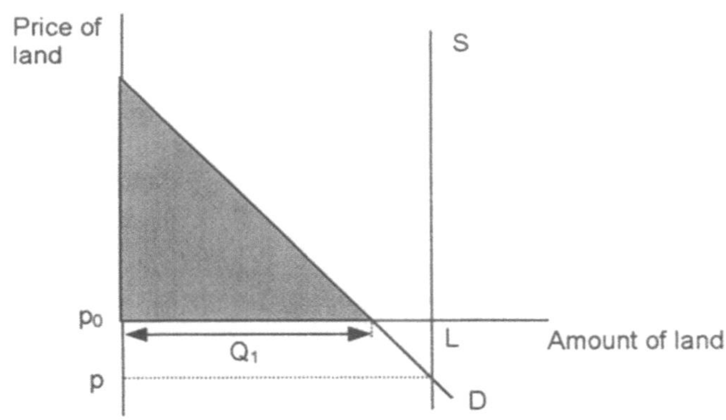

Another explanation of land prices that are not zero when the demand for land is less than the supply occurred in feudal systems where land was owned by one or more landlords. The strong market position of the landlord enabled him to absorb the surplus of the demand curve (representing the willingness to pay) above the equilibrium land price. Thus, even in the case of Figure 6.2 where the equilibrium price of land with competitive supply was below zero, the owner was able to force users to pay. In the case of maximum price discrimination, the total receipts of the landowner would equal the size of the shaded triangle. We should recognise that an increase in demand (an upward shift of the demand curve in Figure 6.1) leads to an increase in the price of land. It does not affect the amount of land used, since the supply is inelastic. This may not be entirely realistic, since when the price becomes high enough, it will pay to expand the supply of land. This can be done in various ways such as by reclamation of land from rivers or seas, for example. As is evident from Figure 6.3, the supply is no longer inelastic in this case: once the price of land is above the critical level of p_0, it pays to supply additional land. When water is free, p_0 is the cost of reclaiming a first piece of land. As land reclamation continues, the additional cost per unit of land increases because the water becomes deeper as the distance from the shore increases. Another possibility of increasing the supply curve is to apply multilevel construction technology (multiple storey buildings, tunnels). Here p_0 can be interpreted as the additional costs of producing a two-storey building compared with a single storey one. This also relates to the substitution of capital for land, a topic that will be discussed in Section 6.3.

Figure 6.3: Elastic supply of land

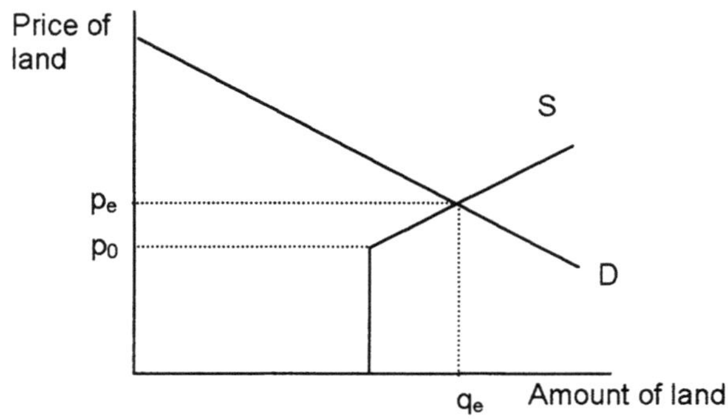

6.2.2 *Various types of land use*

In the previous section we assumed that only one participant was operating on each side of the land market. In reality, many actors operate on the supply and demand sides respectively. Each actor has a demand and supply curve, and determines how much land he/she wants to buy or sell at a certain price. For simplicity, the actors are usually categorised in different groups, such as those that use land for housing, industry, infrastructure, agriculture or nature. This categorisation results in different types of land use. Every type of land use has its own characteristics of demand and supply. Infrastructure, for example, demands parcels of land in such a way that it is possible to build roads. Agriculture needs fertile land, but cannot afford more than the crops yield and nature usually has a predetermined location.

In Figure 6.4, we depict the situation of a land market with two actors that demand land for housing (D_h) and for agriculture (D_a). The demand curve for housing is steeper than the demand curve for agriculture. The steep curve for housing reflects the fact that housing does not need much land and is prepared to pay a high price for a parcel of land. The curve for agriculture reflects the fact that agriculture cannot pay a high price for land, but with sufficiently low prices the demand is large.

The resulting demand on the land market is depicted by the curve $D_h + D_a$. This results in a land price p_e. At this price all of the land is being used. Housing uses the amount of land Q_h and agriculture uses Q_a. This analysis introduces two types of actors, but we still have one single price for land. It is important to notice that the price of land for housing and for agriculture is equal in this simple example. Since the land price is determined by the marginal bidder (who uses the land for agriculture), the users for housing get the land at a price that is much lower than they would be prepared to pay.

In reality we see that there is no single price of land, but the price of land varies between different parcels. The land market is very heterogeneous as the result of the different characteristics of parcels. Government intervention also causes different land prices. Previously we saw that it is possible to identify different (groups of) actors operating in the land market. These actors compete with each other for land, as the supply is limited. Table 6.1 illustrates the different land prices for different functions in the Netherlands in 1996. It can be seen that large differences exist between different types of land use, but also within a certain type of land use.

Figure 6.4: Land market with two participants

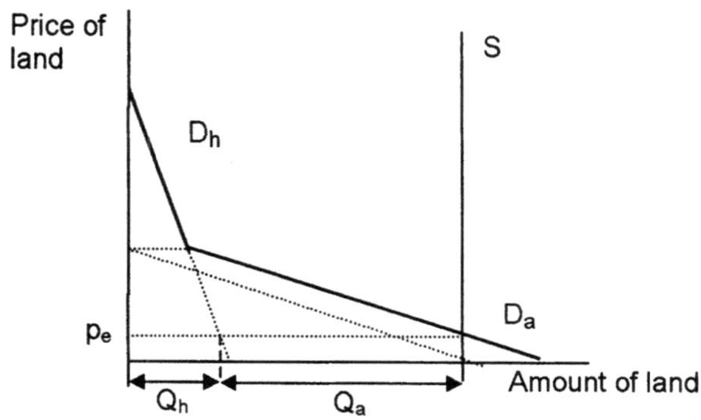

Several factors explain the differences in the prices of land for various types of land use. The first factor is that the prices in Table 6.1 do not relate to identical parcels of land. For example, a distinction is made between serviced and not-serviced land for housing. The price of land for housing often includes the costs of infrastructure of various kinds of public utilities (sewage, drinking water, electricity, telephone, *et cetera.*). In a similar vein, the difference in the price of land for agriculture and nature is that the agricultural land price includes the costs of infrastructure for water management. These additional costs related to infrastructure can also be characterised as costs of transition from one type of land use to another type in the chain: nature to agriculture to housing.

Table 6.1: Estimates of land prices in the Netherlands in 1996

Function	Price (Euro/m^2)	Comments
Nature	0.0 – 0.9	Rough estimate
Agriculture	2.0	Average of different types of agricultural land
Housing	13.2 (2.8 – 22.7)	Average of not-serviced land
	101.2 (49.0 – 196.9)	Average of serviced building land
Companies	54.5 (0 – 226.9 and more)	Large differences between locations

Source: Centraal Planbureau (1999)

Related to this first factor is the fact that land parcels have different properties that cause price differences. More fertile agricultural land delivers more and better crops and needs less fertiliser than less fertile land. It is more expensive to build a house on peat land than on rock. An important property of a parcel of land that causes price differences is its location. This is illustrated in Figure 6.5, which shows a map of the land price for single-family homes by province in the Netherlands. The map shows that if a province is more centrally located in the Netherlands, the average land price (in Dutch florins per m^2) is higher. In the next section we will pay more attention to the spatial dimension.

Figure 6.5: Average land price for single family homes in the Netherlands, 1996

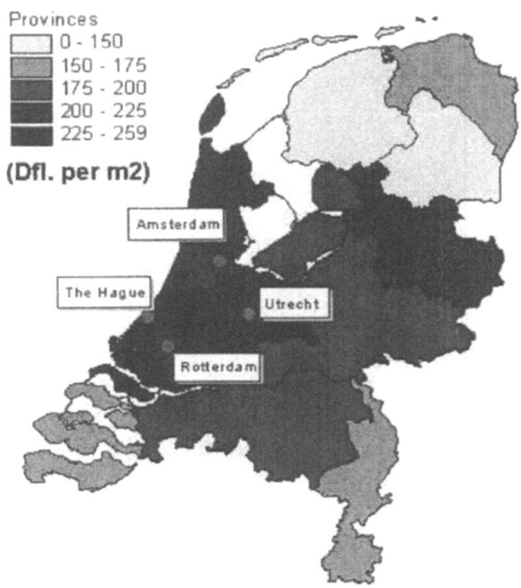

Source: Centraal Planbureau (1999)

Another factor that may explain differences in prices between various types of land use is the existence of price discrimination. The owner of agricultural land who knows that someone wants to buy the land for the development of a housing project may ask a higher price than he would do in case another agricultural buyer buys the land. Once he knows the location of the demand curves he can be sure that the buyer would be prepared, in principle, to pay more than the current agricultural land price. Note, however, that price discrimination cannot be used in all cases. For example, in an entirely competitive market, where owners of land do not collude, the developer will be able to negotiate a low price because he can threaten to go to another owner of land. In addition, the developer may use intermediaries in an effort to hide his intention to use the land for housing.

A final factor to explain differences in prices between various types of land use is government intervention in the form of rationing of land according to various types of land use. In the next subsection we show how government intervention to correct for externalities in land use can explain differences in land prices.

6.2.3 *Externalities and government intervention*
In the case of agricultural land use, the social benefits of externalities may be higher than the direct benefits to the potential user. This can be explained by the fact that society regards agricultural land as being more valuable than the peasant owner does. People enjoy the biodiversity, landscape and open space that it provides. On the contrary, society does not like (new) built-up areas that reduce the amount of open space. Therefore, the social benefits may be lower than the individual benefits in the case of housing. This is also the case with spatial 'spillovers'; the construction of a factory at a certain location will mean that the attractiveness of nearby parcels of land for housing will diminish, thus generating negative external effects in this case.

This divergence between individual benefits and social benefits is illustrated in Figure 6.6 which is equivalent to Figure 6.4 except for the DS_h and DS_a curves. In Figure 6.4, the private demand for housing and agricultural land has been depicted (the dotted curve $D_h + D_a$ in Figure 6.6). The private demand is the demand of a certain actor at a certain price. This actor only looks at his individual situation and does not take into account the effect his behaviour has on other actors. In Figure 6.6, the curve DS_h depicts the social demand curve for housing as below the private demand curve of housing. The social demand curve for agricultural land, depicted by DS_a, is above the private demand curve, meaning that, at a certain price of land, the social demand for housing is lower than the private demand and the social demand for agriculture is higher than private demand.

In the case of a social equilibrium, the amount of land demanded for housing (QS_h) is lower, and the amount of land demanded for agriculture (QS_a) is higher than in the previous case. The equilibrium price of land in this case would be above p_e at ps_e, the social equilibrium price. In Figure 6.1, besides homogeneity, we also assumed a competitive market. In a competitive market, the social equilibrium price will never be reached as the actors in the market behave as 'free riders': they do not take into account the external effects of their behaviour. The land market is, however, far from competitive. Parcels of land have unique properties, so no good substitution between parcels is possible. This and the external effects give governments grounds to interfere in the land market.

Figure 6.6: External effects of land use

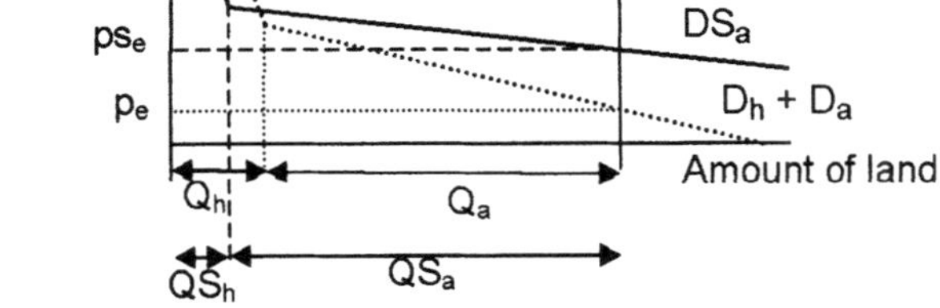

Many possible ways of interfering in the land market exist. Best known are the different kinds of land use planning. In most countries, the land market is segmented because of land use planning, i.e. zoning of the available area into parcels and assigning a type of land use to each. A different price will exist for each segment of the market and this segmentation gives rise to many new questions. For example, in the case of conversion from agricultural land to housing, the price of land increases strongly, so who receives this price increase? From a society perspective, the people who feel the loss of open space should receive some compensation. Another example of a question that arises is why, in some cases, land prices do not go up because of scarcity of land in certain market segments. These questions are a fertile research field for economists.

The questions deal with the fact that the land market is very heterogeneous, as already shown in Table 6.1, and this heterogeneity is reflected in the large number of actors that operate in the land market. Every activity requires land: land is needed for housing, recreation, transportation, production, services, nature, etc. All these actors have different interests that can conflict with each other. For this reason, governments are important actors in land markets. Some further cases of government intervention will be discussed in the context of economic analysis.

A more detailed representation of the effects of the imposition of an upper limit on land for housing by government is shown in Figure 6.7. The negative external effects of housing, the loss of open space, imply that the private willingness to pay for housing is higher than its social value. The demand for agricultural land is assumed to be perfectly elastic, implying a horizontal demand curve. We do not take into account positive externalities of agriculture in this case. Without government intervention, the equilibrium price would be p_e. If rationing would take place, and if it would be based on the marginal external costs of land used for housing, the government would impose QS_h as the upper limit.

Figure 6.7: Land use, taxation and subsidies

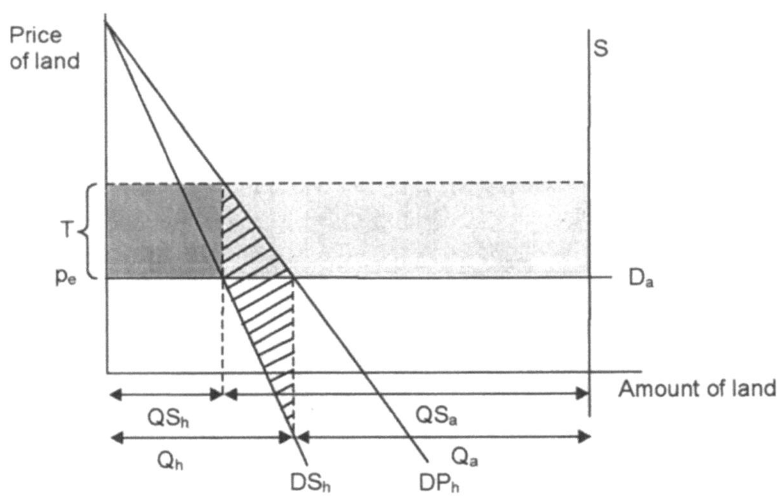

There are basically two ways to achieve this optimal allocation of land. The first is rationing via the direct imposition of constraints on use in the land market. The other way is to use pricing instruments, for example via the imposition of a tax T equal to the marginal external effect. This tax can be imposed at the time of the transfer of land from agriculture to housing, or just as an annual real estate tax. Note that the tax should be only on land used for housing; for agricultural land, it should be zero since we assume that that there are no external costs involved here. The total tax receipts are $T.QS_h$ as one would move from an entirely agricultural area to an area with mixed land use (the dark shaded area in Figure 6.7).

An alternative way to arrive at the same outcome would be the subsidisation of agricultural land use at the level T. This can be done via a single, large subsidy to get agreement with owners of agricultural land that they will not convert it to land for housing, or via an annual subsidy on agricultural activities. The total subsidy to maintain a sufficiently large agricultural area equals $T.QS_a$ (the light shaded area in Figure 6.7).

It is not difficult to see that the amounts of money involved in the taxation or subsidisation would be enormous given the large gap between the price of agricultural land and land used for dwellings (Table 6.1). Even when we correct for the costs of transition from one type of land to the other, this gap will still be substantial. For example, Aalbers *et al.* (1999) estimate that after taking into account the costs of transition, the average price difference in the Netherlands between agricultural land and land for housing is about 40 Euro per m^2 in 1996. Assuming that the Dutch Government has fixed the upper limit of land for housing at the appropriate level, this would imply a negative external marginal effect of built area of this amount. It would imply enormous transfers in terms of taxes or subsidies if this allocation were to be achieved by means of financial instruments. For example, the annual total tax on the built-up area would yield revenues of about 6 per cent of GDP. Similarly, the annual subsidy to owners of agricultural land (and natural areas) who do not convert the land to housing would be about 12 per cent of GDP (Aalbers *et al.*, 1999).

An interesting question is to what extent the above reasoning can be used to estimate the total value of open space. Note that in Figure 6.7, large taxes or subsidies would be needed to prevent a relatively modest reduction of open space from QS_a to Q_a. The marginal cost of open space would equal the amount of T per m^2. The actual external costs of a decrease of open space from QS_a to Q_a would be equal to the size of the area between the private and social demand curve (the dashed area in Figure 6.7). This is obviously much smaller than the total amount of subsidies needed to keep the open space open.

Instead of using taxes or subsidies, governments can put restrictions on the amount of land that can be used for a particular purpose. This is actually more common in practice. Using planning procedures, governments assign a certain function to parcels of land, and in so doing they limit the amount of land that can be used for housing, for example. This situation is depicted in Figure 6.8 where S_h is the supply of land for housing and S_t is the total supply of land. This figure clearly shows the price difference between housing (pe_h) and agriculture (pe_a) that is the result of limiting the amount of land by planning measures.

Note that Figures 6.7 and 6.8 may lead to identical allocations of land between housing and agriculture. Thus, we arrive at the well-known result from environmental economics that there are two ways in which governments may carry out environmental policies, i.e. via prices and via constraints (Baumol and Oates, 1990).

The (above) graphs illustrate that government policies to intervene in the land market in order to correct for externalities may lead to substantial tensions on the land market. The economic interests related to the price differences between open land and built land have become very large. It is not surprising that, especially in densely populated countries such as The Netherlands, land markets are high on the political agenda.

Figure 6.8: Supply restrictions

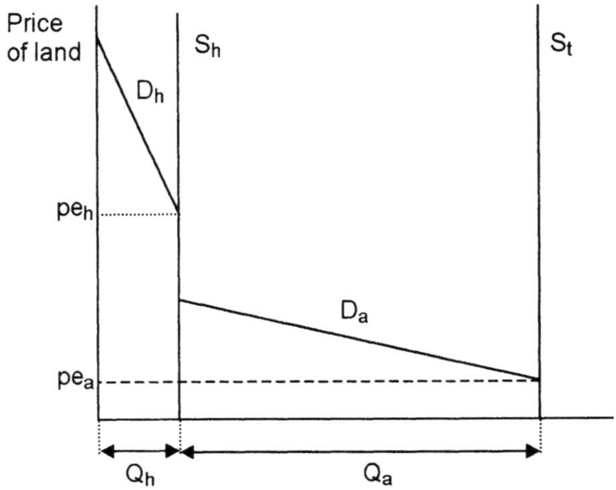

In this section, an analysis of the land market has been given using economic theory. The differences between the prices for land for different functions have been explained. We noted, however, that the spatial dimension is important in such an analysis, but until now we have not explicitly included it in our analysis. This follows in the next section.

6.3 The spatial land market

Many theories have been developed to explain land use patterns from a geographical and spatial economic point of view. Several of these theories are in fact location theories, which try to explain the location behaviour of economic subjects. These theories do not directly explain land use patterns, but they do indirectly as land use patterns are the result of location behaviour. Many land use theories focus on urban land use and the hinterlands of urban areas. With regard to the land market, both the conversion of agricultural land to urban use and topics regarding the structure of urban land use have been the focus of attention (Kruijt and Needham, 1980; Fujita, 1989). In this section a short overview is given of the theories of land use that have a clear spatial component.

6.3.1 *Land use and the location of the market*

One of the first land use models was the well-known Von Thünen model (1826, in Lambooy, 1988) that analyses agricultural land use in relation to the location of a market. Consider a city surrounded by agricultural land. The city is the market, the place where the farmers sell all their products. The agricultural land surrounding the city is homogeneous, i.e. all products can be grown anywhere. The model explains the location of different crops, cattle and forest products by using transportation costs. Suppose the transportation costs increase with distance in a linear way. The selling price of products at the market is fixed because of competition. The production costs are also fixed. In this case, the profit of the farmer for a certain crop is expressed as:

$$P_c = S_c - (C_c + T_c) \quad (6.1)$$

The profit (P) from a certain crop (c) is the selling price (S) minus the production costs (C) and the transportation costs (T). The transportation costs vary linearly with the distance, where t is the transportation cost per distance unit and D is the transportation distance from the production place to the market:

$$T_c = t_c * D_c \quad (6.2)$$

As the farmer does not want to make a loss, he will grow the crop up until the point where the profit is zero:

$$D_c = (S_c - C_c)/t_c \quad (6.3)$$

Different crops have different selling prices, production costs and transportation costs. In Figure 6.9, this is illustrated for two different crops. The crop with the highest profit will be produced 'first', i.e. closest to the city. From distance D onwards, crop 2 is more profitable and will be produced. In this way, concentric circles around a city with different zones of crops are defined. Von Thünen found out that a zone with intensive crop production would exist closest to the city, then a zone of milk production, then a zone of forestry, several zones with grain production in diminishing intensity and finally a zone with cattle. In this model, no attention was paid to land prices. They could be part of the fixed production costs or, more in line with Ricardo's theory, the land rent equals the profit (so the profit will be zero). Closer to the city the land prices will be higher.

Figure 6.9: Land use after Von Thünen

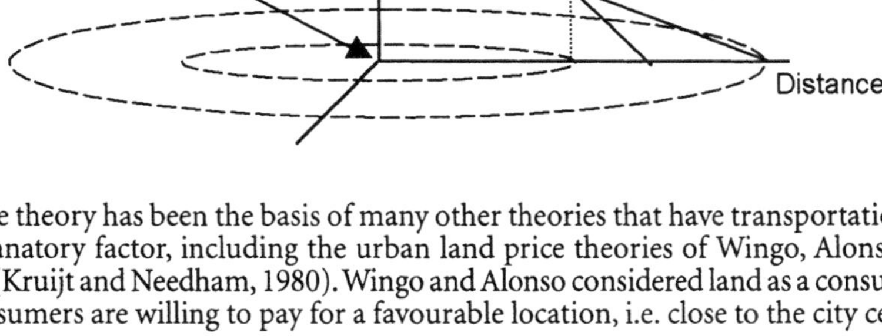

This simple theory has been the basis of many other theories that have transportation costs as an explanatory factor, including the urban land price theories of Wingo, Alonso, Mills and Muth (Kruijt and Needham, 1980). Wingo and Alonso considered land as a consumption good. Consumers are willing to pay for a favourable location, i.e. close to the city centre, to avoid transportation costs. If land prices are low, consumers are prepared to pay more for transportation. Transportation costs and land prices are therefore substitutes. Land is a normal good such that if the price is high, the demand will be lower. At locations with good accessibility, this results in high building densities because the price of land is high and demand for that location will be high. Thus, the demand for land depends on price and location. In this context, Alonso introduced the concept of the bid-rent curve for the demand for land (Kruijt and Needham, 1980; Anas, 1982). Under several assumptions, among them utility maximisation of consumers, this leads to the rent curve depicted in Figure 6.10.

Mills and Hamilton (1994) arrive at the same curve in a different way. In their model, they first focus on the demand for land for production. The land price depends again on the distance to the city centre, but now substitution between capital and land is possible. For example, in order to double production, it is possible to build an identical building next to the original building so the use of land is doubled, or it is possible to build an extra storey on top of the original building. In the latter case, some space in the original building will be lost for staircases, stronger walls, etc. Thus capital can be substituted for land with diminishing returns to scale. As the land rent increases towards the city centre, capital will be less expensive compared to land and capital will be substituted for land. This explains the high capital-to-land ratio (tall buildings) in the city centre. In the model, households maximise their utility regarding the consumption of housing. The value of housing services depends on the amount of uncovered land surrounding the house. Other variables include the costs of travel and the consumption of other goods. In a multi-sector model, the price paid by the highest bidder will be the land rent at a certain location. In this model, the prices again increase strongly closer to the city centre and flatten out towards the suburbs, as shown in Figure 6.10.

Figure 6.10: Relationship between rent and location

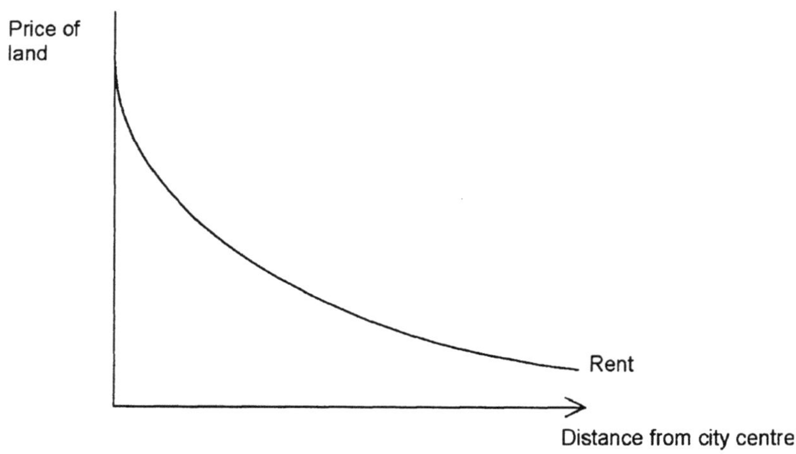

The models described above can be elaborated with, for example, more sectors, decentralised employment or more cities (Mills and Hamilton, 1994). Travel costs remain the most important reason for land use structures. Some people have argued that transportation possibilities have increased so much that travel costs should not be a main factor anymore. However, with problems of traffic congestion, even nowadays travel costs will probably still be a dominant factor.

We conclude that the introduction of the spatial dimension in the analysis of land markets leads to important additional insights. The willingness to pay for land depends on the location of the land and varies strongly among different types of land use. Transport costs and accessibility play an important role in the variations of the willingness to pay for land. The consequence is that land use patterns are far from homogeneous across space. Highly specialised zones may be expected as a consequence of market forces.

Figure 6.11 shows an example of land prices in a monocentric city. The radius of the city equals D_0. It is defined as where the price of land for agriculture and housing is equal. When we compare the agricultural price with the average price for housing we observe a big gap. This result derived in a spatial context differs from the non-spatial context outlined in Section 6.2 where we found that, in a perfect market, the price of land used for housing would equal the price of land used for agriculture. The reason of the difference is that in Section 6.2, land is considered as a homogeneous good, whereas in the present section, the issue of heterogeneity in terms of distance to the centre is addressed. We conclude that the equilibrium condition of price equality for various types of land only holds true at the fringe of the city. Thus, in addition to the factors that explain differences in price levels of land mentioned in Section 6.2, there is another reason related to heterogeneity of land due to differences in accessibility.

A reduction of the size of urban areas in terms of area occupied, for example as a result of planning policy, means that a gap occurs between the agricultural land price and the price of land at the border of the city (Figure 6.11). Without policy intervention the monocentric city would have a diameter of D_0. The policy to save open space leads to a diameter of D_1. Thus, tensions would emerge in the land market at the fringe of the urban areas. Pressures on governments to lift or soften the restrictions on land use at the fringe will be substantial because landowners have much to gain.

Figure 6.11: Land prices in a monocentric city

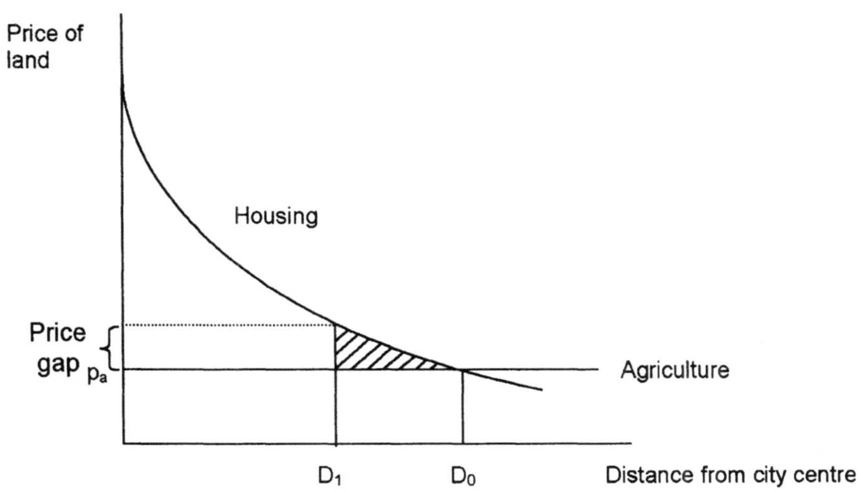

It is interesting to compare this spatial analysis with the non-spatial one in terms of the value of open space. As indicated in the discussion of Figure 6.7, the difference between the price of land for housing and for agriculture would have to be multiplied with the entire agricultural area to determine the total value of open space. In the spatial context of Figure 6.11, we again have to consider the above mentioned price gap but, in terms of subsidies needed for farmers to prevent them from selling the land for real estate developers, the area is much smaller. The area involved in the size reduction from D_0 to D_1 is actually larger than it seems in the figure since the city is assumed to be circular. As a consequence the city area involved is a quadratic function of its radius. In addition, the subsidy needed decreases with distance to the city. As a consequence, the total subsidy needed would be equal to the size of the shaded triangle in the figure.

6.3.2 *Mixed land use*

The simple spatial models of land use that have been reviewed imply that mixed land use does not occur. The highest bidder determines how the land is going to be used, and in the models, the highest bidder just represents one particular type of land use. This leads to a spatial structure with zones of homogeneous types of land use (CBD, housing for single person households, housing for multiple person households, etc). In more complex models, however, heterogeneous land use can be accommodated. There are several reasons for mixed land use. First, transport infrastructure is needed, and this leads to a certain share of all zones being devoted to spatial interaction. Second, public goods such as parks are offered by the public sector and these imply an additional claim for space. Third, households prefer shopping facilities near their dwellings. Also, land intensive employment will be decentralised to avoid the high rents near the city centre. Thus, the original picture with homogeneous zones will soon disappear when mixed land use is recognised and as more sophisticated models are developed.

Of course the degree of heterogeneity will depend strongly on the size of the spatial units considered. With very small spatial units of say 1 m^2, homogeneous land use will occur because for many land use types, a minimum plot size is needed. But when areas are larger than this minimum plot size (areas with a size of more than 250 m^2, for example), it is common to find that more than one land use type is present. This implies a problem when land use is mapped, e.g. in a geographical information system (GIS), in terms of grids of, for example, 100m by 100m. The common practice is to use concepts like 'dominant' type of land use, but this is not satisfactory in the context of patterns with highly mixed land use. A positive exception is land use for line infrastructure. This is relatively easy to represent in a grid context, at least for the major connections such as railway lines and expressways. These line elements can be superimposed on a grid representation of land use. Mapping can become relevant when developing models in a spatial decision support environment. Operational models will be discussed in the next section.

6.4 Operational models

The spatial economic models of the land market and land use discussed in the previous section are derived from theoretical insights and are based on stylised facts which do not always reflect reality. In this section some more operational models are discussed. These models enable one to analyse and forecast land use patterns. The operational models can often be run on computers and can use GIS techniques for the calculations and visualisations. Operational models are usually more comprehensive than the theoretical models discussed above, which can be characterised as partial models. A good land use model should comprise several sectors and systems, like transportation networks, housing market, employment and population (Wegener, 1998). Additionally, the operational models should have a sound theoretical foundation. Besides land market and land use theories discussed above, also theories regarding allocation mechanisms, market forces and probability theory play a role in these models. In this section some operational models will be discussed as an example. We will first look at some models that analyse transportation and land use and then consider cellular automata approaches. For more extensive surveys refer to Wegener (1994; 1998).

6.4.1 *Examples*

Many operational models focus on urban land use and the transportation system. From the 1960s onwards, urban simulation modelling emerged as a major area of research (Anas, 1982). A good urban simulation model must be consistent with economic theory. Some models have been better founded in economics than others. Anas (1982) developed a model that combines urban economic theory with urban modelling. The theoretical foundation lies in bid rent theory, which has essentially been described in the previous section. As bid rent theory cannot explain the dispersion of choices of location within a market segment, probabilistic discrete choice theory is used to extend the model. Walrasian equilibrium analysis is used to model market clearing. An econometric model has been formulated and empirically estimated for Chicago for use in analysing the implications of policy.

Many other models have been developed along similar lines to analyse urban areas. More recently, the models also include environmental issues. Wegener (1998) gives a brief overview of fourteen different models, all of which have been applied to real cities. Regarding their applicability, he notices that if one considers the range of planning problems, the number of issues addressed by the models is rather narrow. Since we are especially interested in questions of whether and how the land market is modelled in these land use-transportation models, two selected models, the RURBAN model and the California Urban Futures (CUF) model, will be explored further.

In the RURBAN (Random Utility/Rent-Bidding ANalysis) model, land use and transportation in a limited metropolitan area are studied (Miyamoto and Udomsri, 1996). The model can be seen as an integrated model of land use and transportation (and the environment). The city is assumed to be a closed system for the land use sub-model, meaning that the supply of land is fixed and extension of the city is not taken into consideration. The area is divided into zones that consist of several sites that are supplied to the highest bidder. On the demand side, several locator groups can be distinguished, each of which chooses the site that gives him the highest utility compared to alternative sites. In order to obtain this site he must bid the highest rent among alternative locators.

The demand for land in the model is derived from random utility analysis and the supply of land is derived from random rent-bidding analysis. Thus the model has two partial equilibrium parts: firstly, the location choice of locator groups based on utility analysis and secondly, the locator choice of zones based on rent-bidding analysis. A general equilibrium can be obtained by means of iteration. The land market can therefore be seen from the perspective of locators and of sites. This simultaneous use of random utility and random rent bidding makes it different from the land use model by Anas described above. Although utility and bid price are closely related (the bid price will depend on the utility a locator obtains from a site), the interesting part of this model is the introduction of two viewpoints: the supply side and the demand side. The model can be integrated in an analysis system with a database, a graphical user interface and integrated policy measures to simulate land use in a real world city.

The CUF (California Urban Futures) (Landis, 1998; Landis and Zhang, 1998) is an urban simulation model that couples a GIS with a profit-maximising model of developer behaviour to project where and when new urban development will occur. The CUF model is work in progress; the second generation of the model is able to deal with multiple land uses and simulate the development effects of transportation investments. CUF-2 is based on the Land Use Change (LUC) model, a series of statistical equations relating observed land use changes between 1985 and 1995, which is calibrated with historical data. The use of GIS in urban activity modelling is a recent development. The availability of data on micro scale and the increasing calculation power of computers and software make it possible to represent an urban area as a collection of individual sites instead of as zones (Landis and Zhang, 1998). This increases the accuracy of the models. The CUF model uses a grid to represent the land use in an area.

The CUF model has four components. The first component is activity projection: a series of econometric equations to forecast future population, households and employment by jurisdiction at ten-year intervals. Employment projection is done by sector. The second component is a spatial database of developable land units in a one-hectare (100 by 100 metres) grid. This database includes ten data layers that describe different land uses. The third component is the LUC sub-model. This is a series of probabilistic equations that relate land use changes in the grid cells to more than 25 site and community characteristics. The fourth component is the simulation engine that calculates site-specific land use change probabilities. These probabilities can be interpreted as bids for (re)development. Different uses can bid against each other for a site.

As the models above, the CUF model uses the bid rent theory and probability theory to describe land use changes. The bid rent theory is modelled in a multinomial logit framework. Land prices are determined as a result of the properties of a parcel and actors bidding for that parcel. However, like most other models, these bid prices have not been calibrated against real world prices. Lack of data remains a problem. *The EuroScanner* model described in detail in a later chapter of this book is in many respects similar to the CUF model as they both consider the total area, without any 'rest' categories. However, as the CUF model still focuses on urban areas, the *EuroScanner* model can cover all land use types in a country.

6.4.2 *Cellular automata*

A different type of land use modelling is that based on cellular automata. Cellular automata models consider a grid in which each data cell contains some data, like the CUF models above. The contents of the cells are however determined by certain behavioural or transition rules which describe how the value of a cell is determined by considering the value of the cell and its the neighbouring cells. The models are based on spatial interactions and are especially useful for simulating dynamics (Câmara *et al.*, 1996).

The behavioural rules are generally quite simple, but the overall system can become quite complex. The theoretical underpinning of cellular automata models is rather limited; the models often behave like a black box. Links with the economic theories of land use as demonstrated in Sections 6.2 and 6.3 are weak. The models can, however, be applied for simulating land use dynamics. Roy and Snickars (1996) have modelled an urban system with cellular automata. The value of a cell is determined by an accessibility function, which determines the attractiveness of a cell for a certain function. The land market is absent in this model.

Another example of a land use model based on cellular automata is the Leef-OmgevingsVerkenner (LOV, Environment Explorer) (for further information, see *http://www.lo.rivm.nl/lov/index.htm*). This is a comprehensive model that covers all land use types in an area. The model contains three scale levels: national, regional and local. On the local level, the allocation of regional land use claims for different sectors and actors is determined by a constrained cellular automata model. The only occurrence of a land market is in the allocation module, in which demand and supply of land are used to allocate land.

6.5 Conclusions

The aim of this chapter has been to give an overview of the land market from a spatial economic perspective. Initially, an economic perspective was taken to analyse the land market as a homogeneous entity, and thereafter the analysis was extended to present a more realistic situation with more participants, externalities and government intervention. However, in reality, the land market is very heterogeneous because parcels of land often have unique properties, locations and demand characteristics. Land use is associated with externalities that lead to the problem that the willingness to pay for land in order to use it for a particular purpose does not fully reveal the associated social benefits. This legitimates government intervention that results in different markets with different prices. Taxation, subsidies and supply restrictions can be used to intervene in the land market.

The land market is pre-eminently spatial. Location is an important property of parcels and externalities are often spatial in nature. Therefore the economic theory of land markets has been supplemented with theories that explain land use patterns. Most theories focus on urban or agricultural land use and arrive at spatial structure with homogeneous types of land use. Since the theories are not very good at explaining the mixed land use that can be observed in reality, more sophisticated models are needed. These more sophisticated models are often operational models, which are mathematical models for analysing and forecasting land use patterns. They tend to be comprehensive as they attempt to model reality and, consequently, they usually focus on a limited area which is in most cases is an urban area. The models are often based on economic bid-rent theory and use probability functions for allocation mechanisms.

A considerable body of knowledge exists to analyse the land market. However, it still seems difficult to make good predictions and analysis of future land use. The land market and land use patterns are complex and need an integrated approach with different disciplines. Sophisticated models can be embodied in information systems to provide decision makers with tools to analyse the impact of their policies. Continuing research is needed to improve the models, understand the functioning of the land market, and improve the way that the implications of policy interventions can be identified.

References

Aalbers, R., Bettendorf, L. and Vollebergh, H.R.J. (1999) Op grond van welvaart, *Economische Statistische Berichten Dossier 'De Grondmarkt'*, l 84(4233): D12-D17.

Anas, A. (1982) *Residential Location Markets and Urban Transportation, Economic Theory, Econometrics, Policy Analysis with Discrete Choice Models*, Academic Press, New York.

Baumol, W.J. and Oates, W.E. (1990) *The Theory of Environmental Policy*, Cambridge University Press, Cambridge.

Câmara, A.S., Ferreira, F. and Castro, P. (1996) Spatial simulation modelling, in Fisher, M., Scholten, H.J. and Unwin, D. (eds) *Spatial Analytical Perspectives on GIS*, Taylor & Francis, London.

Centraal Planbureau (1999) *De Grondmarkt, een Gebrekkige Markt en een Onvolmaakte Overheid,* Sdu Uitgevers, CPB, Den Haag.

Fujita, M. (1989) *Urban Economics*, Cambridge University Press, Cambridge.

Kruijt, B. and Needham, B. (1980) *Grondprijsvorming en Grondprijspolitiek, Theorie en Praktijk*, Stenfert Kroese, Leiden/Antwerpen.

Kruijt, B., Needham, B. and Spit, T. (1990) *Economische Grondslagen van Grondbeleid*, Stichting voor Beleggings – en Vastgoedkunde, Amsterdam.

Lambooy, J.G. (1988) *Regionaal Economische Dynamiek*, Dick Coutinho, Muiderberg.

Landis, J.D. (1998) Simulating highway and transit effects, *Access*, 12: 14-19, University of California Transportation Center, Berkeley.

Landis, J.D. and Zhang, M. (1998) The second generation of the California urban futures model. Part 1: Model logic and theory, *Environment and Planning A*, 30: 657-666.

Mills, E.S. and Hamilton, B.W. (1994) *Urban Economics*, HarperCollins College Publishers, New York.

Miyamoto, K. and Udomsri, R. (1996) An analysis system for integrated policy measures regarding land-use, transport and the environment in a metropolis, in Hayashi, Y. and Roy, J. (eds.) *Transport, Land Use and the Environment*, Kluwer Academic Publishers, Dordrecht, pp259-280.

Roy, G.G. and Snickars, F. (1996) Citylife: a study of cellular automata in urban dynamics, in Fisher, M., Scholten, H.J. and Unwin, D. (eds.) *Spatial Analytical Perspectives on GIS*, Taylor & Francis, London.

Wegener, M. (1994) Operational urban models: State of the art, *Journal of the American Planning Association*, 60:17-29.

Wegener, M. (1998) Applied models of urban land use, transport and environment: state of the art and future developments, in Lundqvist, L., Mattson, L.G. and Kim, T.J. (eds.) *Network Infrastructure and the Urban Environment*, Springer, Heidelberg.

Joost Buurman
Department of Spatial Economics
Vrije Universiteit Amsterdam
De Boelelaan 1105
1081 HV Amsterdam
The Netherlands

Piet Rietveld
Department of Spatial Economics
Vrije Universiteit Amsterdam
De Boelelaan 1105
1081 HV Amsterdam
The Netherlands

Henk Scholten
Department of Spatial Economics
Vrije Universiteit Amsterdam
De Boelelaan 1105
1081 HV Amsterdam
The Netherlands

7 THE ROLE OF REMOTE SENSING TECHNIQUES FOR EUROPEAN LAND USE DATABASE CONSTRUCTION

Henk Kramer and Sander Mücher

7.1 Introduction

The development of sound environmental policies relies heavily on the outcome of environmental models which are significantly influenced by the spatial and thematic accuracy of land cover and land use data. European-wide, georeferenced land use and land cover databases are still scarce. The 10 minutes pan-European land use database of the Dutch National Institute for Public Health and Environment (RIVM), which is largely based on statistical data, was a first step towards meeting the demands of environmental monitoring on a European scale (Veldkamp *et al.*, 1995). In comparison with statistical data, satellite remote sensing has the advantage of collecting up-to-date land cover data with a high degree of spatial accuracy. Until recently, land cover data derived from remotely sensed images has only been available for certain countries and regions in Europe. The CORINE land cover project (CORINE, 1993) is one of the relatively few approaches to European land cover mapping, but the CORINE database does not cover the whole of Europe and is based on satellite images from 1986 to 1997. This hampers the use of the database for time series analysis. The PELCOM database (Mücher *et al.*, 2000) is based on NOAA-AVHRR satellite images from 1997 and is the first 1 km European land cover database based on remote sensing images with acquisition dates from the same year. However, the results of the PELCOM project have indicated that, for monitoring purposes, the use of NOAA-AVHRR data is limited. The land cover changes taking place in Europe do not agree with the spatial detail of the NOAA-AVHRR satellite data. Therefore, in addition to NOAA-AVHRR satellite data, types of satellite data with more accurate detail have to be used, depending on the land cover types, the degree of fragmentation of the landscape and the required detail for monitoring.

The next section of this chapter gives an overview of optical satellite sensors that are available for land use database construction in a multi-sensor approach. This is followed in Section 7.3 by an overview of three existing European databases based on satellite images. Clearly, whilst the remote sensing images provide useful data on land cover, the conversion to land use information is another difficult issue. In Section 7.4, an example of a land use database construction process is presented and some conclusions are drawn in Section 7.5.

7.2 A selection of important satellite sensors

A wide range of optical satellite sensors are available that can be used for land use database construction. The selection given in this section is not a complete list of available sensors but it does give an overview of the spatial and spectral details which are available at the moment. When several sensors with similar characteristics are available, only one is mentioned in this list. Vonders and Clevers (1999) and the CEOS yearbook (1995) give a full description of the capabilities of current and future remote sensing sensors used for land applications.

J. Stillwell and H. Scholten (eds.), Land Use Simulation for Europe, 83–99.

In principle, if a sensor is capable of collecting digital data in the visible and infrared parts of the electromagnetic spectrum, it can be used to classify land cover. For European-wide land use database construction, two additional criteria are important. The first is the spatial coverage of the collected data, the image *extent*. The smaller the extent, the more images are necessary to cover the whole of Europe. The images will probably have been taken at several different dates over a time period, which makes the classification of land cover more difficult because of the changes in vegetation cover during this period. An image acquired in February or March is likely to show a different land cover than an image acquired in August or September, depending on the vegetation dynamics and the present land cover. The second criterion is *detail*, which is directly correlated with extent. Sensors with a wide extent (e.g. 2,400 km) usually have low detail (e.g. 1 km), sensors with high detail (e.g. 10-30 metres) usually cover only a small area (e.g. 60-180 metres).

Satellite sensors measure reflected or emitted radiation in various parts of the electromagnetic spectrum. Table 7.1 shows the characteristics of the Landsat ETM+ sensor, which measures in eight different bands (Lillesand and Kiefer, 1994). The satellite sensors discussed in this section all have comparable spectral characteristics.

Table 7.1: Spectral characteristics of the Landsat ETM+ sensor

1 =	Blue (0.45-0.52 mm) – useful for mapping coastal water areas, differentiating between soil and vegetation, forest type mapping, and detecting cultural features
2 =	Green (0.52-0.60 mm) – corresponds to the green reflectance of healthy vegetation; also useful for cultural feature identification
3 =	Red (0.63-0.69 mm) – useful for discriminating between many plant species; it is also useful for determining soil boundary and geological boundary delineations as well as cultural features
4 =	Near-infrared (0.76-0.90 mm) – especially responsive to the amount of vegetation biomass present in a scene; it is useful for crop identification and emphasizes soil/crop and land/water contrasts
5 =	Mid-infrared, (1.55-1.74 mm) – is sensitive to the amount of water in plants, which is useful in crop drought studies and in plant health analyses; it is also one of the few bands that can be used to discriminate between clouds, snow and ice
6 =	Thermal-infrared (10.40-12.50 mm) – is useful for vegetation and crop stress detection, heat intensity, insecticide applications, and for locating thermal pollution; it can also be used to locate geothermal activity
7 =	Mid-infrared (2.08-2.35 mm) – is important for the discrimination of geologic rock type and soil boundaries, as well as soil and vegetation moisture content
8 =	Panchromatic (0.52-0.9 mm) – is like a black and white photograph; it can be used for spatial enhancement

Source: Lillesand and Kiefer (1994)

The measurements of the sensor are stored in a digital format and can be processed on a computer. For visual interpretation, combinations of three bands can be displayed on a computer screen, using the red (R), green (G) and blue (B) colour guns of the computer screen to create the image. Displaying bands 3, 2, 1 in RGB shows an image similar to an ordinary colour photograph. Displaying bands 4, 3, 2 in RGB shows an image similar to an infrared (IR) photograph. Shades of red in the image indicate the amount of vegetation biomass.

An important parameter that can be calculated from red and near-infrared is the vegetation index (VI). Vegetation has a relatively high reflectance in the near-infrared spectral band and a low reflectance in the red spectral band. Areas with no vegetation (e.g. bare soil) have a low reflectance in both spectral bands and areas with water, snow or clouds have a higher red reflectance than near-infrared reflectance. From this information, the VI can be calculated as near-infrared minus red. Another more often used index is the normalized difference vegetation index (NDVI) where:

$$NDVI = (\text{near-infrared} - \text{red}) / (\text{near-infrared} + \text{red}) \quad (7.1)$$

In theory, the NDVI is an index taking a value between –1 and +1. Negative values indicate water, snow or clouds; a low positive index indicates areas with no vegetation; and a high positive index indicates areas with vegetation. In the remainder of this section, details of sensor characteristics and examples of images from the NOAA-AVHRR, RESURS, IRS, Landsat TM, SPOT and IKONOS satellites are presented.

7.2.1 *NOAA-AVHRR*

The National Oceanic and Atmospheric Administration (NOAA) Advanced Very High Resolution Radiometer (AVHRR) instrument is available from the NOAA-6 to -12 satellite missions. The characteristics of the AVHRR sensor are presented in Table 7.2.

Figure 7.1a shows a part of a NOAA-AVHRR image as received from the Dundee station whilst Figure 7.1b shows a detail of the Netherlands. Land cover classes like forest (f), grassland (g), arable (a), water (w) and urban area (u) can be distinguished from the detailed image in Figure 7.1b. The 1 km resolution of the NOAA-AVHRR sensor only makes it possible to classify large objects. More information can be found at: *http://edcwww.cr.usgs.gov/glis/hyper/guide/avhrr.*

7.2.2 *RESURS*

The Russian RESURS-01 satellite provides information that is bridging the gap between AVHRR's 1 km resolution and LANDSAT TM's 30 metre resolution. Launched in November 1994, this third satellite in the RESURS-O1 series carries the MSU-SK multispectral scanner with a resolution of 160 metres. Table 7.3 indicates the characteristics of the sensor.

Figure 7.2a shows a full scene and Figure 7.2b shows a detail of the Netherlands. The spatial resolution of 160 metres makes it possible to classify land cover with much more detail than with AVHRR data. The scene size is still suitable to perform classification on the European scale. The availability of green, red and near-infrared spectral information makes it possible to classify land cover classes like urban, forest, arable land and grass land. More information can be found at: *http://www.eurimage.com/Products/RESURS_O1.html.*

7.2.3 IRS

The IRS-1C and 1D are identical satellites that are part of the Indian remote sensing program. Table 7.4 indicates the sensor characteristics.

Table 7.2: Characteristics of the NOAA-AVHRR sensor

Satellite	NOAA
Sensor	AVHRR
Spectral bands	Visible Near-infrared Thermal IR (3 bands)
Scene size	2,399 km wide
Spatial detail	1.1 km
Temporal resolution	1/2 day
Available since	1987

Figure 7.1: NOAA-AVHRR image and detail of the Netherlands

(a) Image (b) Detail

Source: Dundee receiving station

Figure 7.3a shows a full scene whilst Figure 7.3b shows a detail of the Netherlands. These images from the IRS-WIFS sensor are comparable in scene size and spatial detail to images from RESURS-MSU-SK. The spectral bands are different, the IRS-WIFS only contains red and near-infrared bands. This limits the number of land cover classes that can be distinguished on the image but provides the information to calculate the NDVI. More information can be found at: *http://www.euromap.de/doc_005.htm*

7.2.4 *Landsat TM*

The Landsat Thematic Mapper (TM) series started with the launch of Landsat-4 satellite in 1982. Since then, Landsat-5 and 7 were launched successfully; the launch of Landsat-6 was not successful. The Enhanced Thematic Mapper + (ETM+) sensor is almost identical to the TM sensor. The thermal infrared band has a higher spatial resolution (60 metres instead of 120 metres) and a panchromatic band has been added (Table 7.5).

Table 7.3: Characteristics of the RESURS MSU-SK sensor

Satellite	RESURS
Sensor	MSU-SK
Spectral bands	Blue
	Green
	Red
	Near-IR
	Thermal-IR
Scene size	600 * 600 km
Spatial detail	160 m
	Thermal-IR 600 m
Temporal resolution	4 days at equator,
	daily at high latitudes
Available since	1995

Figure 7.2: RESURS image and detail of the Dutch flevoploder

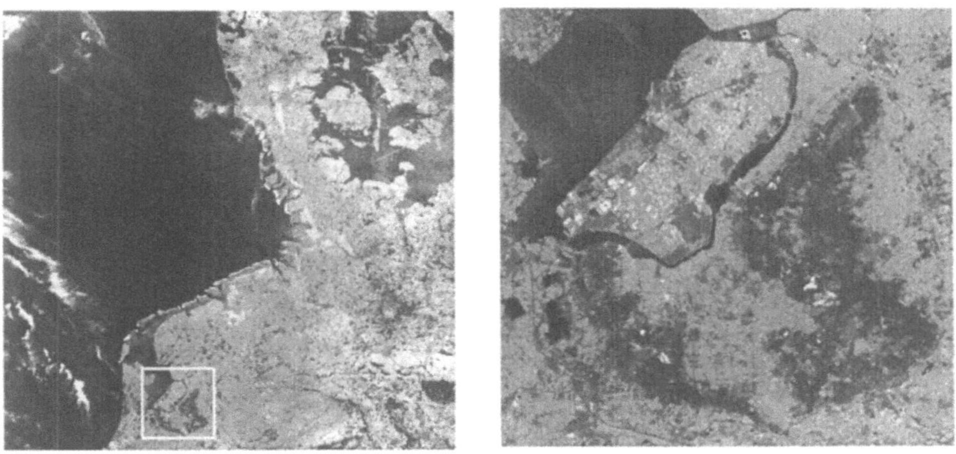

(a) Image (b) Detail

Source: Eurimage

Figure 7.4a shows a full Landsat-5 TM scene for the Netherlands and Figure 7.4b shows shows a detail of the flevopolder. Landsat TM and ETM+ images combine high spatial detail with large scene size. Thus, for example, almost the whole of the Netherlands is captured in two scenes, which are recorded on the same day. The spectral information from the red, near-infrared and mid-infrared make it possible to distinguish many land cover classes, including many different crop types. Images from the Landsat TM have been used to create several land use databases on a national scale (e.g. in the Netherlands and the UK). Modern computer technology (processing power and storage availability) make it possible to create mosaics of Landsat TM images at full spatial resolution for the whole of the world (http://www.geocover.com). More information can be found at: http://landsat7.usgs.gov/.

Table 7.4: Characteristics of the IRS-1C/1D WIFS sensor

Satellite	IRS-1C/1D
Sensor	WIFS
Spectral bands	Red
	Near-IR
Scene size	728 * 812 km
Spatial detail	188 m
Temporal resolution	5 days at equator
Available since	1995

Figure 7.3: IRS image and detail of Dutch flevopolder

(a) Image

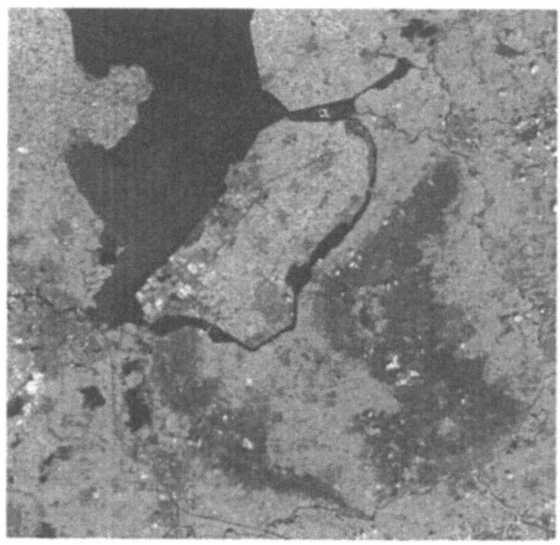

(b) Detail

Source: Euromap

7.2.5 *SPOT*

The SPOT series of satellites are capable of collecting images on the track directly beneath the sensor or to the side of the track of the sensor. This capability of 'off-nadir' viewing makes it possible to collect data from the same part of the earth every three days. These images can also be used for stereoscopic imaging. Characteristics are illustrated in Table 7.6.

Table 7.5: Characteristics of the LANDSAT TM7 ETM+ sensor

Satellite	Landsat TM7
Sensor	ETM+
Spectral bands	Blue
	Green
	Red
	Near-IR
	Mid-IR
	Thermal-IR
	Mid-IR
	Panchromatic
Scene size	180 * 180 km
Spatial detail	30 m
	Panchromatic 15 m
	Thermal-IR 60 m
Temporal resolution	16 days at equator
Available since	1999
	Landsat 4 TM data available from 1982 to 1996, Landsat-5 TM available since 1984

Figure 7.4: Landsat TM image and detail of Dutch flevopolder

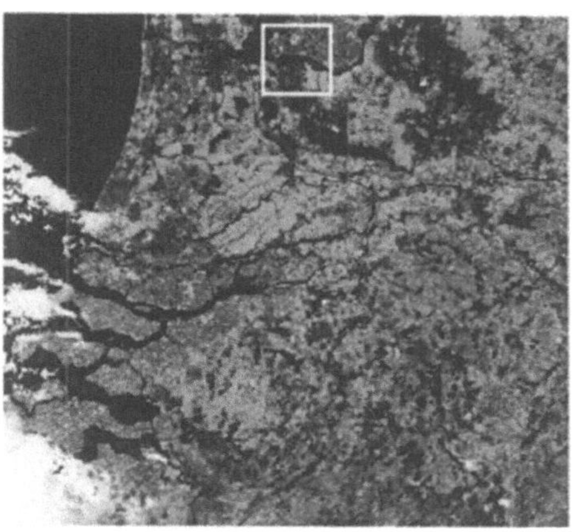

(a) Image

(b) Detail

Source: EOSAT

Table 7.6: Characteristics of the SPOT4 HRV sensor

Satellite	SPOT-4
Sensor	HRV
Panchromatic	10 m
Spectral bands	Green Red Near-IR Mid-IR Panchromatic
Scene size	60 * 60 km
Spatial detail	20 m
Temporal resolution	26 days at equator
Available since	1998 Data from SPOT-1 available from 1986 to 1990, SPOT-2 since 1990 and SPOT-3 from 1993 to 1997

Figure 7.5 shows a full scene of the Netherlands and a detail of the flevopolder. Images from the SPOT-4 sensor are comparable to Landsat TM, only with fewer spectral bands. The detail of SPOT-4 images is higher but its scene extent is much smaller. More information can be found at: http://www.spotimage.fr/home/system/introsat/welcome.htm.

Figure 7.5: SPOT image and detail of the Dutch flevopolder

(a) Image

(b) Detail

Source: Spotimage

7.2.6 *IKONOS*

At the present time, only a few selected IKONOS images are available for demonstration purposes. Information from the infrared band has not been released yet so it is not possible to give an assessment of the capabilities. Sensor characteristics are presented in Table 7.7.

Table 7.7: Characteristics of the IKONOS sensor

Satellite	IKONOS
Sensor	
Spectral bands	Blue Green Red Near-IR Panchromatic
Scene size	11 * 11 km
Spatial detail	4 m Panchromatic 1 m
Temporal resolution	3 days
Available since	1999

Figure 7.6 shows a full scene and a detail from Washington DC. The panchromatic images show high spatial detail and a high detail in grey-level, which makes it possible to identify many small objects. More information can be found at: *http://www.spaceimaging.com/*.

Figure 7.6: IKONOS image and detail of Washington DC

(a) Image

(b) Detail

Source: Space Imaging

7.2.7 *New sensors*

Within the next few years, more satellite sensors with high spatial detail will become available. Orbimage plans to launch its Orbview 3 and 4 in 2001 *(http://www.orbimage.com)*, with similar characteristics as the IKONOS satellite. Spotimage plans to launch Spot 5 in 2001. This satellite sensor will provide 10 metre multi-spectral images and 5 metre panchromatic images. This development ensures the availability of many images with high spatial detail. The European Space Agency (ESA) will launch ENVISAT-1 in 2001. One of the instruments on board ENVISAT-1 will be the MEdium Resolution Imaging Spectrometer (MERIS). MERIS will provide images of land surfaces with a swathe width of 1,150 km and a spatial resolution of 300 metres. They can be used for monitoring vegetation on the global scale. More information about MERIS can be found at: *http://envisat.estec.esa.nl/instruments/meris/index.html.*

7.3 Available European land cover databases

Land cover is strongly related to land use. However, the terms are too often mixed up, with descriptions and definitions being written partly in terms of land cover and partly in terms of land use (UNEP/FAO, 1994). Land cover can be defined as the attributes occupying a part of the earth's surface, such as human artefacts, crops, grassland, forests and bare soil, which can be 'seen' from a distance. Land use, on the other hand, refers to *"man's activities which are directly related to the land"* (Anderson *et al.*, 1976). Examples of such land use classes are industrial areas, grazing land, recreation areas, forestry, fishing areas, etc. These land use classes define the purpose for which the land is used (the land use purpose) (Stomph *et al.*, 1997).

Satellite images do show the land cover. However, the patterns of the land cover are often easily recognised by the human eye as belonging to a certain land use type. For example, 'urban area', which is a land use class, consists of land cover classes like bare soil, buildings, grass, forest, water and infrastructure. But all these land cover classes also exist in the land use class 'rural area'. The way these land cover classes are spatially organised defines a certain area as belonging to the urban area land use class. These structures can be detected by performing a visual interpretation of a satellite image. The use of additional information, such as topographic maps and local knowledge, helps with this interpretation.

Images from Landsat TM and SPOT are often used to create land cover or land use databases. On most occasions, these databases are created within the scope of a project and only for the area which is important for the project, which is usually small. The available knowledge about the project area enables the creation of a land use database and the satellite data is used to get up-to-date information about the area. To create a 'general purpose' land cover or land use database which covers a large area like Europe is much more complicated. The area covered by a Landsat TM or SPOT image is relatively small so many images must be processed and classified. Images from the NOAA-AVHRR sensor cover a large extent but do not contain much detail. This makes the classification of land cover more complicated and restricts the amount of classes that can be classified. In this section, three projects are discussed which have involved the creation of land cover or land use data on a European scale. Each project has its own methodology, spatial detail and nomenclature.

7.3.1 *CORINE land cover database*

The main objective of the CORINE (CoORdination of INformation on the Environment) project, funded by the European Union, is the gathering of coherent information on land cover across Europe and the integration of this information into a GIS. The methodology consists of a computer-assisted visual interpretation of high resolution satellite images, with the simultaneous consultation of additional data, into categories of the CORINE land cover nomenclature. In order to consider the complete spectrum of land cover, a European nomenclature was developed. The legend distinguishes 44 classes (Figure 7.7), grouped in an open three level nomenclature system. The scale of the land cover database is 1:100,000. The surface area of the smallest unit to be mapped is 25 hectares. For line elements, the minimum width is 100 metres (Thunnissen and Middelaar, 1995).

Figure 7.7: CORINE land cover database

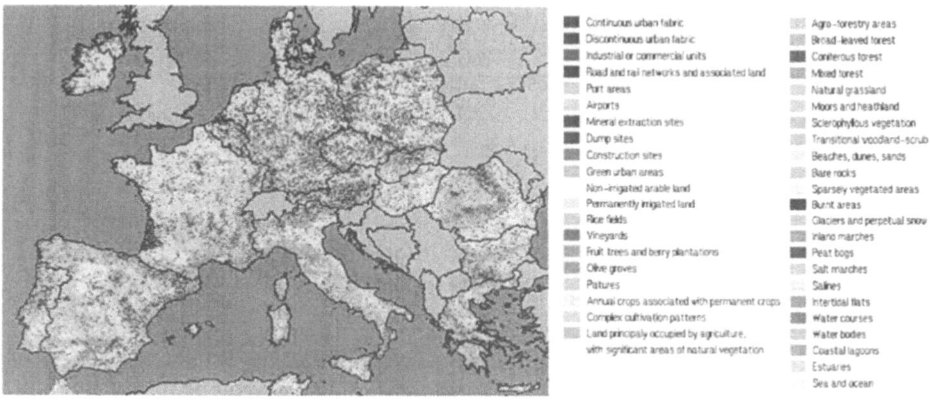

The current CORINE database is based on Landsat TM and SPOT images from 1986 to 1997 and does not cover the whole of Europe. An update of the CORINE database is planned called CLC 2000. For this update, Landsat TM images from 1999 and 2000 will be used, covering the whole of Europe. The update process will be integrated with the update of national land covered databases as far as possible. This will shorten the update time and provide a land use database with a more coherent time span. More information can be found at: *http://etc.satellus.se/*.

7.3.2 *PELCOM*

The Pan-European Land Cover Monitoring (PELCOM) project has been formulated to provide a land cover databases covering the entire European continent with a high spatial accuracy and which can periodically be updated (Mücher *et al.*, 2000). PELCOM is a three year project of a shared cost action under the Environment and Climate section of the European Union Fourth Framework RTD Program. The project started in September 1996 and finished in November 1999. The methodology consists of a stratified, multi-temporal supervised classification of NDVI data with post-classification refinement using multi-source earth science data. The PELCOM project succeeded in the establishment of a 1 km pan-European land cover database (Figure 7.8) with a total average accuracy of 69.2 per cent based on the interpretation of 40 Landsat-TM images over Europe. The smaller and often more fragmented classes had a lower accuracy. The use of NOAA-AVHRR satellite data fails to map such classes as natural grassland, wetlands and permanent crops due to its limited spatial accuracy. More information can be found at PELCOM homepage: *http://www.cgu.girs.wageningen-ur.nl/cgi/projects/eu/pelcom.*

Figure 7.8: PELCOM land cover database

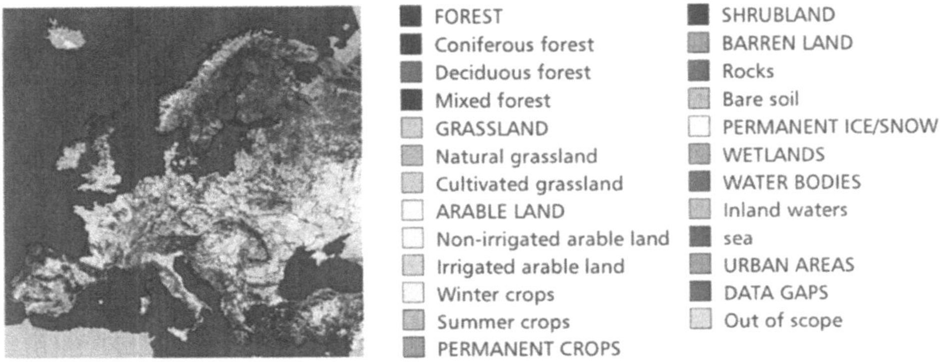

7.3.3 *DISCover*

A third database which covers Europe is the DISCover data set of the International Geosphere Biosphere Program (IGBP) Data and Information System (DIS). This database covers the whole of the earth and consists of 17 land cover classes (Figure 7.9). The methodology consists of an unsupervised classification of monthly NDVI data (derived from NOAA-AVHRR data from April 1992 to April 1993) with post-classification refinement. The global land cover classification was completed in 1997 and the validation was completed in 1998 (Belward *et al.*, 1999). More information can be found at: *http://edcwww.cr.usgs.gov/landdaac/1KM/1kmhomepage.html.*

Figure 7.9: DISCover land cover database

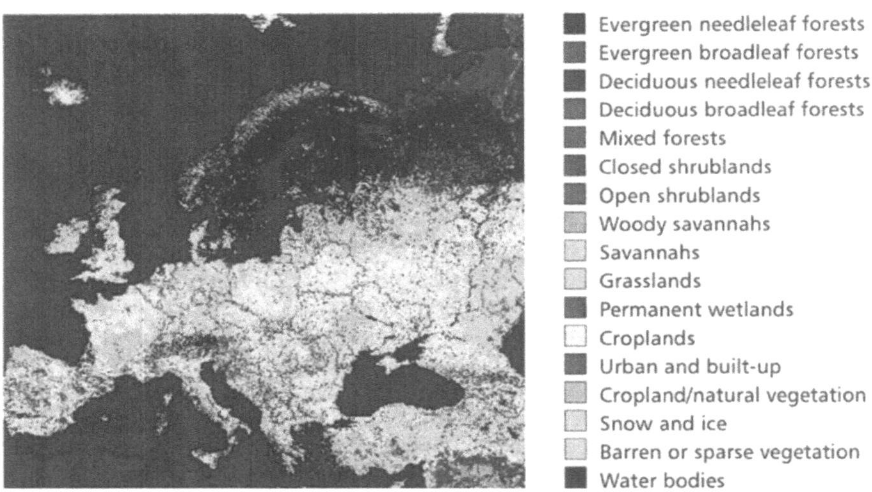

7.3.4 *Datasets in detail*

Figure 7.10 shows details for all three data sets. In each case, the area shown covers the central part of the Netherlands with Amsterdam in the upper left corner and Utrecht in the lower middle. The differences between the data sets are caused by the difference in methodology for the land cover classification and the different satellite images that are used. CORINE is based on Landsat TM images, PELCOM and DISCover both are based on NOAA-AVHRR images. CORINE contains the most detailed information, both spatial and thematically. Both CORINE and Pelcom show the area around Amsterdam and Utrecht as grassland, according to the DISCover database it is cropland. This is also an effect of the applied methodology, DISCover aims at a global classification, PELCOM and CORINE at an European classification. The DISCover database is not intended to be accurate at this scale level.

Figure 7.10: Datasets in detail

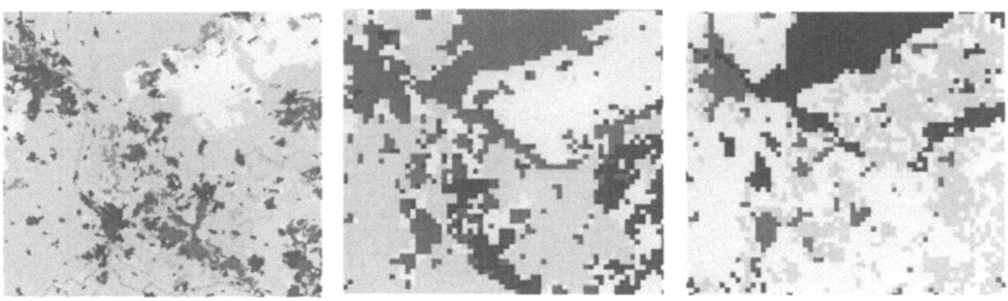

CORINE
Characteristics:
100 m raster
44 classes
Scale 1:100,000

PELCOM
Characteristics:
1 km raster
13 classes
Scale 1:2 million

DISCover
Characteristics:
100 m raster
17 classes
Scale 1:5 million

7.4 Land use database construction

Monitoring and simulation processes often require land use information with land use classes specific to the process. Sometimes these specific classes can be derived from available databases like the CORINE database or created for this purpose from other sources like remote sensing images. In this section, we describe a process for estimating land use for a selected case study area in the Paris region. Historical and up-to-date information about the urban area of Paris is required to monitor the urban land use change. Landsat TM images for 1984 and 1998 have been used for the 'historical' and 'contemporary' databases (Figure 7.11). Additional information from topographic maps and knowledge about the area has been used during the classification process.

A small area from the southern part of Paris (see detail in Figure 7.11) is used to demonstrate the creation of the land use database. The classification process consists of three steps. Firstly a visual interpretation of the satellite images is undertaken to indicate the urban area; secondly, the land cover is classified; and thirdly, the land cover classes are assigned to land use classes. Figure 7.12 shows the visual interpretation results of both images.

Firstly, the urban boundaries are digitized on the 1984 images, based on colours and patterns that identify the urban area. These boundaries then are superimposed on the 1998 image. Additional areas are digitized (indicated with an arrow on the 1998 image).

Figure 7.11: Landsat™ images of (a) Paris; (b) detail, 20/10/1984; and (c) detail, 10/8/1998

Figure 7.12: Visual interpretation of Landsat™ images of Paris

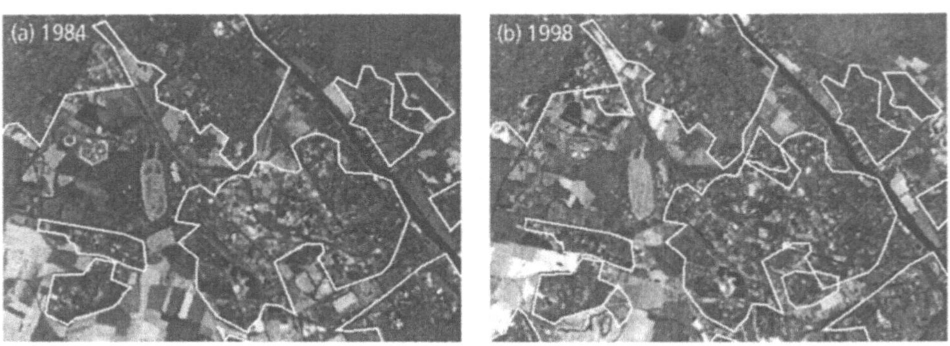

Figure 7.13 shows the results of steps two and three. In step two, the images are classified into six land cover classes (water, bare 1, bare 2, bare 3, vegetation low, vegetation high). Combined with the visual interpretation of the urban area the six land cover classes are assigned to nine land use classes. The classification results from 1984 and 1998 are used to monitor the land use change. Figure 7.14 shows an example of how change detection can be performed with a GIS.

Figure 7.13: Land use classification

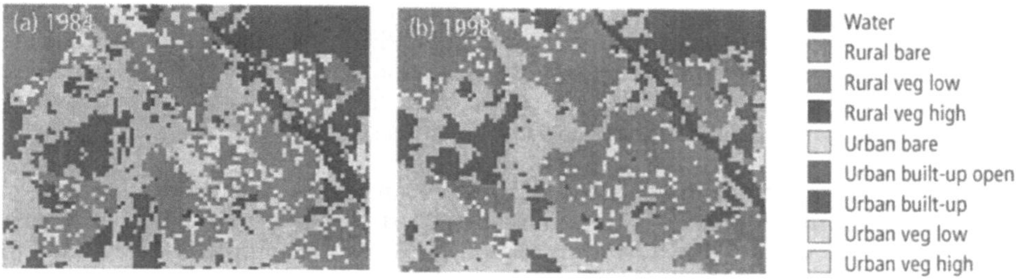

Figure 7.14: Performing change detection using a GIS

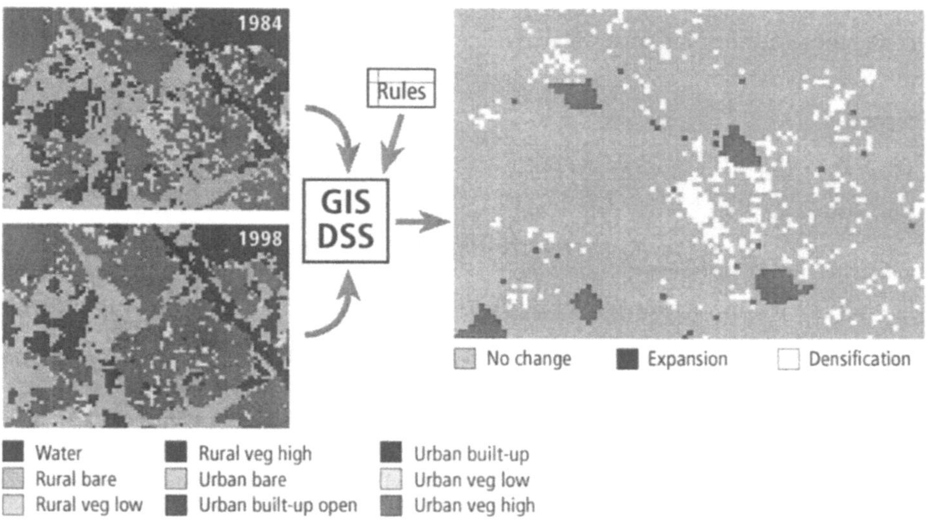

The available land use classes make it possible to detect expansion of the urban area (rural area changed to urban area) and densification of the urban area (bare, vegetated or open built-up urban area changed to built-up urban area). These two types of use change in the future.

7.5 Conclusions

A wide range of optical satellite sensors is available for the creation of land cover or land use databases. Sensors like the NOAA-AVHRR are suitable for use on the global or continental scale, RESURS MSU-SK and IRS-WIFS at the continental scale level, Landsat TM, Spot-HRV at the national or regional scale level and IKONOS at the local scale level. Three land cover databases are available which cover a large part or the whole of Europe.

However, these databases cannot be used for monitoring purposes yet. No European-wide historical databases are available, although an updated version of the CORINE database is planned, and the PELCOM methodology makes it possible to undertake frequent updates of the database. However, use of the PELCOM database for monitoring is restricted to large scale land cover changes. The spatial detail of the PELCOM database does not coincide with the spatial detail of the often fragmented land cover classes in Europe, whilst the CORINE database does provide this spatial detail.

For monitoring projects, tailored land use databases can be created from satellite images. Historical images up to 1982, with the same characteristics as current satellite images, are available for the creation of historical databases. The classification process is tailored to the land use classes that are needed for the monitoring process, thus making the creation of the land use database less complex.

Satellite images have been used in the past for the creation of land cover and land use databases and will provide information in the future. New satellite sensors with higher spatial detail are available and will provide detailed information which will make it possible to create more detailed land use databases in future.

References

Anderson, J.R., Hardy, E.E., Roach, J.T. and Witmer, R.E. (1976) A land use and land cover classification system for use with remote sensing data, *Geological Survey Professional Paper 964*, United States Government Printing Office, Washington, USA.

Belward, A. S., Estes J.E. and Kline, K. D. (1999) The IGBP-DIS Global 1-km Land Cover Data Set DISCover: a project overview, *Photogrammetric engineering and Remote Sensing*, 65(9):1013-1017.

CEOS (1995) *CEOS Yearbook 1995: Coordination for the next decade*, ESA.

CORINE (1993) *CORINE Land Cover – Guide Technique*, Report no. EUR 12585 FR, Luxembourg: Office for Official Publications of the European Community, pp. 144.

Lillesand, T.M. and Kiefer, R.W. (1994) *Remote Sensing and Image Interpretation*, Third edition, Wiley & Sons, New York.

Mücher, C.A., Steinnocher, K., Kressler, F and Heunks C. (2000) Land cover characterization and detection for environmental monitoring of pan-Europe, *International Journal of Remote Sensing*, 21 (6/7): 1159-1181.

Mücher, C.A., Steinnocher, K.T.,. Champeaux, J.L., Griguolo, S., Wester, K., Heunks, C. and van Katwijk, V.F. (2000) Establishment of a 1-km pan-European land cover database for environmental monitoring, In ISPRS 2000 Conference Proceedings, Volume XXXXIII, Part B4, Amsterdam 2000, the Netherlands, pp. 702-709.

Stomph, T.J., Mücher, C.A. and Fresco, L.O. (1997) Environmental impact of land use: a new basis for analysis, *The Land*, 1(2): 129-142.

Thunnissen, H.A.M. and Middelaar, H.J. van (1995) The CORINE land cover database of the Netherlands: Final report of the CORINE land cover project in the Netherlands, *SC-DLO Report 78*, Wageningen.

UNEP/FAO (1994) Report of the UNEP/FAO Expert Meeting on Harmonizing Land Cover and Land Use classifications, GEMS Report series no. 25, Nairobi.

Veldkamp, J.G., Faber, W.S., Katwijk V.F. van and Velde, R.J. van de (1995) Enhancements on the European land use data base, RIVM Report 724001 001, RIVM, Bilthoven, pp. 62.

Vonders, O.W. and Clevers, J.G.P.W. (1999) Multisensor RS Capabilities Land, Report 1: Applications of optical remote sensing satellite sensors, BCRS Report USP-2 98-25, Delft.

Henk Kramer
Wageningen UR – Centrum voor Geo-informatie
Postbus 47
6700 AA Wageningen
The Netherlands

Sander Mücher
Wageningen UR – Centrum voor Geo-informatie
Postbus 47
6700 AA Wageningen
The Netherlands

8 TOWARDS A EUROPEAN SPATIAL METADATA INFRASTRUCTURE TO FACILITATE LAND USE PLANNING

Joana Abreu and Henk Scholten

8.1 Introduction

The development of the 'information society' has been one of the key priorities of the European Union (EU) in recent years. A European Commission White Paper, published in 1993, identified the emergence of an 'information society' as one of the main opportunities to create new jobs and strengthen the global competitiveness of Europe (European Commission, 1993). The responsibilities of the public sector to address the societal impacts of the transition towards an information-based society, include the alleviation of social exclusion and the promotion of policies towards education, training, ethical issues and the protection of privacy. The public sector is the largest producer of information. In each member state, this information is useful for internal policy-making, for administration and for the citizens involved, but also for the private sector to add value to and create new products (Madame, 2000).

The accessibility of data is a key element in the 'information society'. The increasingly digital, borderless community is characterised by access a wide range of good quality services on the Internet, at the click of a mouse. Spatial data services either providing access to the data or to metadata catalogues have started to emerge everywhere with no apparent concern for harmonisation with other systems. It is necessary, therefore, to improve co-ordination of the collection and use of geographic information at a higher level, irrespective of political borders, since many complex spatial problems are not confined to one particular nation or region. Land planners are confronted with major questions concerning the implications of European policies for the modification of land use or the definition of agricultural production quotas across the EU. Thus, the availability of and accessibility to trans-border data is very important and the establishment of geographic data infrastructures to support the dissemination and use of the data is clearly justifiable.

The ability to transfer digital data quickly and easily from user to user and from application to application is the key to much of the success of digital spatial data. Digital spatial data sets are now available from a variety of governmental and non-governmental sources. Spatial data, either in digital or analogue form (such as maps or charts), are used to represent all aspects of physical reality and are of paramount importance in land use planning. Although the value of spatial data is often measured in terms of the costs of acquisition, these costs may be offset by the number of uses to which the data is put (Frank, 1994).

This chapter begins in Section 8.2 with the specification of some of the requirements of an infrastructure for the provision of data on spatial information before setting out some of the problems associated with gaining access to geographic information across Europe in Section 8.3. Existing geographic information infrastructures in Europe and across the world are reviewed in Section 8.4, and the advances that have been made through the European Spatial Metadata Infrastructure (ESMI) project are outlined. In Section 8.5, the benefits of providing direct access to the data through Web mapping technologies are

J. Stillwell and H. Scholten (eds.), Land Use Simulation for Europe, 101–111.

considered and exemplified using examples of emergency service planning. Some concluding comments are contained in Section 8.6.

8.2 The requirements of a spatial data information infrastructure

"*To successfully apply geographic data technologies, users need geographic data that are current, accurate, available over the area of interest, obtainable for as low a cost as is possible, and in a usable form*" (FGDC, 1991). Today, data is increasingly available in digital form from the Internet (Goodchild, 1996) and there is a great demand for rapid or instant access to geographic data or data sources. Users of geographical information (GI) require access to the most up-to-date meta-information that describes the characteristics of the available data as well as to the data itself. At the organisational level, there is growing attention to improve communication between the data users – those who interpret information and make decisions, and the data processors – those who collect and analyse the data. In many management situations, the main obstacle confronting decision-makers is the lack of access to data, be it at a national or multinational level.

A spatial data infrastructure can provide the user with access to data such as road centre line locations, local government boundaries, digital orthophotos, digital elevation models and digital geodetic control networks (Goodchild, 1996) as well as online access to other multiple source GI data and metadata catalogues. Data producers or distributors of remote sensing, topographic, thematic or geo-referenced alphanumeric data should provide (distributed) access to their metadata catalogues or/and to the data itself through the spatial data infrastructure. This is the model of a single access gate on the Internet for spatial geo-referenced data availability. Frank (1994) suggests that the system should be able to provide metadata about:

- *availability*: users will need to know the availability of data, algorithms, products or services, including alternative sources, terms of access, thus limiting the proliferation of data duplication and redundancy. Presently, data duplication occurs as it is faster and easier to collect new data than to search for existing data;

- *cost:* it is difficult for users to determine the worth of information until they actually have it or use it. The value of a particular data set, algorithm, product or service is dependent on the context and needs of each individual user (Rhind 1992);

- *support:* users will want to know what level of support to expect for the data, algorithm, product or service they require (Faulhaber, 1992), such as its update frequency;

- *compatibility:* users will want to know how compatible data, algorithms, products or services are with other data, algorithms, products and services;

- *format:* users will need to know the format of data sets, the acceptable data formats for algorithms and the formats of products and services; and

- *semantics:* users will need to know the concepts underpinning each set of the data, each algorithm, product or service. In some cases, this will be straightforward, as with much imagery data. In other cases, the problem becomes more complex, as when data sets are assembled from multiple observations at differing resolutions and time intervals.

The spatial data infrastructure should also provide access to algorithms and to tools (i.e. software for manipulating data sets) that will enable sophisticated or less sophisticated users either to solve diverse and complex problems or to easily represent information geographically (Frank, 1994). The success of an information infrastructure may be measured by the user's ability to access the right resource, in the right form, at the right time, and at the right cost. However, different users will have different needs that will require different levels of service and sophistication. The spatial data infrastructure should aim at providing a service to all, following the information society principles and needs.

The spatial infrastructure should provide spatial information products (i.e. reports, tables and maps) that users might require to solve many routine problems. These products would be provided by direct or indirect access to spatial database systems coupled with software for manipulation. Finally, spatial information services should be provided. These services might provide automatic data input to car navigation or air traffic control systems or they might simply be *ad hoc* services provided by GIS specialists who would deliver spatial information products or who could even provide spatial data tools. Users would have control over the general nature of the services, but would have very little control over individual operations or products. Users would be interested primarily in the value and appropriateness of these services. Sophisticated applications would attract new users, thus boosting the GI market.

8.3 The problems to access geographic information in Europe

In Europe, there are a number of information services available, each managed by a different European institution. These include:

- the INFormation on Earth Observation (INFEO) system (*http://www.infeo.org*), managed by the European Space Agency (www.esa.int), provides a service to the earth observation community, by allowing the simultaneous search of several data collections as well as tools and algorithms;

- the European Environment Agency (EEA) *(http://www.eea.eu.int)*, which is responsible for setting the standards and a distributed metadata catalogue (CDS) of the environment data produced by its focal points in the EU states;

- the European Thematic Topic Center on Land Cover (ETC/LC), which is an organisation led by the Environmental Satellite Data Centre and works as a consortium of 16 different organisations all over Europe to provide users of land cover data with accurate data that corresponds to needs in a wide range of applications such as environmental monitoring, coastal zone monitoring, flooding, forest fires and acidification; and

- EUROSTAT, which is responsible for GISCO, the EU's statistics system.

These services, selected for illustration, are developed for specific professional groups and have no explicit connections between them. The European Environment Agency and the related European Environmental Information and Observation Network (EIONET) were conceived to provide information to screen, evaluate, validate and process data and information pertinent to the environment. They were also expected to transform this data into efficient information principally for the EU institutions and the member states, and also for the general public at large. Thus, the main goal of the European Environment Agency is to make maximum use of existing data and information and to improve existing capacities in member states and other European institutions to produce reliable and efficient data and information (Steenmans, 1997). Part of the infrastructure in these organisations is their datacatalogue or metadata structure. A visit to their Web sites shows immediately how important this approach is for their search facilities (*http://search.eea.eu.int/compass).*

It can therefore be argued that a European 'information infrastructure initiative' should provide an umbrella service to contain all these different services and others, independent of the target professional group they were initially developed for, thus allowing access of information to a much wider range of different users. Moreover, at European level, there are problems in finding and accessing the relevant information due to linguistic, metaphoric and semantic issues. Linguistic barriers occur when the user is not familiar with the language used on the interface. Metaphoric problems occur when the interfaces do not react as the user expects (Koepel, 2000), forcing the user to adapt to the whims of the technician responsible for designing the interface. Semantic problems occur mainly due to the widespread variation of metadata standards, each interpreted differently by the data producer, which makes comparison and evaluation of the potentially interesting data set very difficult.

Socio-cultural conditions can prevent potential users from accessing the metadata system. A citizen will probably not know which systems are available in a neighbouring country even if the language is not a barrier. This situation can occur in the same country, due to customisation of the systems to target communities. Regionalisation, in geographic and 'cultural' terms, adds a greater degree of complexity to the problem of the potential user of the data, not only because it increases the number of potential usable systems but also because some powerful metadata systems are targeted for specific communities (like the academic community), leaving all other potential users out. Means must be found to improve the coordination of GI at European (and continental) level, as the need for information can have a European dimension, either for land planning, environmental or emergency management purposes (Burrough, 1999).

8.4 Existing geographic information infrastructures

It is not particularly easy to find geographic information for a neighbouring country or region, in spite of the growing number of services available. Some spatial data distributors and users have set up metadata systems to manage their data, but such systems appear to be limited to data sets originating from a single source, or to data sets being used by a select group of users and limited to data sets used within that group. In Europe, several approaches have been followed, depending on whether the initiatives were centrally and government-driven or not. In Portugal, there is a National System of Geographic Information that provides a distributed service which connects the main geographic data producers in the country as well as initiatives by the regions. In Belgium, the federal state supports three different initiatives in the regions of Brussels, Flanders and the Walloon (Madame, 1999). There are several important metadata initiatives mostly nationally oriented. Examples include the National Clearinghouse for Geo-Information (NCGI) in the Netherlands (*http://www.ncgi.nl*) (Beltman and Scholten, 2000), the emerging National Geospatial Data Framework in England and Wales (NGDF) (*http://www.ngdf.org.uk/*), the GIS-Flanders Support Centre in Belgium, the National Land Survey (NLS) in Finland (*http://www.nls.fi/ptk/infrastructure/index.html*) and the Sistema Nacional de Informação Geográfica in Portugal (SNIG) (*http://snig.cnig.pt*) (Gouveia *et al.*, 1997).

Pan-European initiatives include the Geographical Data Description Directory (GDDD) of MEGRIN, which contains metadata of over 19 European mapping agencies, and the Geological Electronic Information Exchange System (GEIXS) of Euro-GeoSurveys which is a geo-science driven network. It is also interesting that some European metadata services, such as the Global Environmental Information Locator Service (GELOS), adopted the Federal Geographic Data Committee (FGDC) metadata standards allowing databases to be searchable through the FGDC system.

This short review highlights the need for a system that could link the metadata services in Europe. The necessity of having a European system is concerned with European diversity. Contrary to the American system, where one government may decide that all institutions have to connect to a central system, Europe is composed of several countries, which have their own governments, languages and cultural characteristics. A European infrastructure has to be responsive to the specifications of the European reality.

In the 1980s, the United States Bureau of Census developed an initiative to create TIGER files (street addresses and census track data), which led to an 'explosion' of new markets that emerged from these files, which have shifted the orientation of NGII. One of the main characteristics of these new files and their use was the fact that topographic data was not used by a large majority of new GI uses in several business domains like banking, insurance, marketing, real estate, transport and traffic, utilities and in the social sciences domains. Led probably by the generalised and increasing use of geographic information in American society, the United States Government mandated the Federal Geographic Data Committee (FGDC) to co-ordinate and develop a distributed search mechanism for digital geospatial data, the National Spatial Data Infrastructure (NSDI) (*http://www.nsdi.fgdc.gov*), in 1994. Using the data elements defined in the Content Standards for Digital Geospatial Metadata, governmental, non-profit-making and commercial participants can make their collections of spatial information searchable and accessible on the Internet. The executive order also urges each governmental agency to adopt procedures to ensure its compatibility with the clearinghouse. The objectives for creating such an infrastructure were to minimise duplication of efforts in the collection of expensive digital spatial data and foster co-operative digital data collection activities. By promoting the availability, quality, and requirements for digital data through a searchable on-line system, a clearinghouse facility would greatly assist in co-ordination of data collection and research activities. It would also provide a primary data dissemination mechanism to traditional and non-traditional spatial data users.

Other spatial data infrastructures have been developed in countries such as Australia, Japan, China, and Canada (GSDI, 2000). To try to coordinate these efforts at a global level, the need for a global infrastructure has been assessed and discussed under the framework of the Global Spatial Data Infrastructure (GSDI, 2000) (*http://www.gsdi.org*). However, the GSDI is the result of voluntary efforts by several national representatives that are responsible for co-ordination of national spatial data infrastructures. The GSDI has recently produced a publication, the *Spatial Data Infrastructure Cookbook* that aims at the dissemination and development of spatial data infrastructures world-wide (GSDI, 2000). However, the need for global thematic spatial data infrastructures, which provide better access to information and networking between professionals is an increasingly important subject. The Global Disaster Information Network (GDIN) (*http://www.gdin-*

international.org) is a joint initiative of the European Commission and the United States, Australia, Asia, and recently the former USSR. As with the GSDI, GDIN is the product of voluntary involvement, and also aims to promote the development of National Disaster Information Infrastructures throughout the world (GDIN, 2000).

8.5 Towards the European Spatial Metadata Infrastructure

According to Madame (2000), and given the strategic importance of exploiting the wealth of information that exists, a considerable number of initiatives and studies have already been taking place over the last few years. These include:

- the 1990 Directive on the freedom of access to information on the environment;
- the two phases of IMPACT projects which have helped in building operational experience on the difficulties of creating market products from (often) public sector data;
- the numerous projects within the 'Interchange of Data between Administrations' and 'Telematics for Administrations' initiatives;
- the GI2000 discussions focusing on the strategic importance of a stable European-wide policy framework on geographic information;
- the three studies on GI-POLICY, GI-META, and GI-BASE commissioned by DG XIII; and
- the projects within the first call of INFO2000 and the Fifth Framework Programme for R&D in which the Information Society will be one of the four vertical actions *(http://www.cordis.lu)*.

At European level, such infrastructure does not exist. However some efforts have been made to assess the needs, constitution and market expansion of the GI and GI community. In 1996, the GI-Meta, GI-Policy and GI-Base studies were undertaken and, in parallel, under the INFO 2000 Programme, the Commission sponsored the *European Spatial Metadata Infrastructure* (ESMI) project which aims to be the technical building block towards the 'European Geographic Information Infrastructure' (Abreu *et al.*, 1998). ESMI has been developed with the philosophy of maintaining the diversification of systems, interface languages and metadata semantics. The gateway aims to connect every metadata provider in spite of differences between the metadata storage medium (database or text files) or the metadata description language (CEN TC 287, ISO TC 211, *et cetera*). As such, the search and retrieve tools connect to distributed sources simultaneously, whether the data is stored in text files (typically eXtendend Markup Language) or in databases. Tests were done using ORACLE and Microsoft Access databases. The problem of different metadata description semantics used by each data producer was solved by mapping the common fields between Dublin Core, CEN TC 287 and ISO TC211 metadata standards. These common fields are used for the interface development and for search and retrieval. Not knowing which language a potential user might use to search the interface proved to be challenging. However, the solution passed not only by the translation of the user interface in different languages, but also by using a stable thesaurus and finally by coding both the thesaurus and the search elements. The search tools used code strings that were at the data producer site translated to the language used to describe its own metadata. As such, a user could choose to search for a data set containing the German word 'autobahn' in Portugal and retrieve Portuguese metadata data sets related with its search term. Therefore, the ESMI concept preserves the cultural diversity in Europe, though successfully providing harmonisation.

The technical infrastructure of ESMI is composed of three main technical components (Figure 8.1). The first component is the Webserver that communicates with both the clients/ users and the data providers. The second component is the Zserver programs that are installed at the ESMI nodes (data providers, national clearinghouses) and the third component is the set of ESMI tools that will allow data providers to become a search node of ESMI. The communication is based on the z39.50 library search standard and CORBA for the OpenGIS catalogue service components (Katwijk, 2000).

Figure 8.1: The ESMI infrastructure

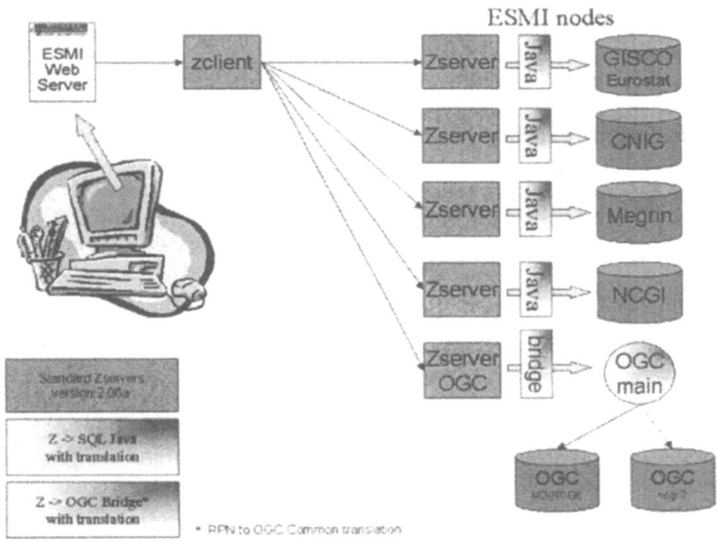

ESMI is a distributed system that allows a data provider to publish metadata in one language but permits access through a common gateway. Since the search programs and scripts and query interface use codes, which are replaced by terms in the SQL query before transferring to the actual metadata database, this makes the ESMI system extremely flexible, for the user and provider. The former is allowed to search using an interface developed in the same language and the latter needs only to map the metadata database to the ESMI terminology, it not being necessary to perform any changes in the storage medium. These distributed systems also provide a good medium for Web mapping applications (discussed in the next section) since, in case the spatial data itself is available, the citizen could easily access it.

Recently, the Commission supported several other projects, which based their developments on ESMI concepts, like the project *La Clef* that included the commercial component, thus allowing producers to directly market its products on the Web. Currently, under the Vth Framework Programme, Information Society and Technologies, the project *European Territorial Management Information Infrastructure* (ETeMII, 2000) aims to evaluate the need for a European geographic information infrastructure by the organisation of discussion fora and workshops on the topics of interoperability, reference data and metadata.

8.6 Providing direct access to the data via Web mapping

Spatial data infrastructures are still basing the locator services in metadata-based systems, either by applying distributed approaches *(http://www.ngdf.org.uk/)* or choosing to store the metadata on central servers *(http://snig.cnig.pt)*. The next step is to allow the user to access multi-source data directly, through the implementation of 'Web mapping' technologies. The Web mapping testbed, which was developed under the framework of the Open GIS Consortium *(http://www.ogis.org)* is the concept of displaying maps in a Web browser that are generated dynamically by a 'mapserver'. Geographic information is stored at the server side and displayed at the client side. So the geographic information has to be transferred somehow from server to client.

The ESMI team undertook some tests using data distributed on two servers in the Netherlands, one in Portugal and one in Spain. The user would therefore access the data portal and select data for Portugal from the Portuguese servers or from Dutch and Spanish servers by highlighting the required server from the menu box shown on the html page (Figure 8.2) that the user is consulting. In this case, data sets available were for natural hazards. The map servers located in each country receive the request, transform the data and send it back to the html page. Each time there are requests for a new map, the system creates the maps (Figure 8.2) and allows the user to zoom in and out and to navigate through the data.

Often, more operational systems that need to provide up-to-date information to the users adopt the new developments more quickly. This is the case in the example of the operation of an emergency service. Emergency situations that affect a country are often seamless or originate from disruptions occurring in a neighbouring country. However, co-ordination of emergency control centres for border countries is often scarce, and data transference is often impossible or done too late. Since the emergency information can be geo-located, the Web mapping tools can easily and immediately overlay the spatial data from multiple sources. In fact, an emergency information network, seamless, which would integrate multi-source data and management tools, would ensure an increase in the efficiency of means, which, in the long term, would mean an increase in the number of lives saved.

The data needed for emergency management is rarely located in a single database, as it is produced by different institutions (not normally the institutions responsible for the emergency management). The operator in the control room will be able to access multi-source data, and to integrate the information in a single interface. To visualise the emergency situation, information may determine the adequacy of response to the situation, as, apparently un-related events might provoke a serious emergency with strong effects. One example of such an emergency was the Chernobyl accident. Another example, at a different level of magnitude, was the devastating flood occurring in Alentejo, Portugal, which grew from several small regional and localised events. The visualisation of these events and the possibility of monitoring its evolution can be crucial in emergency management. Web mapping technologies seem to be optimal to use for complex emergency management, due to the nature of the problem. To overlay and immediately visualise distributed geo-referenced data in a single interface seems to be a complex problem appropriate for a Web mapping solution.

Figure 8.2: Example of overlaying spatial data located on different servers

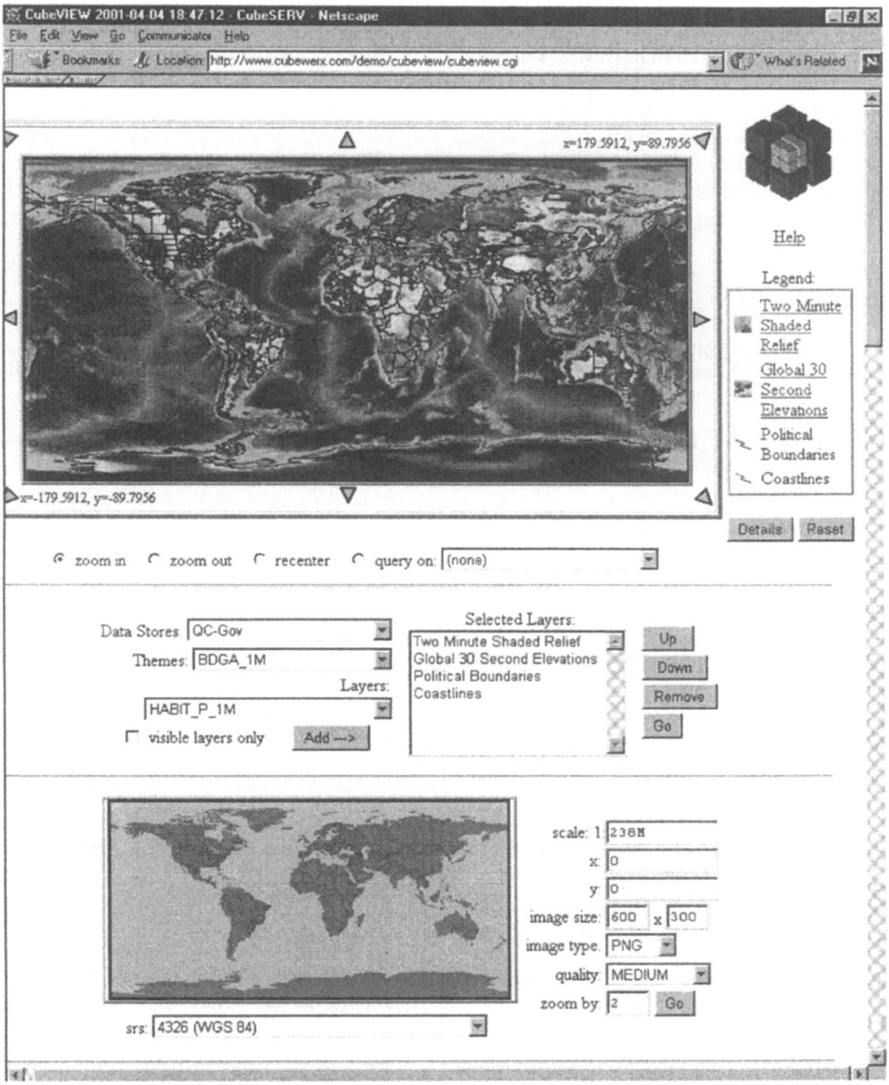

Note: The demo was developed by Cubewerx (www.cubewerx.com) under the Open GIS (www.opengis.com) Webmapping 1 Call for specifications

8.7 Conclusions

Potential geographic information users require access to data sets across Europe in order to provide more efficient government and administration as well as new business opportunities. A European geographic information infrastructure should provide guidelines for metadata standards across the member states and advocate interoperability between systems. It should also promote accessibility to mechanisms, tools and algorithms for data manipulation, queries and retrieval and serve as a single access point to multi-source, heterogeneous, distributed data.

Data availability should also be promoted by the encouragement and harmonisation of data access policies. European citizens should have a means to compare their own national policy with policies elsewhere in Europe. A European geographic information infrastructure is a fundamental building block in the construction of the European market, promoting democratic and transparent access to information by all European citizens.

References

Abreu, J., Bonn, B., Scholten, H., Cabrita, P., Fonseca, T. and Câmara, A. (1998) *European Spatial Metadata Infrastructure, The Geographic Information Gateway on the Internet*, GISPlanet Conference Proceedings, September, Lisbon.

Beltman, R. and Scholten, H.J. (2000) *ESMI: Europa's Toegangspoort tot Geo-informatie*, Jaarboek Geo-Informatie 1999-2000, RAVI, Amersfoort, pp. 146-152.

Burrough, P. (1999). *GSDI and ESDI – Views on Interoperability and Spatial Data Infrastructures in Europe*, (*http://www.gsdi.org/docs/parma.html*).

ESMI (2000) *The European Spatial Metadata Infrastructure*, (*http://www.esmi.org*).

ETeMII (2000) *European Territorial Management Information Infrastructure*, (http://www.ec-gis.org/etermii).

European Commission (1993) *White Paper on Growth, Competitiveness and Employment*, European Commission, Brussels.

Faulhaber, G.R. (1992) Pricing Internet: the efficient subsidy, in Kahin, B. (ed.) *Building Information Infrastructure*, McGraw Hill, pp. 271-295.

FGDC (Federal Geographic Data Committee) (1991) *A National Geographic Information Resource: the Spatial Foundation of the Information Based Society*, PGDE, Washington DC .

Frank, S.M. (1994) Cataloging Paradigms For Spatial Metadata, *PhD Degree Thesis*, NCGIA (National Center for Geographic Information and Analysis), University of Maine, United States, pp.117, (http://www.spatial.edu).

GDIN (2000) *Global Disaster Information Network*, (*http://www.gdin-international.org*).

Gouveia, C. Abreu, J. Neves, N. and Henriques, R. G. (1997) *The Portuguese National Infrastructure for Geographical information: General Description and Challenges for the Future*, GISDATA Conference Proceedings.

GSDI (2000) *Developing Spatial Data Infrastructures: The SDI Cookbook*, Global Spatial Data Infrastructure, (http://www.gsdi.org).

Goodchild, M. F. (1996) The Spatial Data Infrastructure of Environmental Modelling, in Goodchild, M. F., Steyaert, L. T., Parks, B. O., Johnston, C., Maidment, D., Crane, M. and Glendimming, S (eds) *GIS and Environmental Modelling: Progress and Research Issues*. GIS World Books, Fort Collins, pp. 486.

Katwijk, V.V. (2000) ESMI Technical Infrastructure, ESMI Consortium, (*http://www.esmi.org*).

Koepel, D. (2000) GUI's just want to have fun, *Wired Magazine*, 8(10):156-172.

Madame, (1999) *Comparative Evaluation of National Data Infrastructures*, MADAME – Methods for Access to Data and Metadata in Europe, INFO 2000 Project: PUB1108. Deliverable 1, (*http://www.info2000-madame.org*).

Rhind, D. (1992) Data access, charging, and copyright and their implications for geographic information systems, *International Journal of Geographic Information Systems*, 6(1): 13-30.

Steenmans, C. (1997) The European Environment Agency and its European Environmental Information and Observation Network (EIONET), in Baulies, X. and Szejwach, G. (eds) LUCC Data Requirement Workshop, Barcelona, Spain, LUCC Report Series No. 3.

Joana Abreu
Centro Nacional de Informação Geográfica
Tagus Park
Edifício Imovação III
Sala 614
2780 Porto Salvo
Portugal

Henk Scholten
Department of Spatial Economics,
Vrije Universiteit Amsterdam,
De Boelelaan 1105
1081 HV Amsterdam
The Netherlands

PART II: MODELLING ENVIRONMENTAL AND HUMAN SYSTEMS

9 LAND COVER INFORMATION FOR EUROPEAN ENVIRONMENTAL MODELLING

Jean-Paul Hettelingh, Maximilian Posch and Peter de Smet

9.1 Introduction

Land cover maps have been used to assess the sensitivity of ecosystems in support of European policies to mitigate air pollution. Within the 1979 Convention on Long-Range Transboundary Air Pollution (LRTAP) of the United Nations Economic Commission for Europe (UN/ECE), a protocol was signed in 1994 to further reduce the emissions of sulphur oxides, while in 1999, a protocol was negotiated to abate acidification, eutrophication and ground-level ozone. These two protocols were the first to take into account information on the sensitivity of ecosystems, in addition to economic and technical knowledge about the cost and potential of installing pollution abatement equipment in European energy combustion facilities.

A summary of 20 years of international collaboration under the LRTAP Convention to reduce air pollution can be found in UN/ECE (1999). Information about the sensitivity of ecosystems has now also been used to support the European Union's acidification and ozone strategy. More generally, it can be argued that the use of land cover information for the preparation of policy documents has increased both on a European scale to assess prominent European environmental problems (EEA, 1995a; 1999), and on a global scale to assess a variety of environmental issues (UNEP, 1997; 1999). The studies supporting these policy documents use mathematical models to reflect relationships between the causes and the effects of environmental issues. Land cover information generally comes at the end of such 'cause-effect chains' to identify the expected changes from a number of policy alternatives (scenarios). Important environmental themes with relevance to current and future land cover include climate change, air pollution, deforestation, nature and biodiversity, land and food, urban areas and waste management.

This chapter focuses on the scientific and technical support of air pollution mitigation policies in Europe on acidification, eutrophication and tropospheric ozone. This support is of particular interest because it demonstrates the importance of land cover information, illustrates the successful relationship between environmental modelling and policy development, and reflects the importance of international scientific collaboration. Given the emphasis of this book on exploring the potential of current and future land cover assessments, occasional side-steps are made in the chapter to allow the expression of technical (mapping) details or tentative ideas on the further inclusion of land cover and land use in environmental modelling. This means that the description of the use of land cover data is occasionally stretched beyond the focus of the chapter, thus contributing to the formulation and structuring of land cover-related research ideas. Section 9.2 begins with a discussion of environmental modelling at the European scale whilst Section 9.3 introduces the RAINS model which has been used as an assessment tool to support European air pollution abatement policies. The concept of a critical load is defined in Section 9.4 as an estimate of exposure to harmful pollution and the equations for

J. Stillwell and H. Scholten (eds.), Land Use Simulation for Europe, 115–126.

computing the critical load for acidification and for eutrophication using the Steady-State Mass Balance model are described. In Section 9.5, the mapping and assessment of critical loads across Europe is considered and the chapter concludes in Section 9.6 with a reflection on the relevance of land cover information for the assessment of future environmental quality in general and impacts of air quality in particular.

9.2 Modelling the European environment

The support of environmental policies in the EU and UN/ECE increasingly consists of a structured interaction between policy analysts and scientists. The latter use mathematical models to simulate policy options to address environmental issues, such as reductions of SO_2 emissions to reduce acidification, while policy analysts interpret the results and supply inputs to enable the simulation of policy alternatives. This kind of structured iterative interaction between policy analysts and scientists – which also has become an important mechanism in the climate change debate – is generally termed integrated (environmental) assessment (IEA). A unique definition of integrated assessment is not yet available (Tol and Vellinga, 1998).

A general definition of IEA, which has been adopted by the European Environmental Agency (EEA, 1995b), is *"the interdisciplinary process of identification, analysis, and appraisal of all relevant natural and human processes and their interactions which determine both the current and future state of environmental quality, and resources, on appropriate spatial and temporal scales, thus facilitating the framing and implementation of policies and strategies".* IEA involves the analysis and modelling of the chained relationships between sources and impacts in a systematic way as illustrated in Figure 9.1. This 'driving forces'-'pressures'-'state'-'impact'-'response' scheme illustrates how an analysis based on IEA modelling of causes and effects can interact with a variety of policy options.

Figure 9.1: The cause-effect relationships to address environmental issues, including a characterisation of policy options

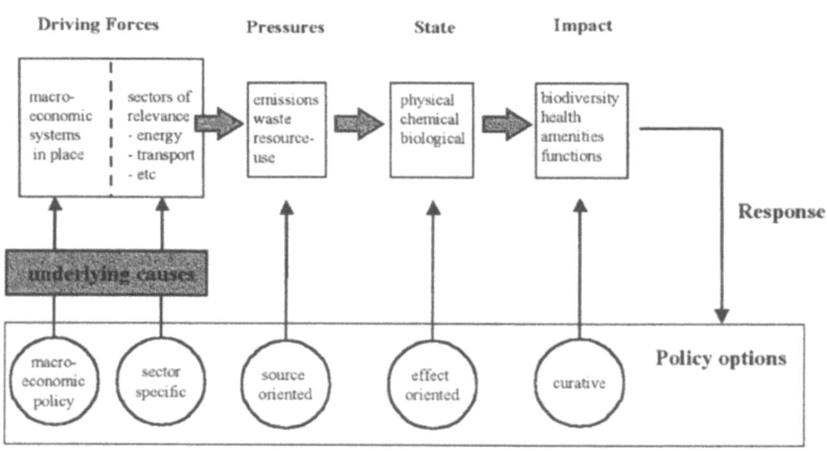

Figure 9.1 helps us to understand cause-effect relationships leading to acidification, for example, that require the identification of fuel combustion as a 'driving force', related emissions, such as sulphur dioxide, in 'pressures', current soil and land cover characteristics in 'state' and finally the assessment of damage to ecosystems, for example, in 'impacts'. Policy alternatives, including the installation of abatement equipment, can be assessed using models such as the RAINS model which simulate relevant cause-effect chains, as will be explained in more detail below. 'Underlying causes' refer to socio-economic or other indirect societal structures that may affect an environmental problem and its mitigation potential. The sequence of 'driving forces' to 'policy response' is generally referred to as the DPSIR framework. It is interesting to develop thoughts on the assessment of geographic information using such a framework. The DPSIR framework could be of use for structuring land cover information and its relationship with land use. For example, urban and regional locations of economic activities (land use) in 'driving forces' could be systematically related to consequences for land cover in 'impacts'. The 'response' would probably be related to intentions of policy organisations in the field of spatial planning. In fact, for many of the environmental issues addressed, in EEA (1995a) for example, it could be a challenging exercise to use the DPSIR framework in the context of geographic assessments in general and of land cover in particular. In the remainder of this chapter the use of land cover as an important part of impact assessment is described in the context of acidification, eutrophication and tropospheric ozone policies in Europe.

9.3 The support of European acidification policies

Scientific support based on IEA was provided to the negotiations of the UN/ECE LRTAP Convention 1994 Oslo Protocol on further reduction of sulphur emissions, the 1999 Gothenburg Protocol to abate acidification, eutrophication and ground-level ozone, and the EU's acidification and ozone strategy.

The Regional Acidification INformation and Simulation (RAINS) model (Alcamo *et al.*, 1990; Schöpp *et al.*, 1999) was used as the integrated assessment tool to support European air pollution abatement policies (see *http://www.iiasa.ac.at/~rains/* for extensive descriptions of the background and applications of the RAINS model). Its structure is shown in Figure 9.2. In terms of DPSIR, the RAINS model does not include an endogenous 'driving forces' module. The energy structure, for example, is exogenous and the model does not include socio-economic variables. The 'pressures' module enables the assessment of emission control options producing output in terms of national emissions and abatement costs. The link to 'state' is obtained with an atmospheric dispersion module which computes sulphur and nitrogen deposition and ambient concentrations of tropospheric ozone (in 150x150 km^2 grid cells) from national emissions of sulphur dioxide, nitrogen oxides, ammonia and volatile organic compounds.

Land use or land cover information is not used in the 'pressures' block of RAINS, but it is embedded in the dispersion module (and in the impact module). The dispersion module of RAINS is derived from a mechanistic atmospheric transport module (Bartnicki, 1999; EMEP, 1999) which requires information on land cover characteristics, such as surface roughness, to establish deposition velocities. Recently, it was noted regarding main problems of regional modelling and mapping of ozone fluxes that *"there is a strong need for improved land cover maps (land use, priority ecosystems, species occurrence and distribution) covering the whole of Europe"* (Simpson and Posch, 1999).

Finally, an impact module of RAINS compares the computed deposition to threshold values of acceptable deposition and acceptable ambient concentration levels, i.e. so-called critical loads and critical levels respectively. An excess of critical loads or levels implies that there is a risk of damage to ecosystems or, with respect to tropospheric ozone, also to human health. No additional information is provided about the kind of damage to natural systems or human health, or its occurrence in time which – formally speaking in terms of DPSIR – implies that the RAINS impact module produces an assessment result in 'state'. Additional information on bio-geochemical, vegetation and human health characteristics could ultimately provide information on the physical damage caused by the excess of critical loads and levels, in which case the assessment result would be part of 'impact' in DPSIR terms. Further elaboration of DPSIR semantics is beyond the scope of this chapter. Behind the RAINS impact module lies a model, bio-geochemical information, land cover data and European scientific collaboration which is addressed in the following section.

Figure 9.2: Schematic overview of the RAINS model

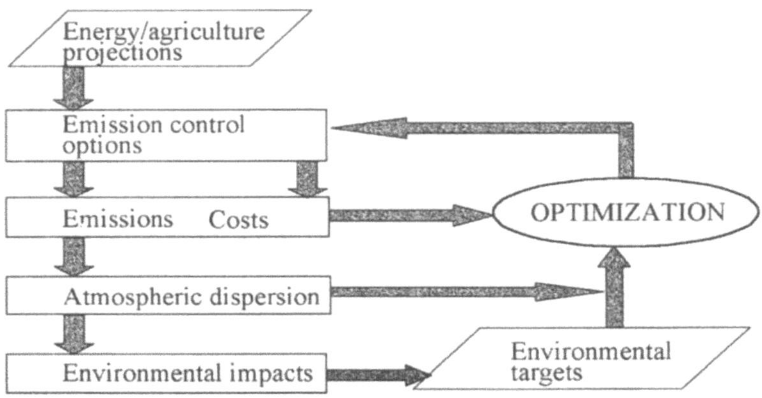

9.4 The computation of critical loads and levels

A critical load is a quantitative estimate of an exposure to one or more pollutants below which significant harmful effects on specified sensitive elements of the environment do not occur according to present knowledge (Nilsson and Grennfelt, 1988). Critical loads are used to assess the effectiveness – in terms of diminishing the risk of acidification and eutrophying damage to natural systems – of emission reduction alternatives which are negotiated in the Convention on LRTAP (Hettelingh *et al.*, 1995). Critical loads are computed using the so-called Steady-State Mass Balance model. This approach assumes a time-independent chemical equilibrium between the production and consumption of acidic compounds. Critical loads are computed for acidification and for eutrophication using the following mathematical formulations, respectively:

$$CL(Ac) = BC^*_{dep} + BC_w - BC_u - ANC_{le(crit)} \tag{9.1}$$

$$CL_{nut}(N) = N_i + N_u + N_{le(acc)}/(1-f_{de}) \tag{9.2}$$

The critical load for acidity (equation 9.1) equals the net input of base cations minus a critical leaching of acid neutralising capacity ($ANC_{le(crit)}$). The net input of base cations consists of sea-salt corrected base cation deposition (BC^*_{dep}) and base cation weathering (BC_w), diminished by base cations which are taken due to the growth of vegetation (BC_u). The critical load expressing the maximum input of nutrient nitrogen (equation 9.2) is based on the nitrogen which vegetation takes up for growth (N_u), nitrogen immobilisation (N_i) and an acceptable amount of nitrogen leaching ($N_{le(acc)}$), e.g., to groundwater. A separate equation for the acidification effect of nitrogen and a more detailed description of the critical load can be found in Posch *et al.* (1997; 1999).

Critical levels are thresholds of ambient air concentrations of air pollutants such as tropospheric ozone, above which direct (i.e. via leaves and stem) damage may occur to crops, forests and natural ecosystems (UBA, 1996). Because of the importance of the relationship between damage, on the one hand, and both the amount and duration of exposure on the other, a measure has been introduced to express the ozone accumulation over time. This so-called accumulated exposure over a threshold of 40 ppb (AOT40 parts per billion) enables the assessment of the total excess (over 40 ppb) ozone concentration to which a plant has been exposed in a relevant (e.g. growth) period. Critical AOT40 values have been formulated for forests (AOT40 = 10,000 ppb-hours during summer) and crops (AOT40=3,000 ppb-hours during spring). Peak hourly values for air pollutants have been formulated both for the prevention of damage to vegetation and by the World Health Organization for human health.

Information on land cover is required for the assessment of uptake and leaching variables. The assessment and mapping of critical loads started in 1990 (Hettelingh *et al.*, 1991a) using a combination of environmental modelling and GIS techniques (Hettelingh *et al.*, 1991b; Meijer *et al.*, 1995). The current use of geographic information in relation to the modelling of critical loads and the excess by atmospheric pollutants is summarised as follows. The Steady-State Mass Balance (SSMB) model is applied to geographic polygons that contain bio-geochemical soil, land cover and meteorological information derived from intersecting relevant maps. An intersection is then made with a grid that is consistent with the domain of the atmospheric model, to allow comparisons between the computed deposition or concentration of air pollutants with critical loads (Figures 9.3 and 9.4).

The grid system used is based on a so-called polar stereographic projection which projects each European point on the Earth's sphere from the South Pole onto a plane perpendicular to the Earth's axis and intersecting the Earth at a fixed latitude (Posch *et al.*, 1999). A stereographic projection is a so-called conformal projection, i.e. an angle on the sphere remains the same after its projection into the plane. This is useful for the modelling of trajectories of atmospheric pollutants that follow wind directions. Each grid cell may contain a variable amount of polygons reflecting ecosystem characteristics, thus leading (using the SSMB model) to a variety of critical loads values. Combinations of sulphur and nitrogen are then identified within a grid cell which protect the same percentage of ecosystems. These combinations form so-called critical load functions. Finally, the RAINS assessment compares sulphur and nitrogen deposited in a grid cell with the critical load functions.

The excess of critical loads has changed over time due to emission reductions that have been agreed in the LRTAP Convention. Figure 9.3 shows the 1960-2010 trend of the area weighted excess of sulphur and nitrogen deposition accumulated over all the critical load functions (Average Accumulated Excess). The excess in 2010 assumes implementation of national emission reductions agreed in the 1999 protocol mentioned above. The result shows the success of emission reduction schemes. After a steady increase of both the magnitude to more than 1,000 eq ha^{-1} yr^{-1} (an excess of 1,000 eq ha^{-1} corresponds to a sulphur deposition excess of 1.6 g m^{-2} or to a nitrogen) and the area of critical load excess between 1960 to 1980, a decrease sets in showing the magnitudes to drop to around 200 eq ha^{-1} yr^{-1} in most parts of Europe. However, the area where critical loads are exceeded still covers a large portion of Europe, mostly due to nitrogen deposition. Nitrogen emissions are difficult to control for a number of reasons including the increase in automotive transport. Figure 9.3 uses a 10-year average of meteorology allowing a clear illustration of the effect of emission variability, not confounded by inter-annual meteorological variability. The light shaded grid squares show areas where critical loads are exceeded by less than 200 eq ha^{-1} yr^{-1}. The dark shaded areas show an excess of more than 1000 eq ha^{-1} yr^{-1}.

Figure 9.3: The Average Accumulated Excess of the deposition of sulphur and nitrogen over critical loads

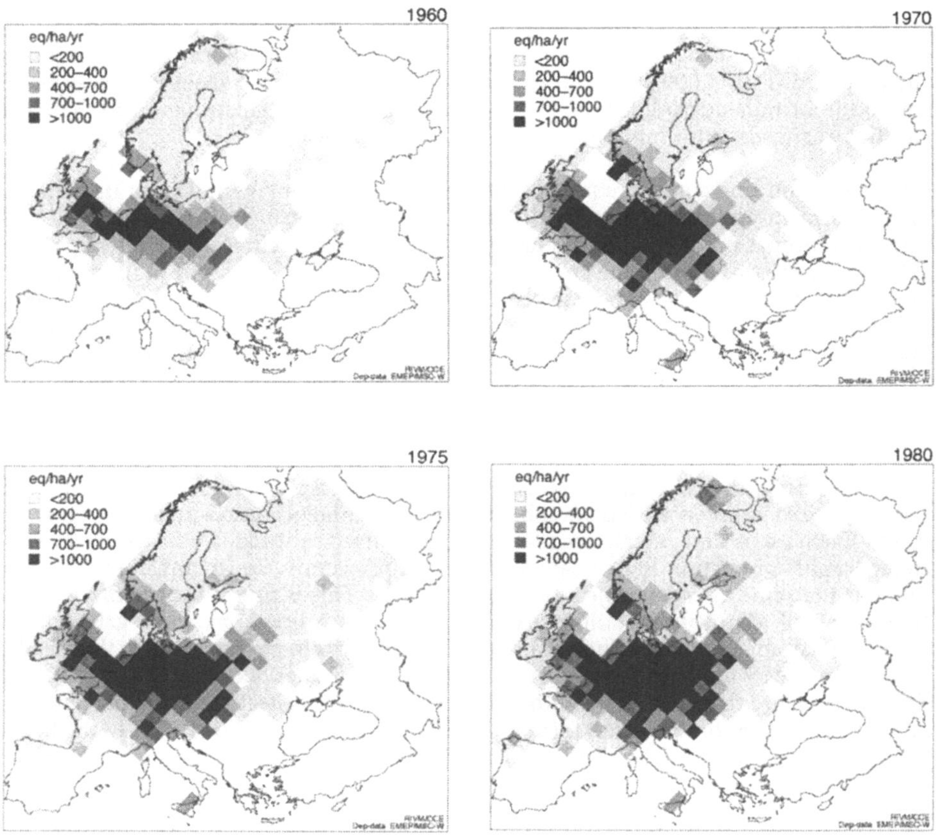

Figure 9.3 continued

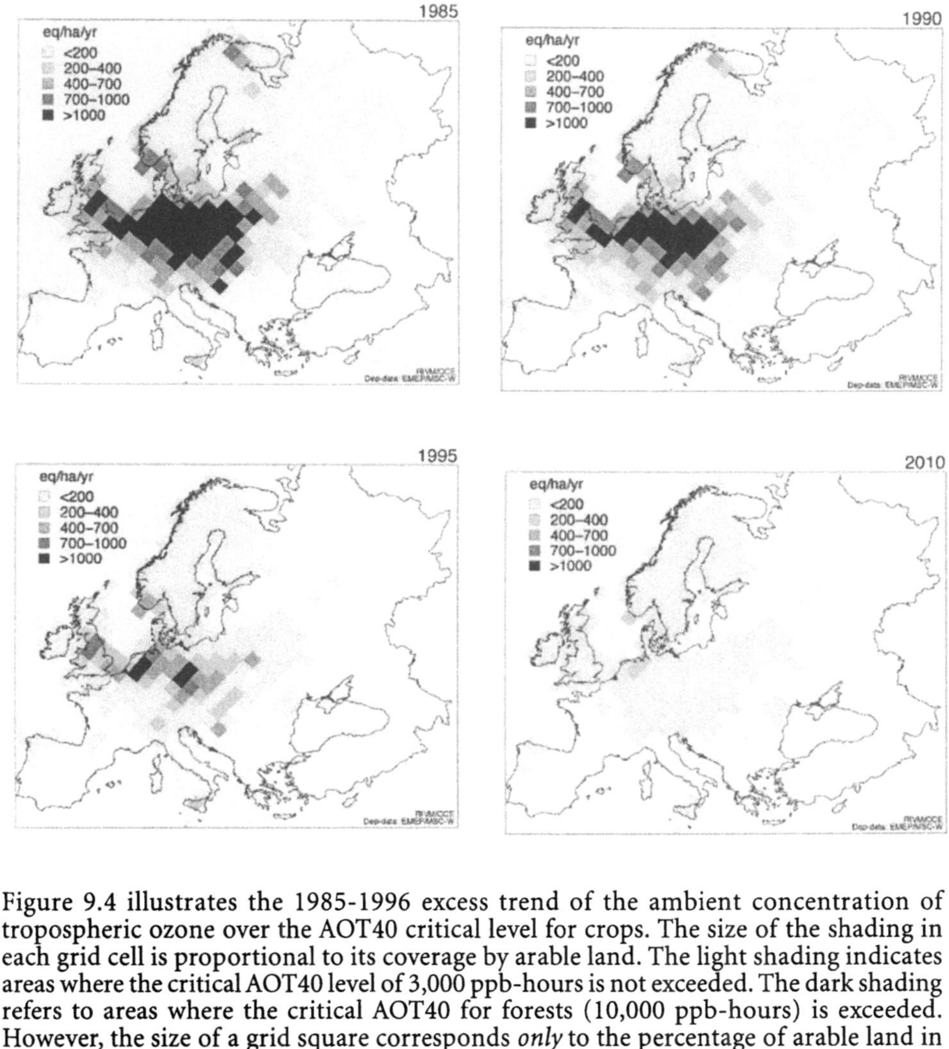

Figure 9.4 illustrates the 1985-1996 excess trend of the ambient concentration of tropospheric ozone over the AOT40 critical level for crops. The size of the shading in each grid cell is proportional to its coverage by arable land. The light shading indicates areas where the critical AOT40 level of 3,000 ppb-hours is not exceeded. The dark shading refers to areas where the critical AOT40 for forests (10,000 ppb-hours) is exceeded. However, the size of a grid square corresponds *only* to the percentage of arable land in that 150x150 km^2 grid cell. Ozone concentration data were provided by the EMEP/MSC-W (Simpson *et al.*, 1997). Maps have been computed using yearly meteorology leading to the display of the confounding effect of the inter-annual variability of meteorological conditions and the variability of ozone precursor emissions. Therefore, in the integrated assessment of tropospheric ozone, using RAINS, a five-year average meteorology is used.The assessment of critical loads and related data involves close international collaboration of scientific institutions in Europe, which is the subject of the next section.

Figure 9.4: The accumulated exposure to ground-level ozone concentrations over a threshold of 40 ppb (AOT40 for crops) in eight years between 1985 and 1996

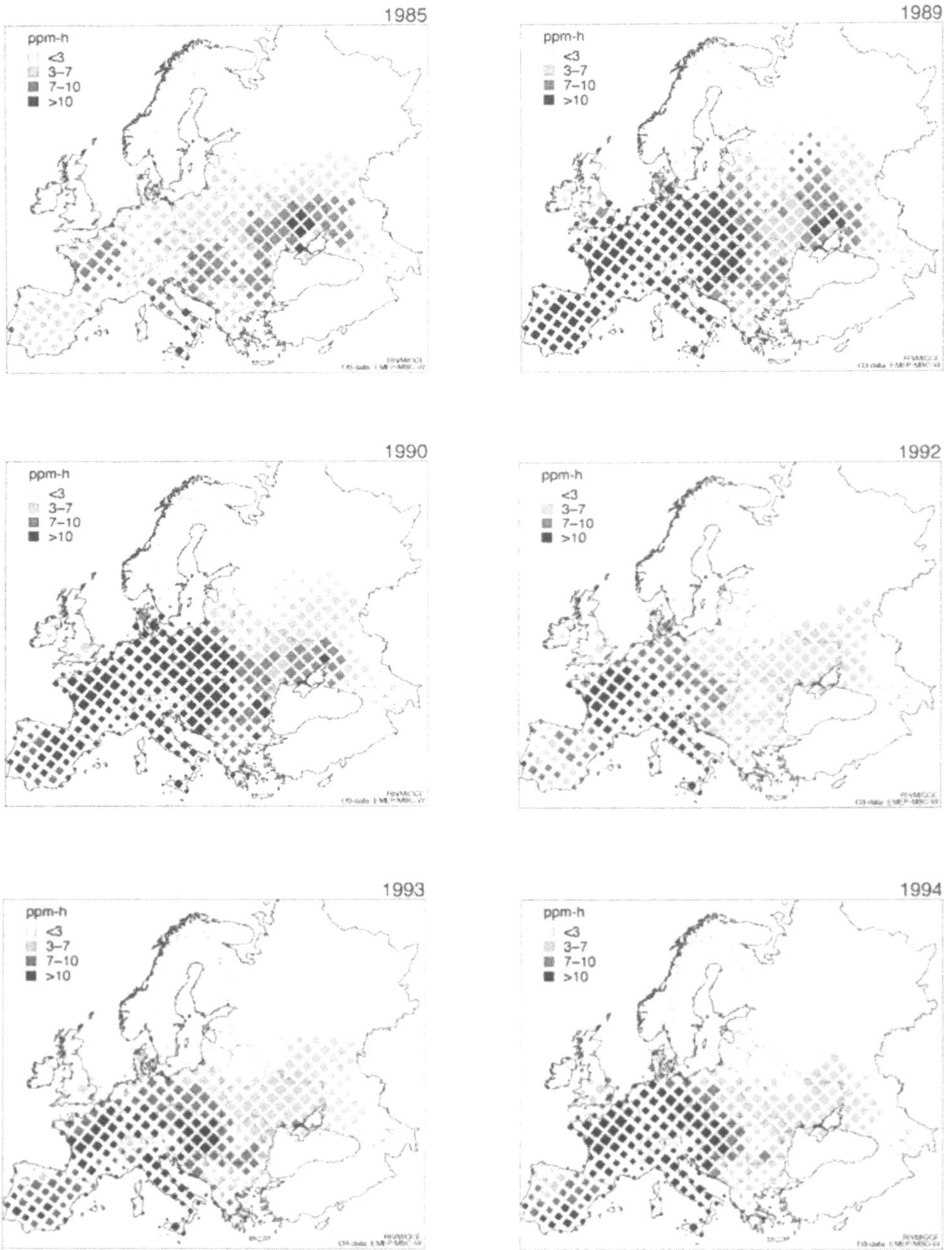

Figure 9.4 continued

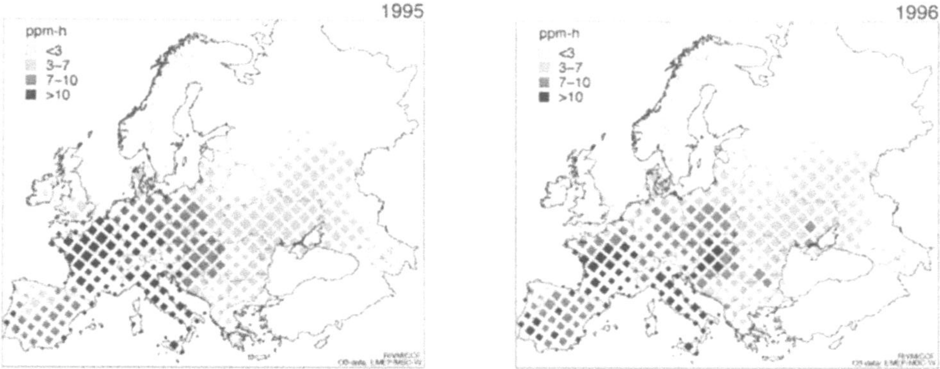

9.5 Scientific collaboration for the mapping of critical loads and levels in Europe

The International Collaborative Programme on Mapping under the LRTAP Convention currently consists of 24 institutions – so-called National Focal Centers (NFCs) – headed by the Coordination Center for Effects (CCE) at the National Institute of Public Health and the Environment (RIVM) in the Netherlands (http://www.rivm.nl/cce). The collaboration aims at the use of scientific methodologies and data for which consensus is obtained within the scientific community. Methods and data are submitted to regular review in (a) dedicated UN/ECE-LRTAP workshops including an annual CCE workshop, (b) UN/ECE-LRTAP reports and other grey literature, and (c) the open literature.

The fact that the collaboration is well framed in the policy structure of the Convention, including a mandate for the activities of the Mapping Programme and the CCE, has stepped up interaction between policy analysts and scientists both on a European as well as on a national scale. The objective of the network is to collect data to enable the assessment of critical loads based on models such as the SSMB – mainly targeting forest soils – or use alternative approaches to obtain critical loads based on other but soil-chemical criteria. Thus critical loads were obtained for a number of ecosystems which vary over countries in Europe (Figure 9.5). It can also be seen from Figure 9.5 that ecosystem coverage varies as well. Denmark, for example, is mapping less than 10 per cent of its country area providing critical loads for coniferous forests, deciduous forests and grass, whereas Norway is mapping almost 100 per cent of its country area addressing surface waters, natural vegetation and forest. The close collaboration with NFCs enables the assessment of critical loads reflecting more than 1.3 million ecosystems covering more than 5.5 million sq. km of Europe.

Information is made available to the CCE for each of the ecosystems shown in Figure 9.5 regarding variables required for equations 9.1 and 9.2, but also providing information allowing the CCE to perform consistency checks. An analysis of critical load and input data variability is part of the responsibility of CCE. Results are described in CCE reports (Posch *et al.*, 1997; 1999) which also publish detailed NFC documentation ensuring visibility of NFCs within the scientific community of the LRTAP. The explicit working relationships in the mapping community of the Convention have also lead to the formation of research consortia applying for external funding. The focus around the common effort of assessing critical thresholds has resulted in the collaboration of scientists from many different fields over the past decade in multi-disciplinary research projects, not restricted to work dedicated to the Convention. The attention of the network for land cover and land use will increase as the Convention will focus more on the verification of the inputs used in the negotiation process, including a further assessment and inventory of possible and actual damage. The latter will require further knowledge of the actual land cover, its historical trend and an assessment of possible future changes.

Figure 9.5: The national distribution of ecosystem types and their areas as percentages of total country area

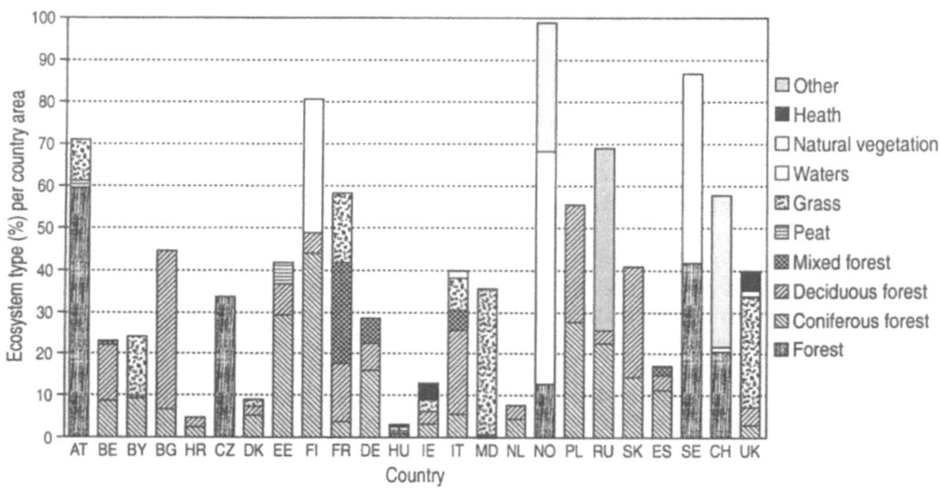

Source: After Posch et al. (1999, p.16)

9.6 Conclusions and recommendations

Information on land cover as well as soil chemical and plant physiological characteristics has successfully been used to quantify critical loads and levels which have been used to assess policy alternatives to reduce emissions of sulphur dioxide, oxidised and reduced nitrogen as well as of tropospheric ozone precursors. Critical loads and levels have been included in an integrated assessment model (RAINS) allowing the evaluation of national emission reductions in terms of cost (to install emission abatement equipment) while addressing the risk of environmental damage through the inspection of the excess of critical loads and levels of European ecosystems. Results of RAINS assessments have been used in the Convention on Long-range Transboundary Air Pollution of the United Nations Economic Commission for Europe as scientific-technical input to negotiations of a protocol on the further reduction of sulphur dioxide (1994) and a protocol to abate acidification, eutrophication and ground-level ozone (1999). In addition, a similar kind of support using RAINS and critical loads/levels was provided to the European Commission in its formulation of an acidification and ozone strategy.

The information on land cover is becoming increasingly important in relation to a large number of environmental problems assessed in both global and European policy frameworks. In a summary of an International Geosphere-Biosphere Programme, it is stated that *"to generate scenarios on future modifications on the Earth system, one has to project how changes in land use are likely to affect land cover in the future using integrated land use change models"* (IGBP, 1999, p.9).

In this chapter, it has been suggested that the so-called driving forces, pressures, state, impact, response (DPSIR) framework used for the systems analysis of causes and effects related to environmental issues in general, could help the structuring of cause-effect related land cover changes. This kind of exploration of relationships between environmental modelling, land use and land cover is recommended as an additional element in the formulation of research projects addressing land cover.

References

Alcamo, J., Shaw, R. and Hordijk, L. (eds.) (1990) *The RAINS Model of Acidification – Science and Strategies in Europe*, Kluwer Academic Publishers, Dordrecht, Netherlands, pp. 402.

Bartnicki, J. (1999) Computing source-receptor matrices with the Eulerian Acid Deposition Model, *EMEP/MSC-W note 5/99*, Norwegian Meteorological Institute, Oslo, Norway, pp. 37.

EEA (1995a) *Europe's Environment: the Dobříš Assessment*, European Environmental Agency, Copenhagen, pp. 676.

EEA (1995b) *A General Strategy for Integrated Environmental Assessment at the European Environmental Agency*, Scoping Study prepared by the National Institute of Public Health and the Environment (RIVM), European Environmental Agency, Copenhagen, pp 35.

EEA (1999) *Environment in the European Union at the turn of the century*, Environmental Assessment Report No. 2, European Environmental Agency, Copenhagen, pp. 446.

EMEP (1999) Transboundary photo-oxidants in Europe, *EMEP Report 2/99*, Norwegian Meteorological Institute (in collaboration with NILU, IER and UiO), Oslo, Norway, pp. 54.

Hettelingh, J-P., Downing, R.J. and de Smet, P.A.M. (1991a) Mapping critical loads for Europe, *RIVM Report*, Bilthoven, The Netherlands, pp 86.

Hettelingh, J-P., Meijer, E.N. and de Smet, P.A.M. (1991b) Mapping acidity in Europe, *GIS World*, 4(3): 42-44.

Hettelingh, J-P., Posch, M. and de Smet, P.A.M. (1995) The use of critical loads in emission reduction agreements in Europe, *Water, Air and Soil Pollution*, 85: 2381-2388.

Meijer E.N., Hettelingh, J-P. and Padding, P. (1995) Mapping critical loads of European ecosystems, in de Lepper, M.J.C., Scholten, H.J. and Stern, R.M. (eds.) *The Added Value of Geographical Information Systems in Public and Environmental Health*, Kluwer Academic Publishers, Dordrecht.

IGBP (1999) Land-use and land-cover change implementation strategy, *IGBP Report 48*, Royal Swedish Academy of Sciences, Stockholm, Sweden, pp 125.

Nilsson, J. and Grennfelt, P. (eds) (1988) *Critical Loads for Sulphur and Nitrogen, Nord 1988:97*, Nordic Council of Ministers, Copenhagen, Denmark, pp. 418.

Posch, M., Hettelingh, J-P., de Smet, P.A.M. and Downing, R.J. (eds) (1997) Calculation and Mapping of critical thresholds in Europe; Status Report 1997, Coordination Center for Effects, *RIVM Report*, Bilthoven, The Netherlands, pp. 163.

Posch, M., de Smet, P.A.M., Hettelingh, J-P. and Downing, R.J. (eds) (1999) Calculation and Mapping of critical thresholds in Europe; Status Report 1999, RIVM-Coordination Center for Effects, *RIVM Report*, Bilthoven, The Netherlands, pp. 165.

Schöpp, W., Amann, M., Cofala, J., Heyes, C. and Klimont, Z. (1999) Integrated assessment of European air pollution emission control strategies, *Environmental Modelling & Software*, 14(1): 1-9.

Simpson, D., Olendrzynski, K., Semb, A, Støren, E., and Unger, S. (1997) Photochemical oxidant modelling in Europe: Multi-annual modelling and source receptor relationships, *EMEP/MSC-W Report, 3/97*, Norwegian Mewteorological Institute, Oslo, Norway.

Simpson, D. and Posch, M. (1999) Report of the Modelling and Mapping session, in Proceedings of the Workshop on Critical Levels for Ozone – Level II, Swiss Agency for the Environment, Forests and Landscape (SAEFL), *Environmental Documentation No. 115*, Berne, Switzerland, pp. 333.

Tol, R.S.J. and Vellinga P. (1998) The European Forum on Integrated Environmental Assessment, *Environmental Modelling and Assessment*, 3(3): 181-191.

UBA (1996) Mapping Critical Levels/Loads and geographic areas where they are exceeded, *Umwelt Bundes Amt Texte 71/96*, Berlin, Germany, pp. 144.

UN/ECE (1999) Strategies and policies for air pollution abatement; major review prepared under the Convention on Long-range Transboundary Air Pollution, *Report ECE/EB.AIR/65*, New York and Geneva, pp. 147.

UNEP (1997) *Global Environmental Outlook, United Nations Environment Programme*, Oxford University Press, pp. 264.

UNEP (1999) *Global Environmental Outlook 2, United Nations Environment Programme*, Earthscan Publications Ltd., pp. 398.

Jean-Paul Hettelingh
Institute of Public Health and the Environment (RIVM)
Coordination Center for Effects (CCE)
P.O.Box 1
NL-3720 BA Bilthoven
The Netherlands

Maximilian Posch
Institute of Public Health and the Environment (RIVM)
Coordination Center for Effects (CCE)
P.O.Box 1
NL-3720 BA Bilthoven
The Netherlands

Peter de Smet
Institute of Public Health and the Environment (RIVM)
Coordination Center for Effects (CCE)
P.O.Box 1
NL-3720 BA Bilthoven
The Netherlands

10 FORECASTING GLOBAL CLIMATIC CHANGE IMPACTS ON MEDITERRANEAN AGRICULTURAL LAND USE IN THE TWENTY FIRST CENTURY

Stan Openshaw and Andy Turner

10.1 Background and context

This chapter is based on three years of research that started in 1996 and was undertaken as part of a European Union (EU) funded project concerned with Mediterranean desertification and land use. This project, known as MEDALUS III, consisted of a wide range of topics aiming to describe, analyse and model various aspects of climatic and environmental change at different spatial scales in the Mediterranean (MEDALUS III Final Report, 1999). One key objective of this work was to integrate socio-economics into physical models of agricultural land degradation. This task could have been tackled in various ways but the approach we adopted involved developing a prototype Synoptic Prediction System (SPS) to model relationships between climatic, physical and socio-economic variables in order to translate potential climate change scenarios into estimates of land degradation risk.

One task that still remains is to package up the various modelling components as a spatial decision support system that can be used to raise awareness of the problems, and develop mitigation strategies to alleviate them. For scientific research on land degradation processes (like MEDALUS III) to have political and public impacts, commensurate with the academic and theoretical importance of the subject, then it needs translating into a form that politicians and a non-technically sophisticated public can understand and appreciate.

There is a growing expectation that global climatic change will have a visible and increasingly fundamental environmental and socio-economic impact in the twenty-first century across many parts of the world. In some regions the effects may well be unnoticed or are irrelevant as they are well within the capacity of existing ecosystems to cope. However, in other regions like the Mediterranean, the environment could be far more fragile. In the Mediterranean, small changes in climate may be sufficient to cause a major impact on the environmental and socio-economic systems that relate to it. The ongoing research challenge is to identify and test ways of predicting climatic change impacts on agricultural land use as a possible basis for raising awareness and creating a framework for action. The hardness of this challenge should not be underestimated but equally there is an increasing urgency to look ahead so that thinking and strategic planning can be proactive rather than purely reactive.

At the start of MEDALUS III there was little published research that focussed on predicting future agricultural land use patterns and most computer models that existed seemed only to be concerned with one or two socio-economic aspects of the problem; for instance, regional employment change or population dynamics. An exception was the work by the CLUE Group (de Koning *et al.*, 1997; Veldkamp and Fresco, 1996; 1997; Verburg *et al.*, 1997). The CLUE (The Conversion of Land Use and its Effects) modelling framework was based on a multi-scale stepwise regression approach that attempted to analyse and model land use and land use change as a function of various socio-economic and biophysical

J. Stillwell and H. Scholten (eds.), Land Use Simulation for Europe, 127–141.

factors. The model was a simple linear continuous time simulation operating at a fairly coarse spatial scale; ranging from 7.5 sq km for Costa Rica to 32 sq km for China. For the EU it was felt that something far more sophisticated but along these lines was needed. The basic requirements of the research in 1996 were that:

- the level of spatial resolution was sufficiently detailed to be useful;
- the key driving factors and process mechanisms that link human and physical environments were explicitly incorporated; and
- that the entire system was driven by climatic change scenarios.

Additionally, the model was required to make good use of the research and model results produced in MEDALUS III.

The principal justification for a modelling approach was that a traditional map-based analysis of indicators had a high risk of failing to recognise the importance of many of the process variables. Additionally, such maps are static and realistically could only be produced and subjectively analysed in detail for small case study regions where our objective was to provide a synoptic overview across the entire EU Mediterranean climate region.

Section 10.2 outlines the structure of a Synoptic Prediction System (SPS) designed to forecast agricultural land degradation. Section 10.3 describes how the various data components were assembled. Section 10.4 describes the neural network based agricultural land use modelling. Section 10.5 describes the fuzzy logic based translation of forecast changes into land degradation risk indicators. Finally, Section 10.6 presents some general conclusions and suggestions for further research.

10.2 Building a Synoptic Prediction System

The objective of integrating socio-economics into MEDALUS III physical models evolved into the task of developing a prototype Synoptic Prediction System (SPS) designed to estimate plausible impacts of climatic change on agricultural land use across the entire Mediterranean climate region of the EU. Designing and developing the system was challenging because many of the theoretically desirable data sets were either not available or did not exist whilst there were significant uncertainties in all the available data. Additionally, system process knowledge was woefully deficient. Virtually all the principal mechanisms for linking the complex dynamics of the physical environment and climate with the associated socio-economic systems were poorly understood.

In essence the underlying and most basic problem was that physical and climatic process models were far better developed than any of the models of socio-economic systems related to agricultural land use. If you asked the question: 'What is the likely effect of climatic change on the hydrology of a specific river catchment or on the speed of erosion of a particular hill slope?', then methods existed that could indicate broad answers under changing land use conditions. However, if you asked the question: 'What is the likely impact of EU agricultural subsidies on the crops growing in the fields on this hill slope or on the agricultural land use of this catchment?', then there were no existing models for the current situation, let alone models which could forecast the effects in the medium to long term. Some novel computer modelling methodology thus had to be devised to satisfy the forecasting objective, cope with the deficiencies in the available data, and manage without good process knowledge.

10.2.1 *A prototype SPS*

The aims of the prototype SPS were threefold. Firstly, to devise a GIS-based model to predict contemporary agricultural land use using climate variables (temperature and rainfall), environmental variables (relating to soil, biomass, elevation, etc.), and socio-economic variables (including population density); secondly, to forecast future agricultural land use under various climatic change scenarios; and thirdly, to translate the predicted agricultural land use changes into surfaces of land degradation risk.

The results of applying the prototype under a given scenario reflect those of the long term weather forecast on which it was based, where regrettably and unavoidably, the geographical details are error prone and the forecasts maybe wrong for many small areas. The results from the prototype SPS are an estimate under a given climate scenario based on subjective interpretations imposed on the analysis.

10.2.2 *Problems*

The principal problem with the prototype SPS outlined here is that the relationship between the climate, environment and land use is not conditioned by the laws of physics or chemistry or biology but involves many ill-defined human factors. In particular, the relationship is (at least) mediated and affected by available technology, market mechanisms, historical tradition, inertia, culture, demography and various economic factors such as land price and subsidy rates. Most of these aspects were not explicitly present in any of the available data that were used.

It would have been preferable to run a process model faster than real time into which crop prices, EU agricultural subsidies, irrigation practices, and each farmer's micro-behaviour were input. However, such a model was beyond technological feasibility and data availability. It is possible that such a model will be built by discovering how to model individual people in artificial world laboratories using a bottom-up approach, but this is probably still years away. Although data for many process variables that were available were explicitly included in the model. At the time it was also hoped that those varaiables that were unavailable were taken into account implicitly by the neural nets that were applied to model agricultural land use. This was probably a most optimistic view, particularly when the forecasts made for 75 years hence assumed a continuation at the same level of all these 'invisible' influences that may or may not have been having a major direct impact.

Unfortunately, there was little that could be done about the constraints of technology, knowledge, data, and research resources. In defense of the prototype SPS, there need be no direct relationship between the overall modelling performance of model sub-components and the quality of the data used.

10.2.3 *Major stages in operationalising the SPS*

The prototype system involved the following seven stages:

Stage 1: Assembling the available data for a common EU-wide spatial framework for the present day;

Stage 2: Building models to estimate population (and other relevant socio-economic variables) for the same spatial framework;

Stage 3: Obtaining and making forecasts of all the variables of interest for about 75 years hence;

Stage 4: Modelling the relationship between climatic (temperature and rainfall), physical environmental (soil, biomass, elevation) and socio-economic variables to predict contemporary agricultural land use;

Stage 5: Forecasting land use for 75 years hence by applying the trained models from stage 4 to classify the forecast values from stage 3 under a climatic environmental change scenario;

Stage 6: Examining and mapping the forecast changes in agricultural land use; and

Stage 7: Interpreting these changes and transforming them into maps of perceived land degradation risk.

10.3 Assembling the data

There were major practical difficulties in assembling the database needed for this project. None of the socio-economic data existed at anything approaching the same geographical scale as the physical and climatic data and a large amount of GIS processing and often modelling applications were needed to represent all the required data from multiple different sources in the chosen spatial framework.

Major differences in data quality and resolution between the available physical and socio-economic data presented a serious obstacle to straightforward integrated agricultural land use modelling. MEDALUS III physical models of land degradation operated at much more detailed spatial and temporal scales than that at which socio-economic data were available. Available socio-economic data related to historically unstable, irregularly shaped zones, whereas physical data were available as grids albeit at a range of different scales. A major reason that much of the existing environmental change research had completely ignored socio-economic systems had already been identified as the lack of socio-economic data at an appropriate level of spatial and temporal detail so as to be directly linked with physical model outputs (Clarke and Rhind, 1991).

10.3.1 *Creating a common spatial framework database*

Appropriate available environmental data was acquired and manipulated into the chosen spatial framework – a geographical projection orientated in terms of latitude and longitude on a regular grid at a spatial resolution of 1 decimal minute (1 DM) using a GIS. A grid was selected as the spatial framework in which to store, manipulate, link and map all the data since it offered flexibility in aggregating upwards and could provide a realistic representation of regional or local variation provided the grid cells were sufficiently small in size. A geographical latitude-longitude projection was chosen to compromise the traditional problems in map projections regarding distance, direction and area distortion caused by the curvature of the earth. A 1 DM resolution (approximately 1 km^2 scale for most of the EU) was chosen as the most appropriate (smallest practically feasible) level of spatial resolution.

A common spatial framework for all the available climatic, physical and socio-economic data was considered essential for integrated modelling to be attempted. Aggregation upwards and interpolation from similar spatial resolution into the chosen spatial framework was relatively trivial compared with interpolation from much coarser levels of spatial resolution. Such disaggregative spatial interpolation was far more problematic and error prone, but some way of doing it needed developing before much progress could be made, see Openshaw and Turner (2001).

10.3.2 *Digital elevation data*

The first data set to be processed was the GLObal Land 1 km Base Elevation source data (GLOBE) (Eidenshink and Faudeen, 1996). This 30 arc-second (0.5 DM) resolution grid in a geographical latitude-longitude projection was imported into a proprietary Geographical Information System (GIS) and aggregated to a resolution of 1 DM as follows. Firstly, the 0.5 DM grid was aggregated from each of the four corners of the origin cell of that grid to produce four 1 DM resolution grids where the value in each cell was assigned the mean value of the four 0.5 DM cells from which it was composed. These four 1 DM resolution grids were then disaggregated back to the 0.5 DM resolution where each cell was assigned a quarter of the value at the 1 DM resolution. Now spatial consistent, the four disaggregated grids at a 0.5 DM resolution were combined to produce a single 0.5 DM grid where each cell was assigned the mean of the four incident grid values. Finally, these averaged measurements were aggregated back to a 1 DM resolution (selecting the point of aggregation so as the aggregate data fit the chosen common spatial framework) where again each cell was assigned the mean value of the four 0.5 DM cells from which it was composed.

This rather convoluted aggregation procedure reduces the spatial bias in the aggregated data in comparison with a more conventional one step aggregation. In this case, the reason for using the mean is because it is easier to compute than the mode or the median and the resulting 1 DM values approximate to the standard distance units of height above sea level. This variable was available for the whole of the EU and was used both for the agricultural land use modelling in the Mediterranean climate region and also for the disaggregative spatial interpolation of EUROSTAT socio-economic data.

10.3.3 *Night-time lights data*

The night-time lights source data is a measure of the frequency of night-time light based on composite satellite images. This (raster) grid of 1 km cells was imported into the GIS, converted into a (vector) polygon coverage and projected from its original Goode Homolosine projection into the required geographical projection using the projection capabilities of the software. The projected polygons were intersected with a polygon coverage coincident with the grid cells of the chosen 1 DM spatial framework. A value was calculated and attached to each intersected polygon by dividing the night-time lights frequency value by the area of the intersected polygon. The intersected polygon values within each 1 DM grid polygon were then added together and the resulting coverage was converted into a grid and clipped to the required size for the socio-economic data interpolation experiments.

10.3.4 *Digital map data*

Two sources of digital map data were used to create predictor variables for the socio-economic data interpolation exercise: Bartholomew's data (1:1,000,000 scale) and the Digital Chart of the World (1:1,000,000 scale). Data from both sources were combined and manipulated to produce various grids in the chosen spatial framework representing the location of, distance from, or density of either sets of geographical features or values of spatial variables as follows.

Firstly, all the various map layers were imported into the GIS, re-projected into the geographical latitude-longitude projection and mapped. Geographical features which seemed to be fairly consistently defined across the Mediterranean region of the EU and Great Britain, and whose location or density were believed to be correlated with population density, were manipulated into location, cost distance and density layers. Cells in location layers were bi-valued, depending on whether the cell partially or completely contained either a selected set of geographical features or a selected set of spatial variable values. Distance layer cells were assigned a value corresponding to the latitude-longitude cell distance to the nearest of a selected set of geographical features (or a selected set of spatial variable values). Proprietary GIS provided the simple functionality for point data, yet lacked functionality for creating density layers from line and area data. Density layers can contain more information compared with location and distance layers. For any particular set of geographical features there are many different density surfaces that can be generated and combined, and from these a huge number of location and distance layers can also be generated.

Proprietary GIS provided a kernel estimated density routine for point features where it was possible to control the range (bandwidth) of the kernel and specify a weight for the kernel height. For various sets of point features the effect of different kernel bandwidths on the density surfaces was investigated. It was found that as the bandwidth of the kernel was increased a larger proportion of the resulting density surface attained non-zero values. In addition, by combining density surfaces made using different kernel bandwidths even more useful density information could be created, and in each case the most useful cross-scale combination was dependent on the distribution of the geographical point features. After extensive mapping and analysis it was discovered that combining kernel density surfaces for overlapping cross-scale ranges of kernel widths by adding their values scaled relative to the size of the kernel which produced them offered the most useful density information for the impending disaggregative interpolation experiments. So, in this way cross-scale density surfaces were created for a various point features including railway stations that were expected to be positively correlated with population density and mountain summits that were expected to be negatively correlated with population density. Turner (2000) describes the creation of cross-scale density surfaces for various types of geographical data.

10.3.5 *Climate data*
Climate data were supplied by the MEDALUS III team from the Climatology Research Unit (CRU) at the University of East Anglia. The base-line data was produced by interpolating measurements from a network of about 50 weather stations across the northern Mediterranean to produce 0.5 DM resolution seasonal average temperature and precipitation totals, see Palutikof and Agnew (1997) for details. The base-line data were based on measurements made from 1970-1979 and were aggregated to produce the desired 1 DM resolution data using the convoluted aggregation procedure described above. In addition to the base-line data for 1970-79, forecasts of future seasonal average temperature and precipitation totals for 2030-39 and 2070-79 based on global climate change models were also made available by the CRU. These data were linearly interpolated with respect to time to produce maps of temperature and precipitation for around about now and 75 years hence. It was felt that the level of spatial uncertainty in these transformed climatic data were matched by equal or greater amounts of climatic scenario forecast uncertainty.

10.3.6 *Soil data*
Soils data came as the soils geographical database of Europe version 3.2 (at scale 1:1,000,000). A classification of 6 soil types was made with the help of a soil expert and each was converted into a location grid. Data concerning physical properties of soil profiles (soil type, rooting depth, soil texture, water regime, slope, and the existence of impermeable layers) were thus combined to produce soil quality layers. This involved coding the expert knowledge of the soil scientist into a set of fuzzy rules. Then via fuzzy inference the soil

quality layers were developed in a similar way to a general land capability classification, but in a simplified way that did not take into account all the physical interactions between the soil and climate (Burrough *et al.*, 1992).

10.3.7 *Biomass data*

Estimates of biomass were provided by MEDALUS III colleagues researching at the University of Leeds. These were outputs from a model which translated temperature and rainfall data via a set of equations and GIS into estimates of potential biomass (Kirkby *et al.*1997; Harris, 1998). The climatic biomass model was based purely on climate data and did not take into account factors like the height above sea level or soil type. The data became available at a 0.5 DM level of resolution towards the end of the project and was aggregated in the same way as the temperature data to produce estimates for the chosen spatial framework.

10.3.8 *Land use data*

Several different land use data sources were used. The soils data contained a classification of dominant and secondary land use; the US Geological Survey (USGS) land use and land cover classifications, including the Seasonal Land Cover Regions (SLCR) classification, (Anderson *et al.* ,1976); the International Global Biosphere Programme (IGBP) classification (Belward, 1996); the Biosphere-Atmosphere Transfer Scheme (BATS) classification (Dickinson *et al.*, 1986); the Global Ecosystems (GE) classifications (Olson, 1994a; 1994b); and the Simple Biosphere Model (SBM) (Sellers *et al.*, 1996).

Multiple land use classification data sets were used because there is both considerable uncertainty regarding the classification of land use and also variation in the details obtained from satellite imagery; for example, the SLCR had 255 different land use classes whereas the SBM had only 10. The land use modelling could have been greatly improved by using the probabilities of each cell belonging to particular land use classes based on the original unclassified satellite data (Moody *et al.*,1996); (Carpenter *et al.*, 1997). However, unfortunately these data were unavailable. For each of the land use classifications, the classes were aggregated into four groups: arable or crop, tree or orchard, shrub, and waste or barren. Each aggregated group was then converted into a location grid. In this way a cell of land could be seen as belonging to more than one land use class.

In addition to the aforementioned land use data, normalised difference vegetation index (NDVI) land cover data were obtained from the USGS. The NDVI data is an indicator of greenness or vegetation abundance derived from AVHRR satellite imagery. It was used to generate additional land degradation risk indicators. These data are available at a spatial resolution of about 1 km^2 and were obtained along with the USGS land use data from the following URL where there is a full description of the data: *http://edcwww.cr.usgs.gov/landdaac/glcc/glcc.html.*

The source NDVI data were obtained in a Lambert-Azimuthal equal area projection and reprojected into the chosen latitude-longitude projection and subjected to aggregation using the methods previously described so as to fit the chosen spatial framework. These data are monthly composites that were aggregated into seasonal values and clipped to the area of interest.

10.3.9 *Population data*

Demographic data at a fine level of geographical resolution were only available for the Great Britain and were estimated using a neural network based interpolation procedure for the rest of the EU. The source data used to train the neural nets were based on the 1991 grid-square population estimates for the UK known as Surpop. The Surpop 0.2 km resolution source data were manipulated into target data for the socio-economic data interpolation experiments in the following way. Firstly, the grid was imported into the proprietary GIS and converted into a polygon coverage in its source Ordnance Survey projection. The polygon coverage was then projected into the geographical latitude-longitude projection of the common spatial framework. These projected polygons were then intersected with a polygon coverage that coincides with a 0.125 DM grid that neatly aggregates to the chosen 1 DM resolution spatial framework. 0.125 DM sized polygons were used since they were approximately the same size as the projected polygons. The intersected polygons were assigned proportions of the population depending on the area of the intersection, then the intersected polygon values within each 0.125 DM polygon were summed and the resulting coverage was converted into a 0.125 DM grid. The total population in the source data was compared with the total population in the transformed data to account for any difference. The convoluted aggregation procedure described previously was then used iteratively until the population values fit neatly into the chosen spatial framework.

NUTS-3 and NUTS-2 resolution population data from EUROSTAT, NUTS-2 population forecasts from the Netherlands Interdisciplinary Demographic Institute (NIDI), 1991 UK Small Area Statistics census data, Italian National Statistical Office (ISTAT) population data for registration zone centroids, the World City Population Database (WCPD), and RIVM's and Tobler's gridded population estimates for Europe were also used in the socio-economic interpolation experiments. These data were manipulated into the common spatial framework using the techniques described above prior to being used in the modelling described below.

10.3.10 *Socio-economic data modelling*

Having assembled a common framework database of available data, the next major challenge was to create similar estimates of socio-economic data. The only readily available socio-economic data for the EU existed for NUTS-3 zones. In Great Britain there are 64 NUTS-3 zones and over 150,000 1 DM cells. So the task was to interpolate data for the 64 NUTS-3 zones in Great Britain to make estimates for the 150,000 1 DM cells using predictor variables that were also available for the Mediterranean region of the EU. Goodchild *et al.* (1993) contains a review of simple spatial interpolation methods that are relevant to creating surfaces of socio-economic data. All these methods suffer drawbacks in terms of the disaggregative spatial interpolation problem involved here. A 'smart' spatial interpolation procedure described in Deichmann (1996) was designed to make population density estimates based on potential surface accessibility relationships of related spatial variables. This idea of using surrogate information to estimate the distribution of socio-economic data was adopted and developed further by broadening the range of input predictor variables and using neural networks to model the relationships, see Openshaw and Turner (2001).

So, the idea was to use the aggregated Surpop population surface to train neural networks to predict population density based on a set of predictor variables. Training/validation data for EU Mediterranean regions were unfortunately unavailable. Consequently the task of making further detailed socio-economic and demographic breakdowns was not worth attemting. It was felt that the EU should organise its basic data resources to a far better degree to tackle this problem.

10.4 Agricultural land use modelling

Land use change and land degradation are complex interacting geographical processes which are themselves intricately related to climatic, physical and socio-economic change. Land use reflects land-capability that depends both on climatic interactions with other physical characteristics of the environment, and also land suitability that is complexly interrelated with various socio-economic factors including the characteristics of market supply and demand.

The task of modelling land degradation is summarised in the following seven steps:

Step 1: Create a contemporary geographical database of climatic, physical and socio-economic variables including agricultural land use.

Step 2: Model contemporary land use and land-cover based on variables in the contemporary database

Step 3: Compile a database of forecasts for those contemporary variables used to predict land use and land cover in Step 2.

Step 4: Forecast future land uses and land-cover by using the models developed in Step 2 to classify the data compiled in Step 3.

Step 5: Translate the estimates of land use change into a land degradation indicators.

Step 6: Create and obtain more land degradation risk indicators.

Step 7: Combine the land degradation risk indicators to create a synoptic scenario based land degradation forecast.

10.4.1 *Predicting contemporary land use and land cover*

Step 2 involved training neural networks to model agricultural land use based on patterns between; soil type, soil quality, potential biomass, average air temperature (in spring, summer, autumn and winter), average monthly precipitation (in spring, summer, autumn and winter), height above sea level, and population density. A list of predictors is given in Table 10.1. Different neural networks were trained to predict arable or crop, tree or orchard, shrub, and waste or barren agricultural land use groups. (Recall that various different aggregated land use classifications were used in effect as a substitute for the lack of unclassified remote sensing data that should have enabled the predictions to be based on probabilities.) The model predictions were used to create maps of predicted agricultural land use where the fit seemed remarkably good; see for example, Figure 10.1 showing predicted waste or barren land and the observed.

NDVI land-cover data modelling was undertaken in two stages: (1) predicting and forecasting the NDVI data based purely on physical and climatic variables and (2) predicting and forecasting also using socio-economic data inputs. The results provided some evidence of socio-economic effects and were used further to generate a land degradation risk indicator that was combined with the land use change based indicator in Step 6.

Table 10.1: Agricultural land use predictors

Predictor	Source
Average temperature in spring	MEDALUS III at CRU-UEA
Average temperature in summer	MEDALUS III at CRU-UEA
Average temperature in autumn	MEDALUS III at CRU-UEA
Average temperature in winter	MEDALUS III at CRU-UEA
Total precipitation in spring	MEDALUS III at CRU-UEA
Total precipitation in summer	MEDALUS III at CRU-UEA
Total precipitation in autumn	MEDALUS III at CRU-UEA
Total precipitation in winter	MEDALUS III at CRU-UEA
Annual climatic biomass potential	MEDALUS III at Leeds
Soil class	Soils geographic database
Soil quality	Soils geographic database
Height above sea level	GLOBE
Estimated average slope	GLOBE
Population in a 2 km radius	MEDALUS III at Leeds
Population in a 20 km radius	MEDALUS III at Leeds
Small & medium sized market density	Bartholomews
Distance to nearest town	Bartholomews
Accessibilty by road and rail	Bartholomews

10.4.2 Forecasting land use and land cover

Forecasting land use and land cover involved applying the trained neural networks from Step 2 to process forecasts of the predictor variables from Step 3. Differences between the contemporary and future predictions were mapped and analysed. Figure 10.2 shows the waste or barren land forecast for around 75 years hence.

Figure 10.1: Observed and predicted waste or barren land areas

Observed

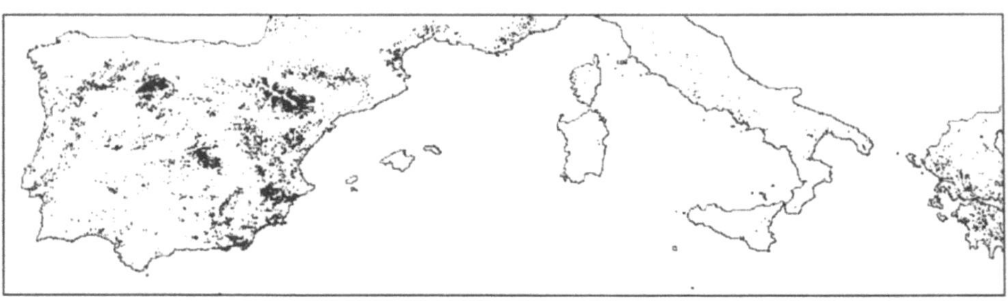

Predicted

Figure 10.2: Forecast waste or barren land areas

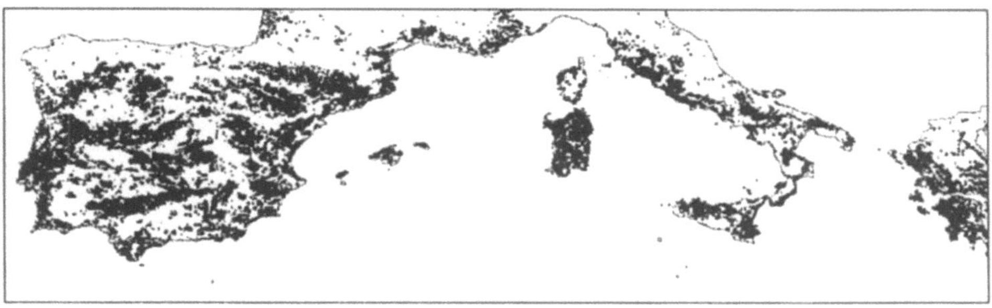

10.5 Translating agricultural land use changes into land degradation terms

The next major step involved translating agricultural land use changes into land degradation terms. This was subjective and the way in which environmental and economic interpretations of agricultural land degradation differ is important. Environmental interpretations tend to be geared towards things like soil quality, biodiversity, biomass, habitat preservation and conservation erring somewhat away from the long term exploitation of natural resources, whereas economic interpretations tend to give greater emphasis to things like crop revenue and land prices erring more towards short term exploitation of natural resources. Consider for example an area of land where only the aggregated tree or orchard land use class has been predicted as a contemporary land use and only the aggregated crop land use has been forecast as a future land use. A more environmental perspective might regard land degradation risk to be high because of likely reductions in biodiversity and biomass and increased soil erosion risk. A more economic viewpoint regarding likely increases in income from crops and potential economic multiplier effects from job creation may contend that the land use change is not degradation.

Fuzzy inference was used to translate the land use changes into land degradation risk surfaces and further combine these with other environmental risk indicators shown in Table 10.2. An illustrative set of the fuzzy rules are given in Table 10.3, and the resulting synoptic scenario based forecast consensus map of land degradation is shown in Figure 10.3. A superficial interpretation might be that Italy is relatively less prone to land degradation than the Iberian peninsular and that Greece is perhaps intermediate under the given climatic change scenario.

Table 10.2: Variables used in the fuzzy translation of land use into land degradation terms

Description	Source
Climatic Erosion Potential (CEP)	MEDALUS III at Leeds
Climatic Biomass Potential (CBP)	MEDALUS III at Leeds
Distance to and density of waste or barren land waste areas	USGS and Soils geographic data base
Seasonal and annual temperature and rainfall	MEDALUS III at CRU-UEA
Predicted and forecast land use	USGS and Soils geographic data base
Predicted and Forecast NDVI	USGS

Figure 10.3: Land degradation forecast

The map shown in Figure 10.3 can be interpreted at a synoptic scale and it is conceivable that such maps could be used by land use planners as decision making aids for allocating funds to implement land improvement schemes. Maps are essentially decision making aids and it is believed that the prototype SPS can be developed to aid spatial decision support relating to land degradation and help formulate policy to control agricultural land use change. With any luck the prototype will encourage informed debate and in the end may provide a means for politicians to justify fund allocation at national, regional and possibly even local scales without the need for a deep understanding of the science involved. The map of forecast land degradation for 75 years hence shown in Figure 10.3 is based on subjective interpretation of land use change related land degradation risk and its perceived importance relative to other land degradation risk indicators as translated via the set of fuzzy rules given in Table 10.3. Given more interpretations it would be possible to attempt to make a consensus. To facilitate this, we hope to develop a web front end to the modelling system and make the analysis interactive.

Table 10.3: Fuzzy rules used to generate a land degradation (LD) output map

1.	IF CEP was very high and increased a lot THEN LD is very high
2.	IF CEP was high and reduced a bit THEN LD is medium
3.	IF CEP was low and increased THEN LD is medium
4.	IF CEP was low and reduced THEN LD is low
5.	IF CBP is low THEN LD is high
6.	IF CBP reduced THEN LD is high
7.	IF CBP is high THEN LD is low
8.	IF CBP increased THEN LD is low
9.	IF temperature in summer was high and increased THEN LD is high
10.	IF rainfall in summer was low and decreased THEN LD is high
11.	IF annual rainfall reduced a lot THEN LD is high
12.	IF density of waste or barren land increased THEN LD is high
13.	IF distance from waste or barren land reduced THEN LD is high
14.	IF Prob(predicted contemporary land use is waste or barren) is low and the Prob(forecast land use is waste or barren) is high THEN LD is very high
15.	IF Prob(predicted contemporary land use is tree or orchard) is high and the Prob(forecast land use is tree or orchard) is low THEN LD is high
16.	IF Prob(predicted contemporary land use is non arable) is high and the Prob(forecast land use is non arable) is low THEN LD is low
17.	IF predicted NDVI reduced THEN LD is high
18.	IF predicted NDVI is low THEN LD is high

10.6 Conclusions

This chapter has outlined a prototype Synoptic Prediction System that provides agricultural land use and land degradation risk forecasts for the twenty-first century. The results reflect the available data and opinions concerning land degradation. The prototype captures the very essence of a GIS based approach to modelling environmental systems; it is visual, it is synoptic, and it is broadly representative. The main aim of this research was to raise awareness of the possible impacts of climate change on land use in the 21st century and suggest tools and approaches that can cope with the uncertainties involved in forecasting land use change. The research challenge remains and the modelling can be greatly improved with further research efforts as better data and process knowledge become available. The results only serve as a preliminary test of methodology; they are too subjective to be used as a benchmark. The prototype SPS appears to provide a useful framework for assessing the possible impacts of climatic change on agricultural land use by linking all the various components in a novel and interesting way. It simultaneously demonstrates what is needed to model the process of land degradation as well as giving some reasoned indication of the likely effects of climate change on Mediterranean agricultural land use.

Acknowledgements

We are very grateful for the contributions of Ian Turton and Tim Perrée of the Centre for Computational Geography at the University of Leeds for the early development of some of these ideas. Many thanks to all that made data available. Thanks to all MEDALUS III colleagues for their help, support and cooperation during and since the end of the contract. The research reported here was supported by EU Grant ENVA4-CT95-0121.

References

Anderson, J., Hardy, E., Roach, J. and Witmer, R. (1976) A land use and land cover classification system for use with remote sensor data, *Professional Paper 964*, U.S. Geological Survey.

Belward, A. (ed.) (1996) The IGBP-DIS global 1 km land cover data set (DISCover)-proposal and implementation plans, *Working Paper No 13*, International Global Biosphere Program-Data Information Service, Toulouse, France.

Burrough, P., MacMillan, R. and van Deurson, W. (1992) Fuzzy classification methods for determining land suitability from soil profile observation and topography, Journal of Soil Science, 43: 193-210.

Carpenter, G., Gopal, S., Martens, S. and Woodcock, C. (1997) Evaluation of Mixture Estimation Methods for Vegetation Mapping, *Technical Report CAS/CNS-97-014*, Boston University Center for Adaptive Systems and Department of Cognitive and Neural Systems.

Clarke, J. and Rhind, D. (1991) Population Data and Global Environmental Change, International Social Science Council Programme on Human Dimensions of Global Environmental Change, Working Group on Demographic Data.

Deichmann, U. and Eklundh, L. (1991) Global digital data-sets for land degradation studies: A GIS approach, United Nations Environment Programme, Global Resource Information Database.

Deichmann, U. (1996) A Review of Spatial Population Database Design and Modelling, United Nations Environment Programme, Global Resource Information Database, and Consultative Group for International Agricultural Research initiative on the Use of Geographic Information Systems in Agricultural Research.

de Koning, G., Veldkamp, A., Verburg, P., Kok, K. and Bergsma, A. (1997) CLUE: A tool for spatially explicit and scale sensitive exploration of land use changes, *Working Paper*, Wageningen Agricultural University, Wageningen.

Dickinson, R., Henderson-Sellers, A., Kenedy, P. and Wilson, M. (1986) Biosphere-atmosphere transfer scheme (BATS) for the NCAR community climate model: NCAR, *Technical Note NCAR/TN275+STR*, Boulder.

Eidenshink, J. and Faudeen, J. (1996) The 1-km AVHRR global land data-set: first stages in implementation, USGS, EROS Data Center, Sioux Falls, South Dakota.

Goodchild, M., Anselin, L. and Deichmann, U. (1993) A framework for the areal interpolation of socio-economic data, *Environment and Planning A*, 25: 383-397.

Harris, T. (1998) Regional land degradation indicators, *Unpublished PhD* Thesis, School of Geography, University of Leeds, Leeds.

Kirkby, M., Harris, T. and Abrahart, R. (1997) Regional Indicators, *Project 3 Report*, MEDALUS III.

MEDALUS III Final Report (1999) *Mediterranean Desertification and Land Use*, Report submitted to the European Commission on MEDALUS III 1996-1999.

Moody, A., Gopal, S. and Strahler, A. (1996) Artificial neural network response to mixed pixels in coarse-resolution satellite data, *Remote Sensing Environment*, 58: 329-343.

Olson, J.(1994a) Global ecosystem framework-definitions, *USGS EROS Data Center Internal Report*, Sioux Falls, South Dakota.

Olson, J. (1994b) Global ecosystem framework-translation strategy, *USGS EROS Data Center Internal Report*, Sioux Falls, South Dakota.

Openshaw, S. and Openshaw, C. (1997) *Artificial Intelligence in Geography*, Wiley, Chichester.

Openshaw, S. and Perrée, T. (1997) Methods for an intelligent continuous surface transformation of socio-economic data for Europe, Extended Abstract for 1st International Conference on Geocomputation, September 1997, Leeds.

Openshaw, S. and Turner, A. (1998) Medalus III project 9.1: GIS based socio-economic modelling:; *http://geog.leeds.ac.uk/staff/a.turner/projects/medalus3/home.htm.*

Openshaw, S. and Turner, A. (2001) Disaggregative spatial interpolation, Paper presented at GISRUK April 2001, Glamorgan.

Palutikof, J. and Agnew, M. (1997) Statistical Downscaling of Model Data, MEDALUS III report from the Climatic Research Unit at the University of East Anglia.

Ross, T. (1995) *Fuzzy Logic with Engineering Applications*, McGraw-Hill.

Sellers, P., Mintz, Y., Sud, Y. and Dalcher, A. (1986) A simple biosphere model (SiB) for use within general circulation models, *Journal of Atmospheric Science*, 43: 505-531.

Turner, A. (2000) Density Data Generation for Spatial Data Mining Applications, Paper presented at GeoComputation September 2000, Chatham.

Veldkamp, A. and Fresco, L. (1996) CLUE: a conceptual model to study the conversion of land use and its effects, *Ecological Modelling*, 85: 253-270.

Veldkamp, A. and Fresco, L. (1997) Exploring land use scenarios, an alternative approach based on actual land use, *Agricultural systems*, 55(1): 1-17.

Verburg, P., de Koning, G., Kok, K., Veldkamp, A., Fresco, L. and Bouma, J. (1997) Quantifying the spatial structure of land-use change: an integrated approach, ITC Journal Special Issue: Geo-Information for Sustainable Land Management.

Stan Openshaw
School of Geography
University of Leeds
Leeds LS2 9JT
United Kingdom

Andy Turner
School of Geography
University of Leeds
Leeds LS2 9JT
United Kingdom

11 DEMOGRAPHY, ECONOMY AND URBANIZATION: A DEMO-ECONOMIC REGIONAL SIMULATION MODEL

Leo van Wissen and Corina Huisman

11.1 Introduction

Since 1997, the Netherlands Interdisciplinary Demographic Institute (NIDI) has been involved in a project aimed at simulating regional demographic and economic dynamics, and the linkages with urban land use in north-west Europe. This project is financed by the Spatial Planning Office (RPD) of the Ministry of Spatial Planning and the Environment (VROM), and will result in a computer simulation model known as *Decores*. The model aims to assist planners and policy-makers to identify the consequences of demographic change at the regional level for regional labour markets and the impacts of regional labour markets processes on regional population dynamics. In addition, it assesses the implications of population and economic growth for urban land use change and those of increased urbanization and congestion for regional allocation of the population.

This chapter aims to provide an outline of *Decores*, to explain the interrelations between each of its components and to illustrate the results of the model by presenting a number of simulation outcomes. Initially, the framework of the model is described in Section 11.2. The modelling system is comprised of four modules – demography, labour supply, economy, urban land use – and each of these sub-modules is explained in Sections 11.3 to11.6 respectively. Section 11.7 contains the results of using the model to simulate economic growth and unemployment in the Netherlands from 1995 to 2025 and to simulate the development of population (by age group), migration, labour supply and urban land use in the Dutch provinces of Groningen and Zuid-Holland. Some conclusions are drawn in Section 11.8.

11.2 Structure of the model

The *Decores* prototype model has been constructed for a system of NUTS-2 regions comprising the territory of three EU member states, the Netherlands, Belgium and Germany. NUTS-2 regions refer to the 12 provinces of the Netherlands and the nine provinces of Belgium. In Germany, the NUTS-2 regions are equivalent to the 38 *regierungsbezirke*, which is one level below the *länder*. At present four modules are specified in the model:

- regional demography, which includes births, deaths, internal and external migration, each classified by age and sex;
- regional economy, which includes a regional production function;
- regional labour market in which labour supply, derived from demography, and labour demand, derived from regional production, are matched; and

J. Stillwell and H. Scholten (eds.), Land Use Simulation for Europe, 143–157.

- regional land use, with a distinction between urban and other land use.

The relationships between the modules and the components within the modules determine the dynamic structure of the model which is presented in Figure 11.1. Starting with the population module, the model works anti-clockwise. Within the population module, births, deaths and external migration are treated first, using parameter values from the most recent regional Eurostat population scenarios (van der Gaag *et al.*, 1999). Internal migration is treated endogenously, since it is dependent on regional unemployment and population density. The population module determines the change in population structure between times t and t+1, using input parameters as well as the unemployment level and population density level at time t.

Figure 11.1: Structure of the *Decores* simulation model

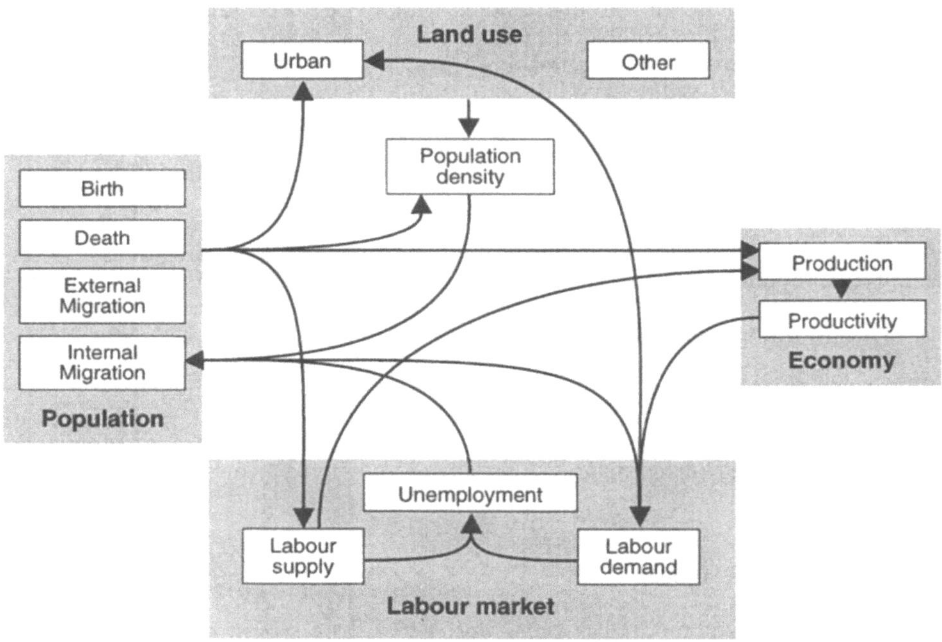

The change in population at the regional level between t and t+1 in turn determines the change in labour supply. A regional production function then calculates the change in gross regional product and from this, using the inverse of the productivity ratio, the change in regional labour demand is projected. Labour supply and demand are matched in the labour market module, which generates unemployment. At present, a commuting module is not implemented. Discrepancies in regional labour supply and demand lead to unemployment. In the first operational version of *Decores* this will also result in changes in the commuting balance of regions. Finally, changes in the population and the number of jobs in the region are translated into changes in urban land, thus completing a cycle of the model between time t and t+1.

Decores is a dynamic model that simulates annual changes in all its components, and changes in one component have an effect on changes in other components. There are a number of reciprocal relationships in the model. Depending on the specification of these relationships, these reciprocal relationships may be stable or unstable. In stable relationships, the model will return to (a new) equilibrium after an exogenous shock, whereas in unstable relationships, a disturbance will move the model permanently out of equilibrium.

An example of a stable relationship is as follows: increase in population → an increase in labour supply → increase in unemployment → decrease in internal migration → decrease in population → decrease in labour supply → and so on. In contrast, an unstable relationship is exemplified as follows: increase in population → increase in regional production → increase in labour demand → decrease in unemployment → increase in internal migration → increase in population → increase in regional production → *et cetera.*

The total effect of both relationships depends on the exact specification of the model parameters. As a result of this specification, the model may be stable or unstable, or stable in certain ranges of parameter values and unstable elsewhere. An analytical description of its properties is infeasible, given the complexity of the structure, and can only be discovered in simulations.

11.3 Demographic module

The starting point for the demographic module was the latest Eurostat regional population scenarios, commissioned by the European Commission and produced by NIDI and Statistics Netherlands (van der Gaag *et al.*, 1999). These long-term regional scenarios were made for NUTS 2 regions in all the countries of the European Union for the period 1995 to 2025. The demographic module in *Decores* differs in one important respect from the projection model used in the Eurostat scenarios in that the internal migration component is endogenous. In a later version of the model, the international migration component will be made endogenous as well. For the exogenous components – births, deaths and (at present) international migration – the baseline scenario of the Eurostat scenarios was used, which describes a more or less average future development, and which is closest to the baseline projections produced by each individual country. The assumptions for births, deaths and international migration are exactly equal to those used in the Eurostat project. The form of the regional demographic model is a separate multi-regional cohort component model for each country.

The four components of (regional) population change (fertility, mortality, international migration and internal migration) are based on the certain assumptions (Table 11.1). Under the baseline scenario, the average number of children a woman will probably have in her life, the total fertility rate (TFR), is expected to stabilise. At the same time, women who postponed childbearing, are having children in their thirties, which results in minor increases in fertility levels. The regional differences are expected to converge slightly. For mortality, life expectancies at birth are expected to continue to increase and regional differences are expected to become smaller in the future under the baseline scenario. Until the year 2000, the average annual increase will be equal to that observed during the past ten years (1985-1995). After the turn of the century, the average increase is expected to slow down. Also, gender differences will become less marked.

Table 11.1: Demographic assumptions according to the baseline Eurostat scenario at national level

	Total fertility rate		Life expectancy Men		Life expectancy Women		Net international migration (1,000)	
	1995	2025	1995	2025	1995	2025	1995	2025
Belgium	1.56	1.80	73.62	79.23	80.23	84.54	1.78	1.40
Germany	1.28	1.50	73.35	77.90	79.73	83.24	5.15	2.38
Netherlands	1.58	1.81	74.62	78.75	80.46	83.75	0.88	2.00

Migration flows into the European Union have been very closely related to specific historical events, particularly in recent years. Examples of events that triggered large migration flows to the EU are the reunification of Germany, the fall of the Iron Curtain and the war in former Yugoslavia. Although migration is hard to predict, it seems unlikely that the high levels of immigration observed at the beginning of the 1990s will continue. According to the baseline scenario, future international migration levels will be lower than in the recent past. Furthermore, international migrants will most likely be more evenly distributed over the regions.

In Decores, there are two options for the internal migration module. In the first or 'consistency' option, the internal migration component is consistent with the baseline scenario in the Eurostat projections. Under the baseline scenario, the level of internal migration as well as the regional differences are expected to remain constant at the base-period level, namely the 1990-1994 average. In the second and more interesting 'no consistency' option, internal migration is endogenous and dependent on regional unemployment, population density and the number of jobs in the region; these are covariates that are endogenous in the model. The form of the internal migration module in this second option is a spatial interaction model.

In the internal migration module of the model, we work with internal out-migration rates, internal in-migration shares and interaction parameters that indicate to what extent migrants from a region differ in preferences for a destination region. It turns out that these interaction parameters are very stable over time (van Imhoff *et al.*, 1997), and that changes in the pattern are mainly caused by the 'edges' of the migration-matrix: total out-migration and total in-migration for each region. In the *Decores* model, these variables are dependent on non-demographic variables and they are not stable over time.

The model formula for the out-migration probability is expressed as:

$$Prob(Outmig) = exp\ \{Z\}\ /\ (1 + exp\{Z\}) \tag{11.1}$$

which is a binary logit formulation for grouped data. The linear sum, Z, is calculated as:

$$Z = \chi_1+\chi_2 Unem+\chi_3 Dens+\chi_4 Dens^2+\chi_5 LD \tag{11.2}$$

where *Unem* is equal to the regional unemployment figure relative to the national level percentage of the labour force, *Dens* is equal to population density (persons/km^2), $Dens^2$

the density squared, and *LD* the regional share of the number of jobs. The parameters χ_1 to χ_5 of equation 11.2 were estimated using data for the Netherlands in the period 1990-1996. Pooled data was used as well as data for separate years. Models were estimated based on cross-sectional as well as on first differences. The results of the first difference method using the period 1992-1994 were the most satisfactory (t-values in parentheses):

$$Z = 5.49*10^{-4} + 0.019Unem - 5.84*10^{-3}Dens + 2.27*10^{-6}Dens^2 + 19.51LD \qquad (11.3)$$

(0.04) (2.04) (-2.44) (1.85) (2.52)

R^2 (adj.) = 0.52

The results indicate that if unemployment increases, so does out-migration. The relationship with population density is non-linear and U-shaped (Figure 11.2). Population density in the Netherlands has a maximum value of 1,200 inhabitants per km². In other countries, occasionally higher densities can be found. The estimated coefficients show that negative consequences of congestion will lead to high increases of out-migration at very high density levels (2,000 inhabitants per km²). The turning point however is at 1,300 inhabitants per km².

Figure 11.2: Relationship between population density (inhabitants/km²) and out-migration probabilities estimated for Dutch provinces, 1992-1994

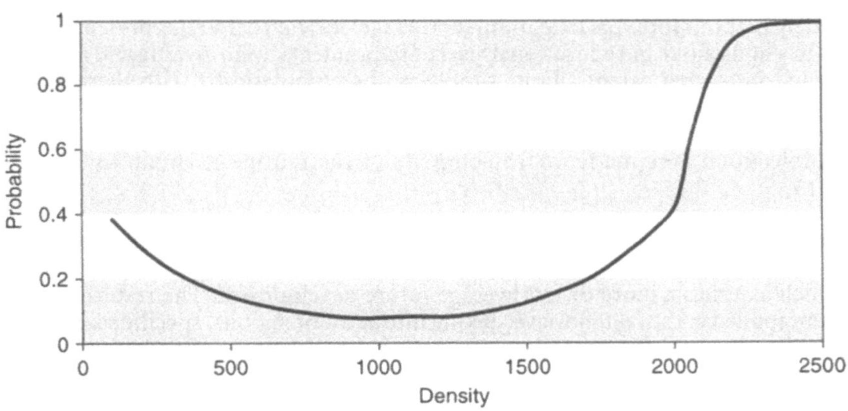

The model for in-migration shares has the following form:

$$P_i = exp\{Z_i\} / \Sigma_k \, exp\,\{Z_k\} \qquad (11.4)$$

which is a multinomial logit model, that guarantees that $\Sigma_i P_i = 1$. The form of *Z* is equal to that of in-migration:

$$Z = \lambda_1 + \lambda_2 Unem + \lambda_3 Dens + \lambda_4 Dens^2 + \lambda_5 LD \qquad (11.5)$$

This model was also estimated using cross-sectional as well as first differences for the period 1990-1996 in the Netherlands. The most satisfying results are produced using cross-sectional data for 1994:

$$Z = 0.098 - 0.089Unem + 4.81*10^{-4}Dens - 6.25*10^{-7}Dens^2 + 7.52LD \qquad (11.6)$$
$$\qquad\qquad (0.28) \qquad\qquad (0.66) \qquad\qquad (-0.41) \qquad\qquad (2.09)$$
R^2(adj.) = 0.56

Most of the λ coefficients are not significant, but all are of the right sign. These results indicate that the higher the unemployment rate, the lower the in-migration. In addition, the relationship between density and in-migration shares is a dome-shaped curve. For relatively low density values, the attractiveness increases with increasing density. After a certain threshold value, which is different for each region, the pull effect diminishes. Using out-migration rates and in-migration shares, a matrix of flows was estimated using historical flow information about origins and destinations of internal migrants.

11.4 Labour supply

Labour supply is calculated by applying regional participation rates to the (projected) population. In *Decores*, activity ratios are exogenous to the model and were taken from existing labour supply scenarios prepared for the European Commission by NIDI (van Dalen *et al.*, 1999). These labour supply projections were also based on the Eurostat baseline projections which form the basis for the demographic module of *Decores.*

In the study aimed at formulating meaningful labour supply scenarios, key factors that may influence labour force participation, such as age, sex, marital status, presence of young children in a household and educational level of respondents, were investigated and insights were gained into what extent labour supply varies systematically with these factors. A number of assumptions about future developments in these key factors were formulated in a number of scenarios, and using the established relationship with activity ratios, labour supply projections were made for 15 countries of the European Union for the period 1996-2015.

These labour supply scenarios predict participation ratios (and consequently labour supply) at the national level for all 15 EU countries. In *Decores*, the base line scenario was used, which assumes a more or less average future development. The resulting national ratios were applied at the regional level, taking into account regional specific scaling factors, based on the observed age-group specific regional participation figures in 1995. From 2015 onwards, participation rates are assumed to be constant. In Figure 11.3, the 1995 female participation rates for the three countries are depicted.

11.5 The economic module and labour demand

Modelling labour demand can be made a very complicated task. This is even more true at the regional scale. This route was not followed in *Decores.* Instead, a simple production function was specified, relating relative growth in gross regional product (ΔGRP) to relative growth in regional exports ($\Delta Export$), regional population (ΔPop) and regional labour supply (ΔLS). This formula was proposed by Ledent (1980). Here, exports and population are two final demand categories, and labour supply is an indicator of the size of the production system, or intermediate demand. The function is a simple linear equation, and, since all variables are specified in relative change terms:

$$\Delta GRP(t) = \beta_1 \Delta Export(t) + \beta_2 \Delta Pop(t) + \beta_3 \Delta LS(t) \qquad (11.7)$$

The β coefficients are elasticities, since all variables are specified as growth rates. ΔPop and ΔLS are endogenous and produced by the model itself. $\Delta Export$ is exogenous. The elasticities have to be estimated using empirical data. Figure 11.4 indicates the growth rate for export on a national level in the period 1991-1994. The negative value for Germany in the period 1992-1993 is remarkable and no doubt has to do with the unification of the country.

Figure 11.3: Female participation rates by age for Belgium, Germany and the Netherlands, 1995 (national level)

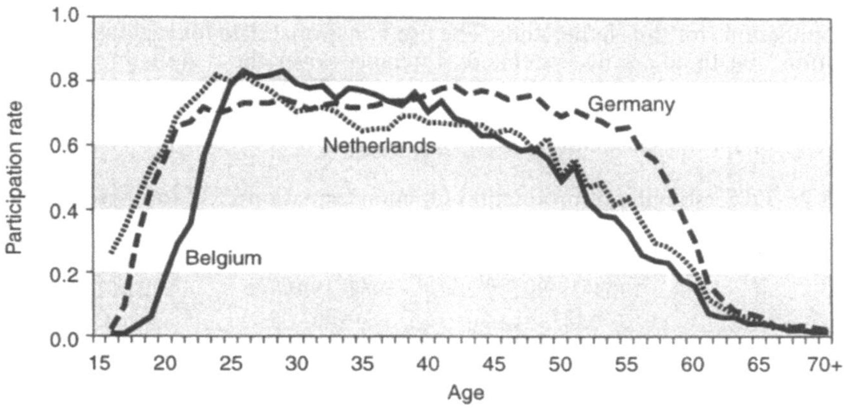

Figure 11.4: Annual and average export growth in Germany, Belgium and the Netherlands, 1991-1994

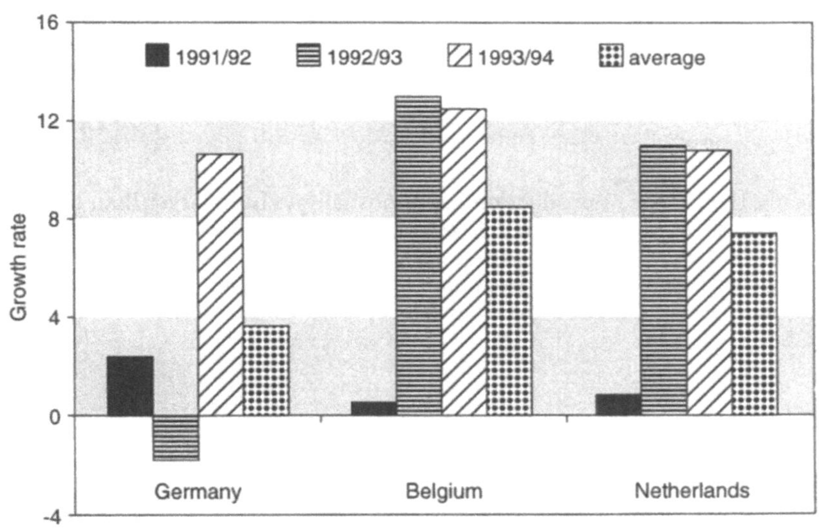

Since regional time series of exports were not available, their values could not be based on region-specific data. To obtain an idea of the relative importance of each term at the national level, the model was estimated from national data of the Netherlands, Germany and Belgium. The results for the Netherlands and Belgium are relatively similar (Table 11.2). The elasticity of population is larger than 1. Economic growth may be more related to the growth in the number of consumption units, or households, which is also larger than the growth of the population. The elasticity of the production system (indicated by the labour supply) is around 0.5. The export elasticity is slightly lower. The German results are different, but these were estimated from a much shorter time series, that includes the unification of Germany. Based on these results, we used the estimated Dutch elasticities in the simulations for the Netherlands. The use of national data for regions rests on the assumption that there are no systematic differences over the regions in terms of the coefficients of the production function (equation 11.7). Although this assumption will certainly not hold in empirical testing, we have at present no additional information to include region-specific elasticities.

Table 11.2: OLS-estimates of production function (equation 11.7) for three countries

	Germany 1984-1996 n =12*		The Netherlands 1961-1996 n=36		Belgium 1962-1996 n=35	
Parameter	Coeff.	t-value	Coeff.	t-value	Coeff.	t-value
ΔExport	0.24	3.90	0.30	5.89	0.39	6.52
ΔPopulation	0.39	0.52	1.17	1.90	1.29	1.38
ΔLabour supply	0.13	3.16	0.45	2.60	0.56	0.82
R^2	0.72		0.73		0.64	

* *1991 excluded*

Labour demand is derived from the production function. Annual change in labour demand is calculated as:

$$\Delta LD(t) = \alpha\, \Delta GRP(t) \tag{11.8}$$

Historically, the change in production has almost always been larger than the growth in employment, so α is smaller than 1. α is the job elasticity with respect to regional production. In the seventies, it was around 0.8, in the eighties around 0.95, and since 1990 it is around 0.97 (van Wissen and Huisman, 1999). This means that the annual increase in productivity ($1/\alpha - 1$) has become smaller over time. In the model, a value of 0.95 was specified.

11.6 Urban land use

Changes in urban land use (*Urban*) are modelled using the following formula:

$$\delta\, Urban(t) = \gamma_0 + \gamma_1\, \delta\, Pop(t) + \gamma_2\, \delta\, LD(t) \tag{11.9}$$

where the operator δ specifies the growth of a variable x in absolute terms: $\delta x(t) = x(t+1)-x(t)$. Note that this is different from the operator Δ, which results in a growth *rate*. The coefficients, γ, denote the change in urban land use per inhabitant of the region and per employed person. They are estimated using time series of Dutch data. The estimated model is:

$$Urban = \underset{(0.81)}{1.53} + \underset{(3.82)}{1.63 \cdot 10^{-4}}\, \delta Pop(t) + \underset{(1.23)}{2.16 \cdot 10^{-4}}\, \delta LD(t) \qquad (11.10)$$

$R^2 = 0.78$

Each additional person in a region adds 163 m^2 urban land to the region, and each new job adds 216 m^2 of the urban land to the region respectively. These values were used in the model. The change in urban land is assumed to reduce the areal agricultural land. Negative change in the areal urban land is not possible.

In order to evaluate the impact of each of the parameters of the model on the results, a number of sensitivity analyses were performed. These analyses showed that the most important parameters are the export elasticity β_1 and the population elasticity β_2. The other parameters have less profound effects on the performance of the model. Of course, the picture may be different for each region. In regions with a high unemployment rate, changes in the unemployment parameter will have larger effects than in regions with low unemployment.

11.7 A number of illustrative outcomes of *Decores*

A number of simulation outcomes can be used to show the type of results that may be obtained with the model. However, a number of remarks should be made at the outset. First, scenario results for population and labour supply are based on existing scenarios for the European Union. With the exception of the internal migration module, which is endogenous in the model, these values are exogenous and fixed. Therefore, population and labour supply only react to developments in the economy or labour market through the internal migration module. Other mechanisms, primarily changes in activity rates are not incorporated in the model, which makes the labour supply side in *Decores* probably less responsive to changes in the economy than may occur in reality. Second, the uncertainties pertaining to the economic and labour demand module are probably much larger when compared to the population and labour supply module. We will come back to some of these mechanisms in the concluding section, but an important characteristic of the model results is that they are based on simulations for the Netherlands, using the estimated elasticities β_1-β_3 for the production function and a job elasticity α of 0.95. As described in the previous section, these values reflect the observed values of the Dutch economy in the recent past. Nevertheless, based on these coefficients, a significant labour shortage will appear within a few years.

In Figure11.5, economic growth and unemployment for the Netherlands are shown in two simulations ('estimated' and 'adjusted'). The first simulation uses the estimated elasticities of production function (β_1=0.30, β_2=1.17, β_3=0.45) and job elasticity (α=0.95), together with observed export growth in the period 1995-1998, and assumed export growth diminishing from 2.5 per cent in 1999 (the average annual growth rate for the Netherlands in the period 1991-1998) to zero in 2013 and afterwards. This scenario results in moderate to small positive economic growth, but large labour shortages after 2000.

Figure 11.5: Unemployment (left) and economic growth rate (right) in two simulations

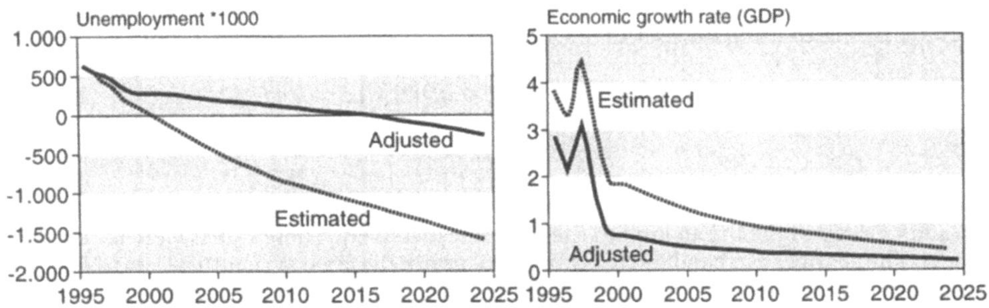

We can compare the outcomes of this model with observed economic growth and unemployment until 1999 using adjusted elasticities (β_1=0.30, β_2=0.60, β_3=0.30) and job elasticity (α=0.80). Economic growth in the period 1995-1998 is about 1 per cent overestimated. In reality, economic growth varied between 2.5 and 3.5 in this period. By the end of 1999, unemployment in the Netherlands had dropped to a level of 200 thousand, which is only slightly higher than the value predicted by the model of 135 thousand. If this trend is continued, large imbalances will occur in the labour market, even in a scenario with export growth rates decreasing to zero.

In reality, a number of economic mechanisms will mitigate the labour shortage resulting from a shrinking labour supply: increased activity rates (not implemented in this version of *Decores*); increased productivity (lower values of α in *Decores*); increased immigration (not implemented in this version of *Decores*); and increased imports and reduced exports (lower values of elasticities β_2 and β_3 as well as lower export growth).

Some of these adjustments are implemented in the second simulation shown in Figure 11.5 ('adjusted'). Here, the elasticities have values β_2=0.60, β_3=0.30, and α = 0.80. Moreover, export growth after 1999 is zero. This results in a simulated economic growth rate between 2 and 3 per cent in 1995-1998, which is close to observed values, and decreasing until 0.2 percent in 2025. Unemployment almost halves between 1995 and 1999, which is again close to observed figures, and reaches zero around 2017, after which it becomes negative. Even in this moderate growth scenario, characterized by zero export growth, large increases in productivity and, compared to the past, a much larger reliance on imports the labour market will be very tense in the next decades.

The adjusted model is the basis of a number of illustrative simulations reported below. In Figures 11.6 and 11.7 we show a number of results of the model, at the national and the regional level of the Netherlands. First, in Figure 11.6 the simulated population development of the Netherlands is depicted. The age distribution of the population is shown in Figure 11.6. Here, the population development in three age categories for the Netherlands clearly shows the process of ageing in the next decades. The three components of population change are also shown in Figure 11.6: birth, death and international migration. The latter component is the most important component of population change in the next decades.

Figure 11.6: Population development in the Netherlands 1995-2025 according to *Decores* basic scenario, by age category (left) and components of change (right)

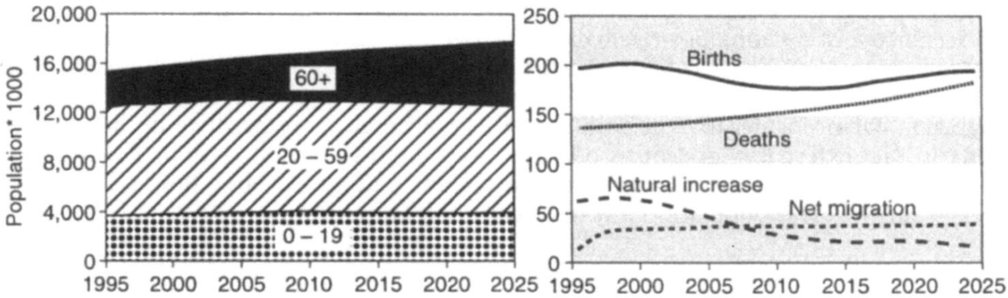

The development of labour supply and labour demand over time is shown in Figure 11.7. Labour demand is based on the adjusted parameter values of the economic module. This clearly shows the narrowing gap between labour supply and labour demand, obviously resulting in lower unemployment rates, as discussed above.

Figure 11.7: Labour demand and supply in the Netherlands 1995-2025 according to *Decores* basic scenario

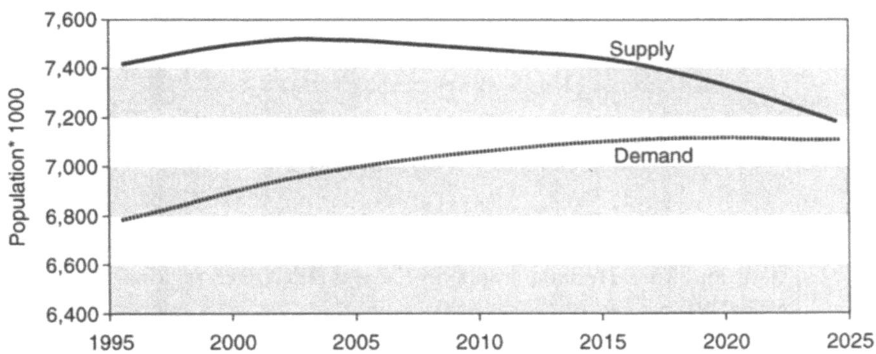

A number of regional results are presented in Figure 11.8. Here, the results for two regions are shown: Zuid-Holland, the largest and most urbanized region, that includes the conurbations of Rotterdam and The Hague, and Groningen, one of the smaller and rural regions, in the north. Results are presented for two scenarios: *EU*, which is consistent with the EU regional population scenarios, and *Decores*, having endogenously determined internal migration.

In Figure 11.8 the development in the regional population of Groningen and Zuid-Holland is depicted for the two scenarios. Clearly, the *Decores* scenario, with endogenous internal migration, results in a higher population development for Groningen and a negative population growth for Zuid-Holland. This is caused by the negative impact of population density in Zuid-Holland, having a strong outward push effect on internal migrants. Thus, in *Decores*, a strong deconcentration takes place as a result of the negative effects of population density. Figure 11.9 shows the large difference between the scenarios in terms of internal migration. Groningen has a negative net internal migration component in the EU scenario, but a strong positive balance in *Decores*, which even grows over time. For comparison, we have included the external migration balance as well, which is small but positive for Groningen, and is similar for both scenarios. In Zuid-Holland the internal migration balance is negative in both scenarios but much more negative in the *Decores* scenario. Net external migration is positive throughout the whole period. This picture reflects the common pattern of large urban regions that attract large flows of external immigrants, but lose population due to internal migration.

Figure 11.8: Total population development 1995-2025 in Groningen and Zuid-Holland in *EU* and *Decores* scenarios

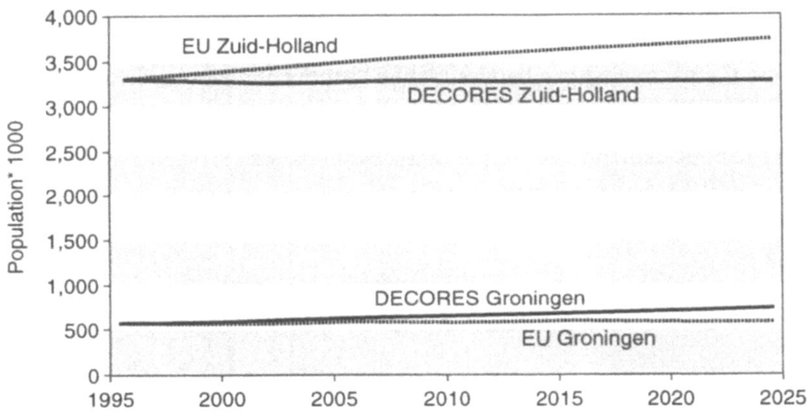

Figure 11.9: Internal and external migration development 1995-2025 in Groningen (left) and Zuid-Holland (right) in *EU* and DECORES regional population scenarios

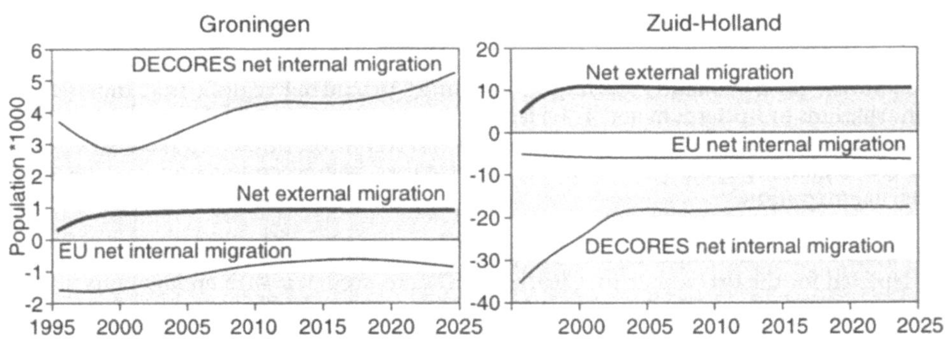

Figure 11.10 shows the effects of these developments for the age distribution in both provinces. The *Decores* scenario leads to a relatively younger population in Groningen, having relatively more young people, more people in the working age category and less people in the retired age category, than in the *EU* scenario. The reverse is true for Zuid-Holland, with the exception that here also in the *EU* scenario the number of 60+ is larger than in the *Decores* scenario.

Figure 11.10: Population development 1995-2025 by three age groups for Groningen (left) and Zuid-Holland (right) in *EU* and *Decores* scenarios

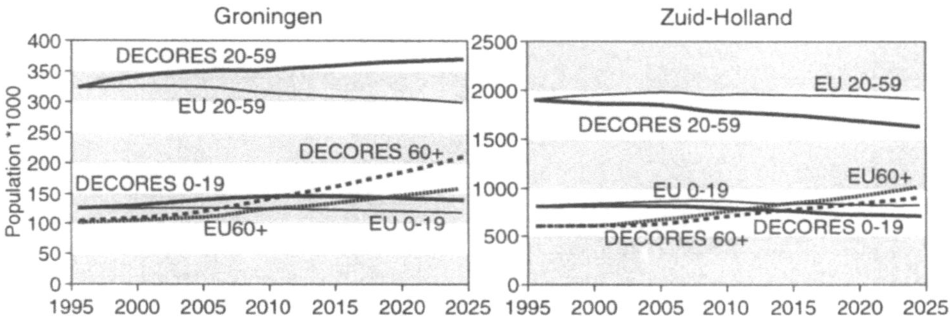

The differences between both scenarios in terms of labour supply and demand are shown in Figure 11.11. For Groningen, in the *EU* scenario, labour demand will exceed supply before 2020. In the *Decores* scenario, supply and demand will almost match in 2025 in this province. In Zuid-Holland, in the *Decores* scenario, labour supply reduces drastically, and around 2010 demand exceeds supply. In the *EU* scenario, supply will remain stable until 2015 and in 2020 the number of jobs will increase above the number of employed. Thus, in both scenarios in Zuid-Holland, labour shortages will occur after 2010, but there is roughly a time difference of up to 10 years between both scenarios. These labour shortage problems will be larger in Zuid-Holland than in the peripheral region of Groningen.

Figure 11.11: Development in labour supply and number of jobs 1995-2025 in *EU* and *Decores* scenario in Groningen (left) and Zuid-Holland (right)

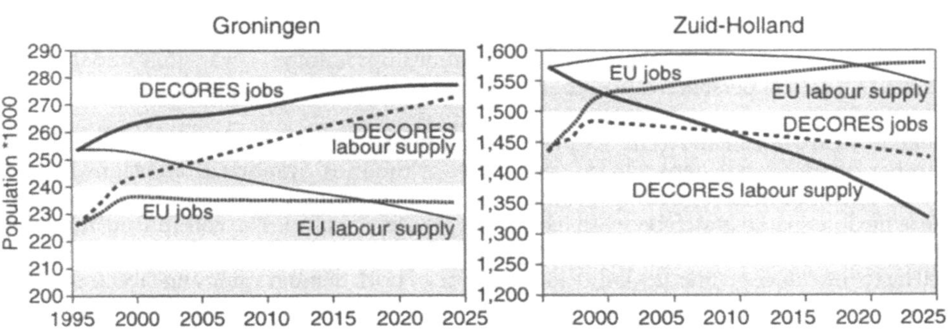

Finally, in Figure 11.12, the consequences of population and economic change for urban land use are depicted. The difference between both scenarios in terms of urbanization is clear. In the *Decores* scenario, the urban area in Groningen will grow substantially, whereas in the *EU* scenario, growth occurs in Zuid-Holland and not in Groningen. This reflects the strong deconcentration effect of the congestion parameter in internal migration.

Figure 11.12: Development 1995-2025 of urban land (in km²) in Groningen and Zuid-Holland in *EU* and *Decores* scenarios

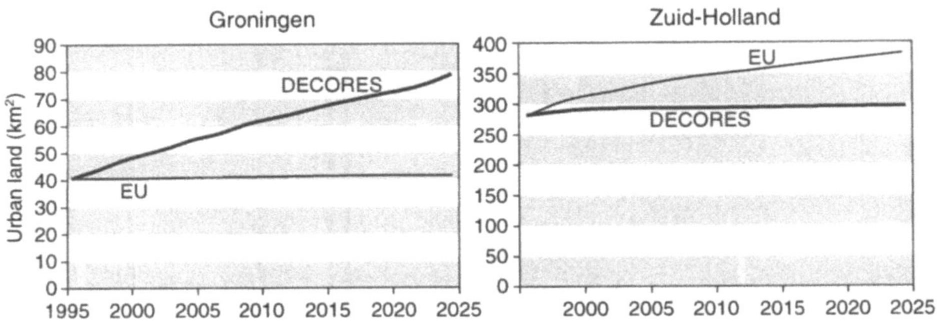

11.8 Conclusions

Decores aims at integrating the most important societal sectors that shape the spatial and regional landscape. The model structure presented links regional demography, economy, labour market and urban land use dynamics in a unified framework. In this effort, increased insights into and knowledge of often complex interrelationships are more important than quantitatively correct results. Confronting demographically determined labour supply with simple models of labour demand shows potential problems in the regional labour market. In addition, the consequences for urban land use of the added demographic and economic changes is helpful for prospective regional planning.

Some of the components are more detailed than others. The demographic module and the labour supply module are more elaborated than the other modules. An important characteristic of the model is the reciprocal relationships between a number of components, e.g. between demography and the labour market, or between land use and demography. However, the exact nature of these reciprocal relationships is not yet always fully understood from a theoretical perspective and more research in this direction is necessary.

Some links are not included in the present version of *Decores.* In subsequent research, a number of additional steps will be taken. One of the most important additions to the model structure involves adding a commuting module to the system, thereby allowing a more flexible spatial match between labour supply and demand. The commuting module will have an effect on internal migration and job relocations as well. Another improvement will be to introduce greater flexibility in the choice of both demographic and labour supply parameters and make the regional allocation of international migration endogenous (i.e. based on economic and other indicators). In addition, border migration between the three countries is an important process that should be taken into account explicitly. Finally, more attention should be given to obtaining more data for each of the countries involved, and to parameter estimation.

Acknowledgements

This chapter is based on research sponsored by the Netherlands Spatial Planning Office (RPD) of the Ministry of Housing, Spatial Planning and the Environment (VROM).

References

Dalen, H. van, Huisman, C. and Imhoff, E. van (1999) Arbeidsaanbod in de Europese Unie, convergentie of divergentie, *Demos*, April 1999, NIDI, The Hague.

Gaag, N. van der, Wissen, L. van, Imhoff, E. van and Huisman, C. (1999) National and regional population trends in the European Union, *Working Paper 3/1999/E/No. 8*, Eurostat.

Imhoff, E. van, Gaag, N. van der, Wissen, L.van, and Rees, P (1997) The selection of internal migration models for European regions. *International Journal of Population Geography*, 3: 137-159

Ledent, J. (1980) A demo-economic model of interregional growth rate differences, *Geographical Analysis*, 55-67.

Wissen, L. van, and Huisman C. (1999) *Decores* An integrated demo-economic regional scenario model of northwest Europe, Version 1, 22 June 1999, *NIDI Working Paper*, The Hague.

Leo van Wissen
Netherlands Interdisciplinary Demographic Institute (NIDI)
Postbus 11650
2505 AR Den Haag
The Netherlands

Rijksuniversiteit Groningen (RUG)
Faculteit der Ruimtelijke Wetenschappen (FRW)
Postbus 0100
9700 AV Groningen
The Netherlands

Corina Huisman
Netherlands Interdisciplinary Demographic Institute (NIDI)
Postbus 11650
2505 AR Den Haag
The Netherlands

12 INTERREGIONAL MIGRATION AND LAND USE PRESSURE

Beatrice Eiselt, Nicla Giglioli and Robert Peckham

12.1 Introduction

A recently completed project carried out by the European Commission's Joint Research Centre on behalf of the European Statistical Office (EUROSTAT, 1997) set out to explore the possibility of using European socio-economic data available in the REGIO database to model interregional migration flows. The aim was to look for correlations between migration and other socio-economic statistics available at European level, and to assess whether they could be used in simulations related to the European Spatial Development Perspective (ESDP).

The work made use of spatial interaction models (otherwise known as gravity models) and, after an initial assessment of the data quality and availability, it followed the classical steps of exploratory analysis, model formulation, model fitting and finally testing the use of the model for prediction. A GIS-based tool for visualization of the actual, modelled and predicted migration flows was also developed as an aid to exploring and understanding the migration flows, and as a means to visualize the results of model predictions.

Some useful correlations were found, in particular between migration and GDP, and between migration and unemployment rate. Since one of the motivations for change in land use is the migration of people, and this can be expected to be influenced by economic driving forces, this work can give some indications on how it may be possible to link these factors in future simulation models for land use change.

The description begins with a very short outline of data and software in Section 12.2. Clusters of regions are then identified on the basis of four variables in Section 12.3 and regression models of interregional migration are calibrated in Section 12.4. An optimum model is used to generate simulated migration projections in Section 12.5, where a GIS-based visualization tool is used for mapping the flows according to user specification. Section 12.6 offers some suggestions as to how the modelling work might be developed so as to provide inputs to models aimed at simulation of land use pressures and changes. Some conclusions are drawn in Section 12.7.

12.2 Data and software

The REGIO database provides information on agriculture and forestry, demographic statistics, economic accounts, energy statistics, Community labour force survey, research and development, transport statistics, and unemployment in European Union regions. The present work has made use of the demographic statistics at NUTS-1 and NUTS-2 levels (NUTS = Nomenclature of Territory Units for Statistics) which include tables of annual average population by sex, and interregional migration matrices. The tables of

J. Stillwell and H. Scholten (eds.), Land Use Simulation for Europe, 159–170.

GDP and unemployment rates at NUTS levels 1 and 2 from the economic statistics were also used. However, as these data are compiled by assembling and harmonising data available from national organisations, which have different levels of availability and time coverage, this has lead to some gaps in the data which restrict the range of countries and time spans to which the modelling can presently be applied. These considerations of data availability have restricted the tests so far to concentrate on Germany, Italy and Spain for time periods within 1990-1994.

The software used for the model fitting and statistical analysis was SPSS (version 8.0) while ARC-VIEW was used for the GIS and visualization as this is now the standard desktop GIS product chosen for use in the European Commission. The link between the attribute data in REGIO database and the polygons representing administrative boundaries (available from the GISCO database) is conveniently established using the NUTS codes.

12.3 Cluster analysis

The objectives of the cluster analysis are two-fold. First of all, clustering is used as a descriptive tool to get an overview of the general situation in each of the three countries. Secondly, it is used for classification, as the grouping of the regions into homogeneous clusters helps in the visualization of the results. Since the regions can be effectively divided into classes, the flow patterns can be displayed by cluster, or by cluster-to-cluster, rather than by region.

The cluster analysis was based on the values of total migration inflow divided by population, total migration outflow divided by population, GDP per inhabitant, and unemployment rate (percentage of unemployed out of total workforce). These variables are relative and are hence not influenced by the population size of the regions. By using these variables, the clusters should represent the economic, employment and mobility variations in each of the countries.

The regions used are the 20 NUTS-2 regions of Italy in 1994, the 16 NUTS-1 regions of Germany in 1993 and the 18 NUTS-2 regions of Spain in 1994. For each cluster, the most representative region, i.e. the region closest to the average value, is determined. By looking at the characteristics of the representative region of each specific cluster, information about the group of regions is implied. For each country, the latest year with a complete data set was selected.

When interpreting the clusters, it is important to keep in mind that the variables included in the cluster analysis not only reflect the economic and employment situation in each of the regions. The fact that in-migration and out-migration intensities or rates are included adds a less intuitive dimension to the clustering. It reflects more general attractiveness/repulsiveness features of the regions and the general propensity to migrate of people in the regions. It is expected that people tend to leave poor regions or regions with high unemployment rates and tend to be attracted to rich regions or regions with low unemployment rates. However, there are clearly other and more complex reasons for people migrating.

The clusters for Italy are shown in Figure 12.1. Cluster 1 agglomerates most of the northern and central regions, having high GDP/inhabitant and rather low unemployment rates. For these regions, the inflow is greater than the outflow, indicating that they are attractive regions. Cluster 2 includes regions with high GDP/inhabitant and low unemployment rates, and the biggest in-migration and out-migration rates. Again, the inflow is greater than outflow. Cluster 3 is composed of some of the southern regions characterised by rather low GDP/inhabitant and high unemployment rates, while the migration is high and the outflow is bigger than the inflow, indicating that they are net losers. Finally, cluster 4 is comprised of the poorest southern regions with the lowest GDP/inhabitant and the highest unemployment rates. Here, migration is low with higher outflows than inflows.

Figure 12.1: Region clusters and representative regions in Italy

The clusters for Germany are illustrated in Figure 12.2. Cluster 1 represents the regions of West Germany with relatively high GDP/inhabitant and low unemployment rates. Berlin is an exception, with its relatively high unemployment rate, very high GDP per inhabitant and greater outflows than inflows. The in-migration and out-migration flows of the other regions in the cluster are medium but generally the inflows are greater than the outflows. Cluster 2 represents former Eastern Germany and is characterised by very high unemployment rates and the lowest GDP/inhabitant. The outflow is bigger than the inflow. Cluster 3 represents regions with very high migration rates both into and out of the regions concerned.

Figure 12.2: Region clusters and representative regions in Germany

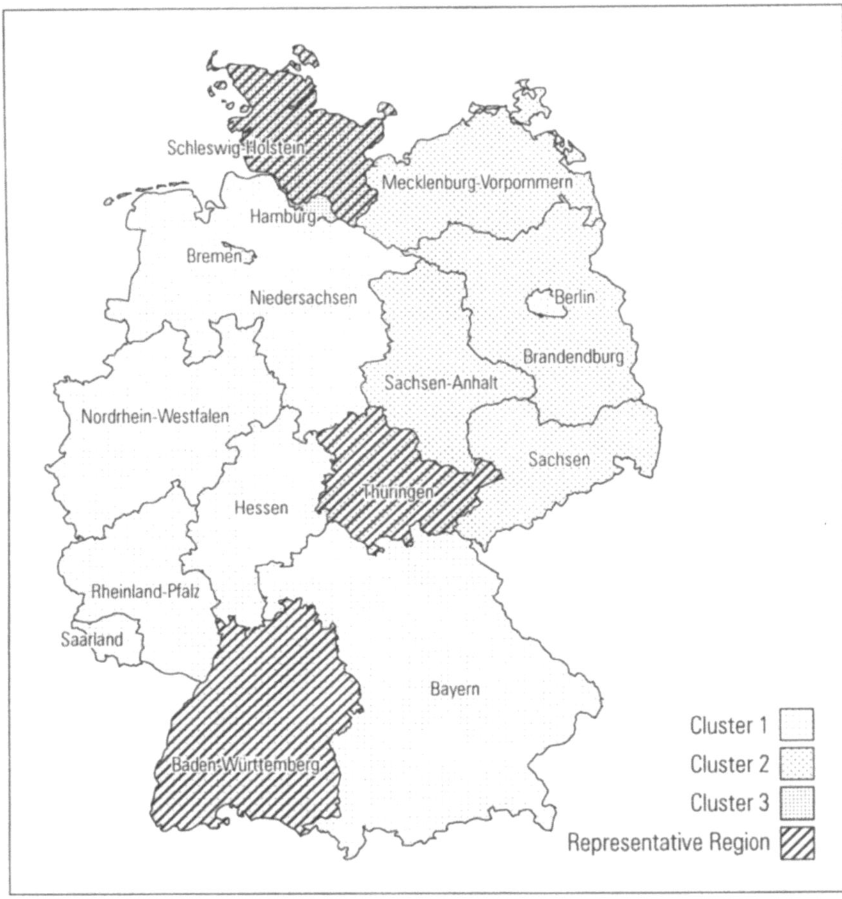

A similar cluster analysis was also performed for Spain. Overall the clusters obtained in all three countries agree with our perception of the political/economical/social situations. For each country the total migration in the country (note that since we only deal with internal migration, arrivals and departures are equal) has been plotted against age to get an idea of mobility of different age groups. These typically indicate significant peaks in the age range between 20 and 40 years suggesting migration for reasons of employment. Furthermore, in order to appreciate the differences between the clusters, the net rate versus age has been plotted for each cluster. An illustration of the results for Italy is shown in Figure 12.3 indicating pronounced peaks in the vicinity of 30 years. Unfortunately these considerations cannot presently be exploited in the modelling part, since the migration matrices are not available for age groups in REGIO.

Figure 12.3: Age structure of flows for clusters in Italy

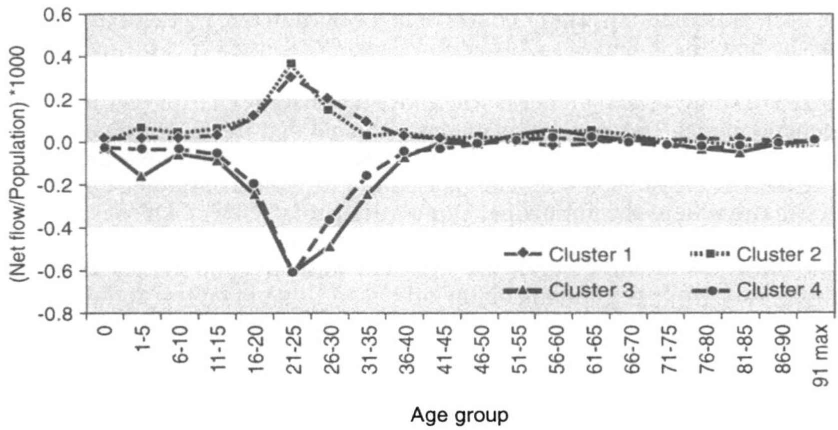

12.4 Migration modelling

For describing and explaining interregional migration, spatial interaction models have been explored, as they have been used by a number of previous studies with some success (e.g. Fotheringham and O'Kelley, 1989; Stillwell and Congdon, 1990; Sen and Smith, 1995; Nijkamp amd Reggiani, 1999; Bailey and Gatrell, 1995; Rees *et al.*, 1996). These models have components representing origin repulsiveness, destination attractiveness, and the distance decay effect. These components can either be abstract parameters to be estimated or observed explanatory variables.

The general spatial interaction model has the following form:

$$Y_{ij} = \mu_{ij} + \varepsilon_{ij} \tag{12.1}$$

where Y_{ij} refers to the migration flow from i to j, and ε_{ij} is an error term, and

$$\mu_{ij} = \alpha_i \beta_j e^{\gamma d_{ij}} \tag{12.2}$$

where α_i are calibrated parameters that characterise the propensity of each origin to generate flows, β_j are calibrated parameters that characterise the attractiveness of each destination, and γ is a calibrated distance deterrence parameter, where d_{ij} refers to the distance between i and j. From this model, more detailed models can be developed by substituting the origin repulsiveness and destination attractiveness parameters with a function of one or more explanatory variables, such as, in our case, total GDP, GDP per inhabitant and unemployment rate. Such a model is desirable because it has explanatory power.

Models developed from the general model commonly used are those where the logarithm of the flow is a linear function of the explanatory variables or a linear function of the logarithm of the variables (double logarithmic model). Another possibility is modelling the flow divided by population, either of origin or of destination or of both. When fitting the models, two different distributions of the data can be assumed, the Poisson assumption and the normal assumption. The goodness of fit for the different models can be evaluated considering both the R^2 and the adjusted R^2 values.

In the present study, several models of the above type have been explored, including the most general model. The correlation analysis implied that the most influential factors on migration are population and total GDP. The normal distribution assumption turned out to be the more reasonable one. The explanatory variables considered for both origin and destination were the following: population, total GDP, GDP per inhabitant, unemployment rate and distance. It was expected that GDP per inhabitant and unemployment rate would describe the economic push and pull influences generating migration flows, while population, on the other hand, would represent the importance of the pool of people. The results using distance between regions and the three explanatory variables for origin and destination, however, were not completely satisfactory.

Given the high correlation between total GDP and total inflow and outflow, the GDP per inhabitant was substituted by total GDP and population was taken out of the model as an explanatory variable. Instead, in some of the models, it was used as the dependent variable in the sense that the flow was divided by population (origin, destination and both). For models including these variables, reasonable results were obtained using the double logarithmic model. Adjusted R^2 values were generally higher.

The following model was found most suitable for Italy, Germany and Spain:

$$\log Y_{ij} = c + \alpha_1 \log GDP_i + \alpha_2 \log UNEM_i + \beta_1 \log GDP_j + \beta_2 \log UNEM_j + \gamma \log d_{ij} \quad (12.3)$$

where Y_{ij} is the migration flow from region i to region j, d_{ij} is the distance between i and j, and γ, α_1, α_2, β_1, and β_2 are empirical coefficients to be estimated. GDP_i, GDP_j, $UNEM_i$, and $UNEM_j$ refer to total GDP and unemployment rate of origin and destination variables respectively.

The adjusted R^2 value is above 70 per cent for all three countries for both years. The dependent variable, migration flow, is not divided by population in this model. The dependent variables are total GDP, unemployment rate and distance. For Germany, it was not possible to find a good-fitting model with the variables available from the REGIO database. Note that the total GDP is a variable that also captures the effect of the population size of the respective regions, as the dependent variable is not relative to this population size. An important aspect is the stability of the model over time, especially if the model is to be used for simulation or forecasting. The adjusted R^2 value is stable for all the models and the parameter estimates also show reasonable consistency.

Thus, a range of different log linear models have been considered in this section. Comparing the fits of all these models, it is found that several models yield good fits for the countries and years considered. However, a model with real world explanatory variables is more useful for simulation purposes than a model with lumped parameters; based on this criterion a single model could be selected that includes total GDP and unemployment rate as explanatory variables.

12.5 Forecasting and simulation

It is interesting to explore how well this model works for forecasting and simulation. To investigate this, the model calibrated on data for 1990 for Italy and Spain, and 1991 for Germany, was employed but using input data from the later year (i.e. 1994 for Italy and Spain, and 1993 for Germany). In other words, the model with parameters calibrated using data for the earlier period is used to forecast the flows in the later period by using the known values of total GDP and unemployment rate. The predicted flows thus obtained were then compared to the observed flows of the later year using linear regression. Note that the actual observed and predicted flows were compared, not the logarithms of the flows.

For Italy 1990, the regression of the expected flows (1990) on the observed flows (1990) gives an R^2 value of 0.784, which means that the model has an acceptable fit. Using the model of 1990 to forecast the flow in 1994, and regressing the predicted flows against the observed flows in 1994, gives an R^2 of 0.568. For Germany (NUTS 1), the respective R^2 values were 0.74 and 0.656 and for Spain 1990, they were 0.746 and 0. 552. The R^2 values are high for all three countries, which gives a good confidence in the usefulness of the models for prediction. Figure 12.4 illustrates the comparison of modelled and observed flows for Germany in 1993.

Figure 12.4: Comparison of observed and predicted flows for Germany, 1993

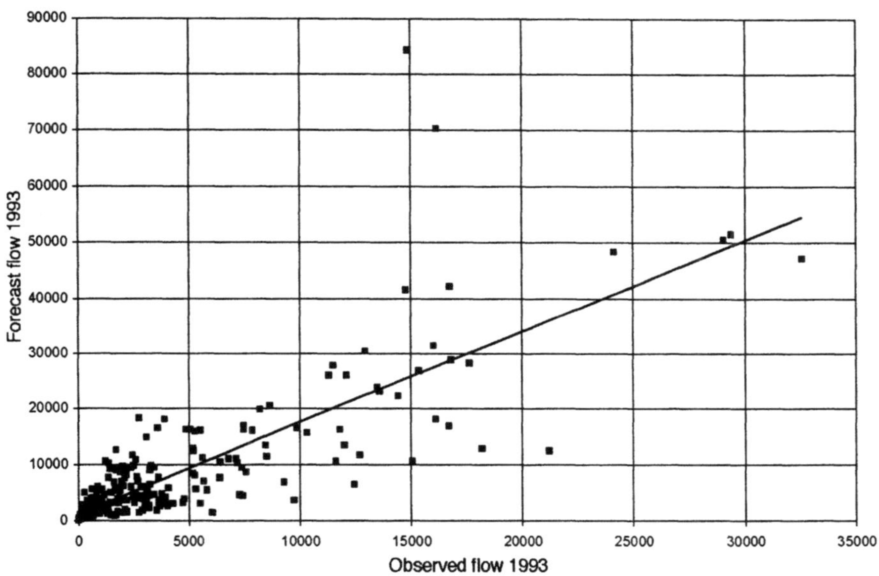

In order to be able to visualise actual migration data, a GIS-based visualisation tool was developed. The user is able to select a country, a region and a year and to select the type of data for visualisation. The latter can either be observed data, modelled flows or predicted flows. Furthermore the user can choose the type of flow (inflow, outflow and net flow) and the type of visualisation (colours, arrows). If the type of data selected is the predicted flow, then the parameters for prediction have to be set (year of prediction, expected change in GDP and unemployment). Figure 12.5 illustrates the Dialog Editor box which is used in the GIS to select the required visualisation.

Figure 12.5: The Dialog Editor in the GIS based visualization tool

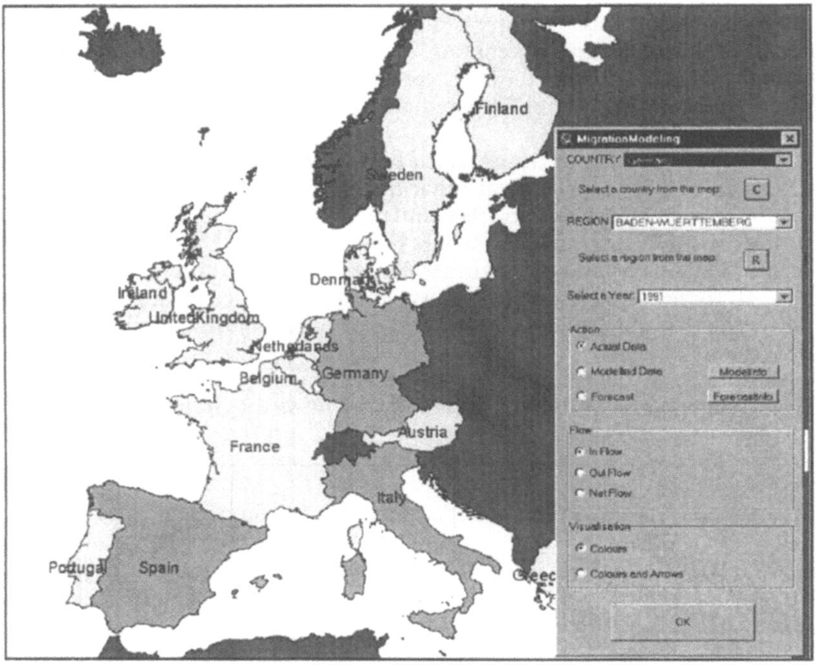

Visualisation of the flows is achieved either by colouring the regions according to the magnitude of the flows to or from a selected region, or by using arrows. The arrows can be generated by using the selected region as a reference and thus visualising direction (direction of the arrow) and magnitude (size of the arrow) of the flows as illustrated in Figure 12.6.

12.6 Application to simulation of land use pressure

The results described above indicate that the model can be used to build a simulator for future migration flows. In other words, it could be used to support a scenario analysis. Once a suitable model has been identified for a country, it can be explored; i.e. one can change the explanatory variables, total GDP and unemployment rate, and see how they influence the migration flows. Since one of the motivations for land use change is population change, the migration models could provide a link between the driving economic force (GDP) and land use change. It is therefore worthwhile to explore how this kind of modelling could be used as an input to simulations oriented to land use pressures and changes.

Figure 12.6: Visualisation using arrows of modelled net flow for Saarland, 1991

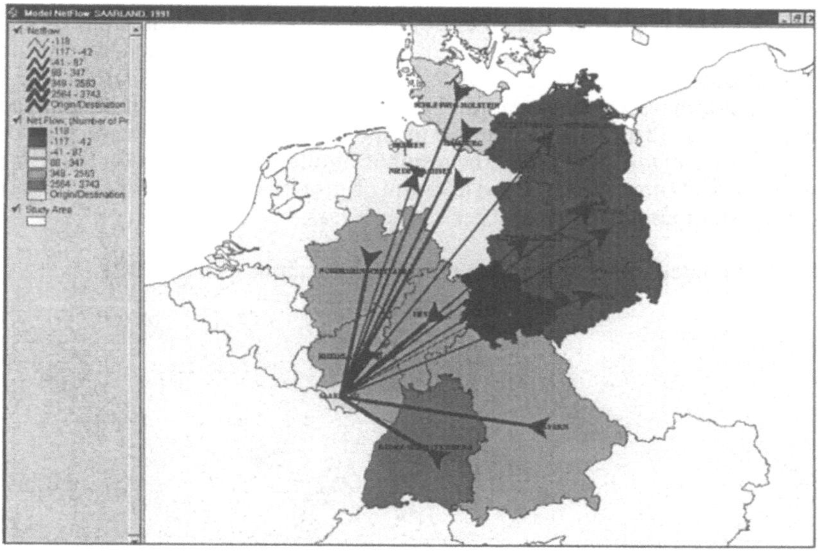

One way to investigate the link between population and land use change would be to measure the extent of the urban areas in several major European cities at two different times with a time interval of 5 or 10 years (this is now feasible using satellite images) and to examine the correlation between urban area and population. If a useful relationship could be established, and calibrated, between population increase and urban area increase, then with the aid of the migration modelling described above, it would be possible to link the three aspects: GDP → migration → land use change for the purposes of simulation. It is worth noting in this context that in many European countries total population is now stable or even declining, so migration is likely to be more important for population changes than increases through the birth rate.

A project is currently under way to measure changes in land use for a number of European cities. This is the project MURBANDY for Monitoring URBAN DYnamics (*http://www.riks.nl/RiksGeo/projects/murbandy/*). The so-called 'CHANGE' component of this project has the objective to implement an Earth observation-based procedure for monitoring land cover changes in urban and peri-urban areas. The changes in the spatial extent and urban structure over the past 40 years are being measured in 15 European cities. At the time of writing, published results from this project are not yet available, but a personal communication has indicated that changes in urban area have been detected and do display a relationship with population. The results to be delivered by this project should therefore provide clear indications as to whether the kind of modelling links outlined above will be feasible.

A useful way of linking migration and land use changes for purposes of explanation and simulation would be to devise an index of land use pressure based on values of population and urban area. The first and most obvious choice would be to use population density, obtained by dividing population by urban area (e.g. for a province or region). However, there may be situations where this value is high, but stable, i.e. not changing with time and not adequately reflecting pressure on land use. Considering that, in a dynamic situation, pressure is likely to be felt in terms of changes, another possibility would be to divide change in population by urban area. A test of this concept was made for Italy by creating an index, obtained by dividing net flow by the total urban area (obtained by summing up the grid cells classed as urban in the CORINE land cover data) for the regions. The index was normalised to have a maximum absolute value of 1. The index can be mapped and colour coded so as to distinguish regions feeling positive and negative pressures as a result of migration flows. Figure 12.7 shows the result of simulating this index for the year 2000, based on the model fitted for 1994 and assuming a growth in GDP of 5 per cent per year with a decline in unemployment of 2 per cent per year.

Figure12.7: Simulated index of land use pressure for regions of Italy, 2000

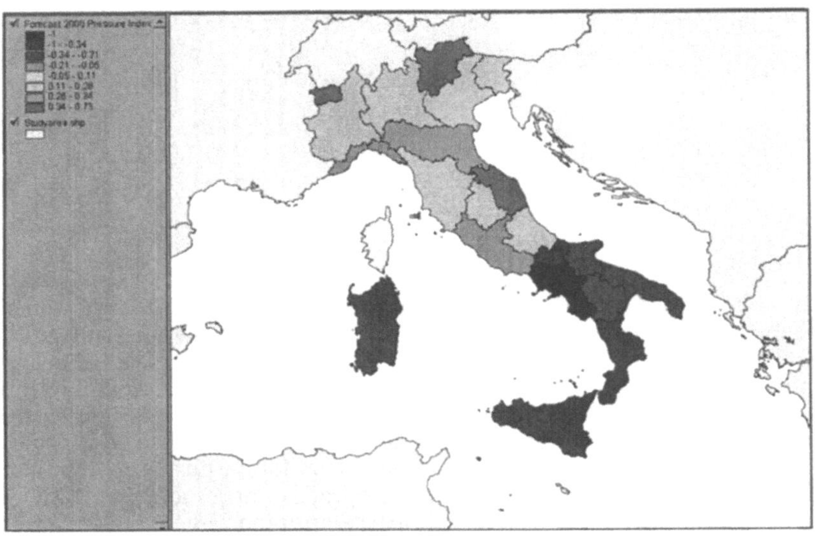

It should be stressed that this is a first test simulation designed to illustrate the concept and should not be used in relation to any policy discussion. However, it clearly preserves the north-south trends already identified for Italy and illustrates how some of the regions in the north could be identified by such an exercise as being possibly susceptible to higher levels of land use pressure from migration flows.

12.7 Conclusions

The work described on interregional migration modelling has attempted to explore the question of whether the data in the REGIO database, which is the official data assembled at European level, could be used in connection with spatial interaction models for explanations and simulations of migration flows. The results showed that some useful models can be calibrated with this data, using GDP and unemployment as explanatory variables, though there are limitations imposed by problems of data completeness and quality. Improved results could be expected if the data were to include age distribution of the flows and their education structure.

A suggestion has also been made for an index of land use pressure due to migration. This has been defined by dividing net flow for regions by total urban area as obtained by summing the areas of urban cells in the CORINE land cover data. Using this index, a method of linking the two important driving forces of GDP and unemployment, via simulated migration flows, to a simulation of land use pressure has been illustrated. It has also been pointed out that it should soon be possible to calibrate models linking urban expansion and population change by using results from satellite images taken at different times.

Acknowledgements

Sincere thanks are due to Tine Sorensen and Andrea Saltelli for their support and contributions to the work on interregional migration modelling.

References

Bailey, R.C. and Gatrell, A.C. (1995) *Interactive Spatial Data Analysis*, Longman Scientific & Technical, England.

EUROSTAT (1997) *Case Study A – Spatial Analysis of Interregional Migration in correlation with other Socio-economic Statistics*, Lot4 – Ref. 97/S140 – 88522/FR.

Fotheringham, S. and O'Kelly, (1989) *Spatial Interaction Models: Population and Applications*, Kluwer Academic Publishers, Dordrecht.

Nijkamp, P. and Reggiani, A. (1999) *Interaction, Evolution and Chaos in Space*, Springer, Berlin.

REGIO database – Available on CD from European Statistical Office.

Rees, P.H., Stillwell, J.C.H., Convey, A. and Kupiszewski, M. (1996) *Population Migration in the European Union*, John Wiley & Sons, Chichester.

Sen, A. and Smith, T.E. (1995) *Gravity Models of Spatial Interaction Behavior*, Springer, Berlin.

Stillwell, J.C.H. and Congdon, P. (1990) *Migration Models: Macro and Micro Perspectives*, Belhaven Press, London.

Beatrice Eiselt
Punto Energia c/o Regione Lombardia
Via Stresa 24
20125 Milano
Italy

Nicla Giglioli
The European Commission
Joint Research Centre
Institute for Systems, Informatics and Safety
APPST Sector
TP 361
21020 Ispra(VA)
Italy

Robert Peckham
The European Commission
Joint Research Centre
Institute for Systems, Informatics and Safety
TP 650
21020 Ispra(VA)
Italy

PART III: EVALUATION AND VISUALISATION METHODS

13 TOWARDS A SUSTAINABLE FUTURE OF CITIES IN EUROPE: AN EVALUATION OF SUSTAINABLE CITY INITIATIVES USING MULTICRITERIA DECISION SUPPORT METHODS

Adele Finco and Peter Nijkamp

13.1 Setting the scene

Concerns about quality of life are not just a pressing phenomenon of modern times. There are records on environmental decay in periods of ancient history, in both urban environments and landscapes, and many medieval cities used to have some sort of environmental regulation (Banister *et al.*, 1999). What has made environmental concerns more pronounced in the second part of the twentieth century, is the large scale occurrence of environmental decay. There is no country or area that is not directly or indirectly suffering from environmental externalities. This quantitative extension of environmental problems has several backgrounds including the increasing number of people and a more internationally interlinked network economy with intensive trade and transport (the 'ecological footprint'). But there are also qualitative and structural factors that are responsible for the so-called 'new scarcity', such as the use of synthetic and non-biodegradable materials, the shift from local to global environmental change or the transition towards more mobile lifestyles. Against this background the notion of sustainable development has become *en vogue* (for a broad survey of relevant issues see Van den Bergh, 1996). Space and time play a critical role in this concept, since through these two dimensions, it is possible that environmental externalities are dispersed, either to other areas or to future generations.

In the debate among economists regarding measures for coping with environmental externalities, the standard therapy for solving market failures, i.e. Pigouvian taxes, has become rather popular in recent years (witness the discussion on eco-taxes, for example). However, others advocate alternative policy approaches such as tradeable permits, standard setting or even prohibitions. In practice, we have seen a portfolio of different policy measures reflecting a compromise between different political-economic viewpoints. In most policy and scientific discussions on sustainable development, we observe nowadays the need for a broad evaluation of environmental issues, in which economic, social and environmental motives, characteristics or attributes play an intrinsic role (Figure 13.1). It is now recognized that sustainable development is not a given endpoint-state of a compound economic–social–ecological system. On the contrary, it is essentially based on complex trade-offs, related to priorities attached to intergenerational equity, weak and strong sustainability, absolute and relative delinking, local versus global sustainable development (including ecological footprints) (see also Van den Bergh and Hofkes, 1998 and Van Pelt, 1993).

In recent years we have witnessed also an intense debate on the so-called 'delinking hypothesis', which means that in a balanced growth situation, it may be possible to realize a rise in income per capita accompanied by a simultaneous rise in environmental quality. The debate is particularly induced by the so-called 'factor 4' and 'factor 10' discussions initiated by the Wuppertal Institute (for a survey see Veen-Groot and Nijkamp, 2001). In the macro-economic literature, this issue is reflected in the Environmental Kuznets Curve

J. Stillwell and H. Scholten (eds.), Land Use Simulation for Europe, 173–192.

(EKC) debate, in which it is assumed that an inverse U-shaped curve for the relationship between environmental degradation and income per capita may exist, in particular if a growing income stimulates environmental awareness and hence effective environmental policies (for a broad survey of the literature and empirical tests, see a recent study by de Bruyn, 1999).

Sustainable development, as sketched in Figure 13.1, means a balanced development in a triangular force field. The multidimensional judgement of sustainability conditions is necessarily based on political choices, as there is – in the presence of externalities and other market failures – no way to create unambiguous trade-offs among diverse welfare constituents. For example, in the area of land use and landscape protection, one might make a distinction between environmental functions which may be evaluated on the basis of their use value and environmental functions related to option values (such as bequest values or conservation values). Only part of these functions can be translated into the 'measuring rod of money'. This has far reaching implications for environmental planning at all spatial levels, including the city level.

Figure 13.1: The triangular concerns of sustainable development

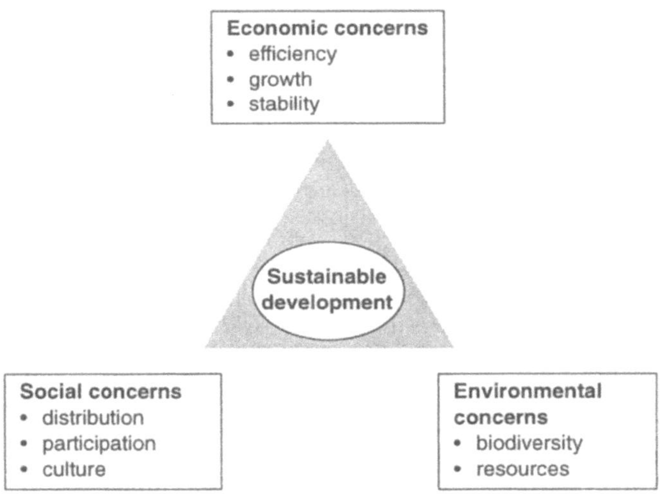

The central question addressed in this chapter is now the following. Given the complexity of the concept of sustainable development, is it possible to employ modern methodological tools for conflict management, in particular multicriteria methods, which may serve as decision support techniques for the evaluation of sustainability initiatives at the local level? The chapter is organized as follows. First, the relevance of sustainable city planning is highlighted (Section 13.2), followed by an exposition of evaluation issues in urban sustainability policies (Section 13.3). To test the validity of the approach and to demonstrate its feasibility, a representative case study on urban sustainability strategies is described, viz. the Italian city of Cremona (Section 13.4). The results are presented and interpreted in Section 13.5, while the chapter is concluded with some retrospective and prospective remarks in Section 13.6.

13.2 Planning for urban sustainability

Our world is becoming increasingly urbanized – with all the advantages and disadvantages associated with that process. The geography of the twentieth century exhibits an intensified trend towards an urban way of life in a modern society. Despite suburbanization – and sometimes de-urbanization tendencies – the city appears to remain the nucleus of a developed economy. It is undoubtedly true that the economies of density and scale are decisive factors for city formation. Clearly, there are also dis-economies as witnessed by congestion and environmental decay. Nevertheless, the positive features of the city still appear to outweigh the negative externalities, as the city is an extremely efficiently organized geographical entity. This does not only hold for the economy in terms of scale and scope advantages, but also for the use of scarce space and environmental resources (for empirical arguments, see Capello *et al.*, 1999).

The general concern about urban environmental decay clearly causes much interest in negative environmental externalities, but it ought to be recognized that also various positive externalities may exist. For example, in a study on urban externalities, Stanghellini and Stellin (1996) distinguished various types of externalities (Table 13.1). In the same vein, Camagni *et al.* (1998) developed a classification of both positive and negative externalities as a frame of reference for finding and judging strategies for sustainable cities (Table 13.2).

Table 13.1: Externalities in the urban area

EFFECTS			EXTERNALITIES POSITIVE	EXTERNALITIES NEGATIVE
Technological	Source	Production	Integration of production sectors	Emissions/pollution
		Consumption	Accessibility to environmental, cultural, architectural resources	Pollution Congestion
	Type of use	Private	System of private waste (resources) recycled	Traffic pollution Waste
		Public	Environmental urban public goods (e.g. green areas)	Derelict areas
Monetary			Increasing value of suburban rural land	Increasing cost of basic resource scarcity

Source: Stanghellini and Stellin (1996)

Apparently, the general and popular concept of sustainable development was not sufficiently focused to serve as a benchmark for urban policy. And as a consequence, we have witnessed in recent years an increasing interest in the concept of the *sustainable city*, which has inspired many authors including Atkinson *et al.* (1997), Banister (1999), Banister *et al.* (1997), Breheny (1992), Haughton and Hunter (1994), Nijkamp and Perrels (1994), Selman (1996), Capello *et al.* (1999), Pearce (1999), Satterthwaite (1997) and Wallner *et al.* (1996). As our world is moving towards an urban world (with an urbanisation rate exceeding approx. 70 per cent), it certainly makes sense to consider cities as focal points of sustainability policies. From a management perspective, cities are also anchor points for administrative and policy action, while they may also be in a better position to induce citizen's participation and public support. And finally, from a research perspective, cities are clearly demarcated statistical units that may be able to offer the necessary empirical data for policy analysis.

Table 13.2: Positive and negative external effects of the interaction between the different environments in a city

	Interaction between economic and physical environments	Interaction between economic and social environments	Interaction between social and physical environments
Positive external effects	• Efficient energy use • Efficient use of non-renewable natural resources • Economies of scale in the use of urban environmental amenities	• Accessibility to qualified housing facilities • Accessibility to 'qualified jobs' • Accessibility to social amenities • Accessibility to social contacts • Accessibility to education facilities • Accessibility to health services • Diversification of options	• Green areas for social amenities • Residential facilities in green areas • Accessibility to urban environmental amenities
Negative external effects	• Depletion of natural resources • Intensive energy use • Water pollution • Air pollution • Depletion of green areas • Traffic congestion • Noise	• Forced suburbanisation due to high urban rents • Social frictions on 'the labour market' • New poverties	• Urban health problems • Depreciation of historical buildings • Loss in cultural heritage

Source: Camagni et al. (1998)

Some authors have interpreted a sustainable city in a rather narrow sense by addressing predominantly urban form (for instance, in a compact city perspective, see Jenks *et al.*, 1996) or urban transport (in relation to energy consumption, see Newman and Kenworthy, 1989). It should be noted however, that the urban environment is a multi-faceted phenomenon ranging from 'hard' pollution indicators to 'soft' quality-of-life indicators. The urban environment is, in addition, the playground of many conflicting interests, institutionally, sectorally and geographically, so that the concept of the sustainable city is an interesting test case for the notion of '*civitas*' or civil society (Selman, 1996). This chapter will, therefore, adopt a broad perspective on urban sustainability.

From an environmental perspective, urban sustainability has also played an important role in recent discussions on urban spatial configurations, such as 'the resourceful city', 'the green city', 'the garden city', 'the ecological city', 'the edge city' and 'the virtual city'. Thus far, no unambiguous concept has emerged and, in reality, we observe a parallel development of various contemporaneous urban policy concepts (Nijkamp and Perrels, 1994). Although it is likely that environmental quality problems may become more severe with urban size, there is no clear evidence that urban size as such causes environmental decay. According to Orishimo (1982), it is not the sheer city size, but rather the implied land use, the transport systems and the spatial layout of a city which are critical factors for urban environmental quality. In conclusion, and in the spirit of Figure 13.1, a sustainable city should offer the foundations for a long-lasting balance in the economic, social and environmental force field of the city.

One important issue is whether the spatial scale of the city should be limited to the urban boundary or whether the broader geographical spillovers would also have to be taken into consideration. This question of the so-called ecological footprint has been extensively discussed in the literature (Wackernagel and Rees, 1996). It leads essentially to a distinction between internal and external urban sustainability (Nijkamp and Opschoor, 1997). In this context, strong urban sustainability would emerge in the case of both internal and external urban sustainability. Consequently, urban sustainability has also important geopolitical dimensions, which are partly related to land use and partly to other functions of the city (including its surroundings). This is also clearly reflected in the great many initiatives following the adoption of Agenda 21, which require a great variety of policy initiatives and strategies. Clearly, the environmentally sustainable development of a city can only be attained by initiating appropriate policy strategies (on land use, quality of life, urban amenities, for example). Actual initiatives in cities world-wide differ strongly in the adoption and implementation of such strategies, because each city has its own specific geographical, political and environmental setting. Nevertheless, there is a need for general integrative concepts and evaluation procedures which can serve as guidelines for many cities undertaking sustainability initiatives. A broad survey of such concepts can be found in Selman (1996), while an overview of policy strategies can be found in OECD (1995).

The general lessons from the above mentioned survey are that policies addressing sustainable development of cities should cover multiple fields like rehabilitation, land use, transport systems, energy management, architecture and conservation policy, and cultural policy. Local authorities will have to share those tasks with other actors in the urban space (including the private sector). Lack of commitment among stakeholders to develop an effective and sustainable urban development policy will reinforce urban sprawl and will expand inner city problems to a much larger area. On the other hand, environmentally-benign policies may attract new investments, favour urban employment, and hence contribute to an increase in quality of life.

Any policy focused on the achievement of sustainable development in space and time requires clearly an enormous political effort. The Rio Summit of the United Nations on Environment and Development (1992) has expressed this in clear terms. To attain these high ambitions, it is necessary to establish – from an international perspective – local sustainability actions and initiatives. The local Agenda 21 presents many such ideas and commitments based on a balance between local economic, social and sociological interests of cities. A further translation in the European setting can be found in the European Charter from Aalborg (1994) in which partnership programmes are foreseen (Selman, 1996).

Local authorities have the possibility to exert both a direct and indirect influence on the successful conditions for urban sustainability. The question whether ultimately a given urban development is sustainable or not is, of course, co-determined by the targets set by policy-makers. Clearly, there is not a single unambiguous urban sustainability measure, but there is a multitude of quantifiable criteria that may be used in an empirical test. A necessary condition for implementing an effective planning system for urban environmental management geared towards maintaining sustainability is the development of a system of suitable urban environmental indicators (for a first attempt, see OECD, 1978). Such indicators, which should represent a balance between the necessary quality of information and the costs involved, would have to be related to economic, social, spatial and cultural dimensions of the city. The OECD list contains elements which are decisive for urban environmental quality and which would have to be included in such an indicator system. Examples include housing, services and employment, ambient environment and nuisances, and social and cultural concerns. However, precise empirical evidence on urban environmental quality and on its implications for both household and firm behaviour is not always available in such an indicators system.

Nevertheless, it is a *sine qua non* for a proper judgement of successful urban sustainability policies that an identification and measurement of relevant indicators takes place. Illustrative listings of such indicators can be found in the so-called Dobrìs Report (Stanners and Bourdeau, 1995). In a more analytical way, also the OECD (1994) has developed the so-called PSR (pressure-state-response) approach, while the International Institute for the Urban Environment (IIUE, 1995) has proposed the so-called ABC (area-basis-core) indicators list. Such approaches can be very helpful in identifying the driving forces of urban sustainability, while they may also be extremely helpful in pinpointing the relevant criteria to be considered in comparing alternative urban sustainability plans, e.g. by using multicriteria analysis. This approach, which will form the basis of our empirical application, is described concisely in the next section.

13.3 Decision support methods for urban sustainability policy

The previous sections have pointed out that urban sustainability – as a policy concept – is not an unambiguous state of affairs, but a multi-faceted phenomenon, fraught with conflicts and uncertainties. As mentioned, the notion of a sustainable city comprises a great variety of (sometimes) conflicting dimensions, such as economic, social, land use, ecological and transportation interests, among which a balanced compromise has to be found by urban policy-makers (Banister, 1999). Conflict resolution is, of course, a political action, but presupposes proper knowledge on the pros and cons of alternative choice possibilities. From an economic perspective, this would ideally imply that all foreseeable costs and benefits of a planned initiative would have to be assessed. Policy analysis offers, in general, an assessment and evaluation framework in the public sector with the goal to increase the efficiency and effectiveness of government decisions, seen from broad and multidimensional perspective. In public decision-making, normally a wide range of decisions has to be made without a clear reliance on the market system. This is partly caused by the nature of choices in the public sector (with emphasis on multi-actor participatory democratic modes of decision-making) and partly by the complexity of government projects (with long-lasting and often uncertain implications). And it is indeed increasingly recognized that decisions based on market forces alone do not necessarily lead to optimal results. In the context of urban policy, structural market failures as well as unexpected external factors may need a policy system that ensures an improvement of socio-economic and environmental conditions.

In the past, several methods have been developed and applied in policy analysis, in which a market evaluation played a prominent role. The most well-known example of such a market evaluation method is based on *cost-benefit analysis* (as an operational application of welfare theory). This method forms the foundation for many policy assessment methods and has been successfully applied in many case studies in the public sector. Despite its great many merits, it is increasingly recognized in modern policy analysis that is also has severe limitations, because not all relevant welfare implications of public sector initiatives can be expressed using the 'measurement rod of money' (Nijkamp *et al.*, 1991).

Cost-benefit studies seem to be most applicable and appropriate if the decision concerns a well-demarcated and *a priori* precisely defined project which does not generate many unpriced or qualitative externalities. If, however, the decision concerns a more general policy programme (of which the details and even sometimes the major features are unknown), then the translation of the impacts into precisely measurable and quantitative consequences and subsequently into monetary figures is often rather problematic. Similarly, if a public investment is likely to generate a wide diversity of social costs (e.g. landscape destruction, loss of safety, health effects, loss of biodiversity or rare species, destruction of archaeological sites), it is often a heroic research task to come up with reliable figures which are broadly accepted in the policy area. This does not mean that cost-benefit analysis would have to be discredited; but it would have to be complemented with more appropriate evaluation tools.

As a response to the shortcomings of conventional evaluation studies, a great diversity of modern assessment methods has been developed over the last 10 years in order to extend the range of and to provide a complement to conventional cost-benefit analysis and to offer a perspective for procedural types of decision-making in which various quality aspects are also incorporated. Many of these methods simultaneously investigate the impacts of policy strategies on a multitude of relevant criteria, partly monetary, partly non-monetary (including qualitative facets). They are often coined *multicriteria methods* and are also known as multi-assessment methods.

It is noteworthy that, in the past few years, an avalanche of assessment and evaluation studies has been undertaken in the regional, transportation and environmental field, but unfortunately, an integral study and a systematic comparison of findings of previously undertaken assessment studies has proven to be difficult due to different analytical approaches and differences in presentation. Several problems underlying a decision-making process in a urban sustainability context have to be coped with, such as:

- the relevant urban information or available data always contain a component of uncertainty;

- the data or information may be stored in different statistical data bases that may be difficult to access, manipulate, compare and study;
- a large set of often conflicting urban objectives or targets has to be taken into account;
- the decision-making process itself might be influenced by power relations or selfish motivations of urban stakeholders; and
- a decision-making process has to take place within the shortest time possible to avoid countervailing effects from the side of various urban actors.

The above considerations highlight that the policy objective of urban sustainability is not unambiguous, but may lead to conflicting actions. Is it possible to devise a decision support system that is able to shed clear light on this difficult issue? The answer to this question may be given by referring to conflict management techniques development in the multi-assessment literature. The existence of multiple objectives means that, in a given societal setting, the best possible alternative or policy has to be determined that also creates sufficient public acceptability or at least social feasibility. In other words, the basic question is: how to determine the optimal policy?

It is generally accepted nowadays that most decisions – certainly in the public domain – can be typified as multiple objective or multicriteria problems (Janssen, 1991; Nijkamp *et al.*, 1991, Beinat and Nijkamp, 1998). This means that an optimal – or most acceptable – compromise alternative from a set of competing alternatives has to be identified which best satisfies a number of – often conflicting – objectives or decision criteria. Another complicating factor is that usually in the public policy domain, besides a set of quantitative criteria, qualitative criteria must also be taken into account in a multi-actor decision-making process. Examples are the interest of the biotic and a-biotic environment, the protection of school children, accessibility conditions of the elderly generation, or the risk of criminality in public transport.

As mentioned above, cost-benefit analysis has severe shortcomings when it comes to an operationalisation of intangible aspects, so that this theoretically elegant method has often limited applicability. In most public policy evaluation studies, especially the assessment of environmental impacts turns out to be troublesome, since all advantages and disadvantages of policy options would have to be translated into a common monetary unit. Hence, incommensurable criteria of an unpriced and intangible nature cannot be included a decision-making procedure based on a standard cost-benefit analysis. Furthermore, in the current policy practice in many countries there is hardly any applicable and meaningful way of including distributional impacts on welfare (e.g., through a weighting system for different groups) into policy evaluation. Therefore, a proper decision-making tool is needed that is able to handle qualitative and intangible information in a proper way. Consequently, for our analysis of urban sustainability initiatives, we will resort to multi-criteria analysis (MCA) as a modern decision support method in the public sector.

MCA derives its strength from the fact that it is, in principle, able to handle qualitative, quantitative and mixed data on distinct choice possibilities in decision-making. Although there is a variety of multicriteria methods, many approaches are based on pairwise comparisons. The techniques employed may then be different, but the main principles are still the same. This applies, or example, to concordance methods, the Saaty method, the Electre method and the Regime method (Nijkamp *et al.*, 1991). For our case study in Section 13.5, we will use a rather simple method, coined VISPA. MCA has become is a modern guiding tool for the choice among a set of finite and *ex ante* demarcated discrete alternatives (projects), which have to be evaluated in relation to a finite number of criteria, for which each distinct alternative presents a performance index or score (Finco and Nijkamp, 1999). The selected case study area is introduced in the next section.

13.4 The case study area

Our aim is to investigate whether modern decision support methods, such as multicriteria techniques, are helpful for an evaluation of urban sustainability strategies. For this purpose, we have selected the Italian city of Cremona as the case study area and this section aims to provide an empirical framework for sustainable town planning with a view to the identification and achievement of a balanced development scenario. Cremona is a small to medium sized town (about 72,000 inhabitants) in the Lombardia Region of northern Italy, situated in the Po plain. Its economy is traditionally dominated by the agricultural (especially dairy-farming) and agro-food sectors which have gained high productivity levels and play a crucial role in the Po area. Furthermore, the economic structure of the town is characterised by a late but strong industrialisation process that took place in the 1960s and 1970s. As a result, the city is currently facing various sustainability challenges, such as brownfield development, a restructuring of the local economy to a service orientation, an international network infrastructure and a weak but potentially promising image. Thus, there is need – and scope – for a new perspective in which local policy-making would have to take the lead.

The Local Plan, the *Piano Regolatore Generale* (PRG), which is at present developed together with the Provincial Territorial Plan, creates an important opportunity to build a lively and attractive town. However, the Council still has to make a multi-faceted choice among different – often conflicting – development plan options, which are all important but are also bound by the town budget restrictions. In order to identify the relevant Local Plan options and the key objectives to be met, it turned out to be necessary to start from an exploratory study on the urban (local) use of its territory and of the urban social and economic dynamics of the city. This first phase has proven to be essential to test the feasibility of various urban sustainability projects and to select the most significant objectives/indicators. The analysis addressed both socio-demographic issues and economic development issues (agricultural, industrial and service sectors).

This initial investigation has brought to light some interesting information. For example, the proportion of total population that were elderly in 1996 turns out to be rather high – more than twice as high as the younger age cohort. In addition, the economy appeared to be dominated by commercial and bank activities (51 per cent of urban GDP), while in the past decade, industry and tourism have shown a decline. Although rich from an economic and cultural point of view, Cremona is lagging behind in terms of an endogenous growth of its local economy. This town clearly needs a new stimulus for its socio-economic and environmental profile, and more dynamism in general. A dynamic and competitive urban system has to face various crucial issues regarding the maintenance of its market position, including in particular:

- a sufficient carrying capacity of its environmental and economic base (soil, natural and manufacturing resources);

- a rich supply of multidimensional public services (e.g., an advanced local system of training/information); and
- a better provision with sophisticated communication and interaction networks.

Apart from a mapping of the above issues at hand, it was also necessary to trace the prevailing policy directions and priorities of different stakeholders in the city. The main priorities brought to light by our policy analysis relate to the need for sustainable development (i.e., a development process that safeguards and regenerates environmental and architectural resources, and pays due attention not only to suburbs, but also to the entire urban territory). These priorities are represented by the following main policy trajectories of the city:

- protection of green areas;
- design of a balanced mobility plan;
- recovery of the existing urban area (derelict areas, brownfield development);
- strengthening and exploitation of the commercial system; and
- cultural and tourist promotion.

Urban environmental quality in Cremona appears to depend on both natural and social elements. The critical factors of successful urban quality of life policy are the low pollution of the air-water-soil system, the favourable condition of the built environment, and the protection of green areas. The sustainability questions to be addressed in Cremona are certainly not exclusive or specific, but exemplify the complex policy issues faced by many cities. Hence, the case study of Cremona may be regarded a typical case for policy analysis of any sustainable city.

13.5 Multi-criteria analysis for the Local Plan of Cremona

A good evaluation technique must be easily implementable by those involved in policy-making and should support efficiency in the policy analysis. One of the benefits of MCA is its flexibility and its user-friendliness. In our case study, the MCA is used as a decision support system for policy-makers with the aim of choosing one or more promising projects in the context of the preparation of the new Local Plan of the city of Cremona. Extensive and time consuming field work was necessary to identify all relevant courses of action, all important policy criteria or all expected consequences of all actions. This led to a comprehensive database on many aspects of the city, including qualitative information.

The study phases for the Cremona case study, which are more or less standard for any MCA, involved:

1. choice of the alternative courses of action for urban sustainability;
2. choice of the indicators for guiding urban sustainability;
3. weight assignment to criteria (on the basis of policy simulation);

4. matrix construction and score assignment for the impact table;
5. software implementation on the data set; and
6. evaluation of policy options.

The selection of n finite alternatives and of k indicators leads ultimately to a *k by n impact assessment matrix*, from which the indicators/criteria are policy objectives to be Pareto-maximised (Table 13.3). The selection of the project alternatives was based upon a consultation process involving local stakeholders (main actors and policy-makers, economic actors, labour unions, entrepreneurs, professional associations; social groups, *et cetera*). An important information source was the use of a systematically designed survey questionnaire through which not only data were collected, but also important policy criteria could be formulated. This questionnaire was prepared in order to identify the choices that each relevant group thinks to be desirable or necessary. Such ideas were then tested and compared with the policy-makers' opinions.

In the original urban sustainability framework for Cremona, a total set of 35 choice alternatives was distinguished. These have been reduced for the ease of presentation to 12 representative options in our study application. This choice is typical of a strategic planning approach. Clearly, the method of MCA allows also to make a cluster for land use policy (in particular, derelict areas or brownfields). In the final assessment matrix, it is then possible to have the most important alternatives envisaged and evaluated. The financial aspects of each choice option were considered in advance, so that only alternatives with a financial feasibility were pre-selected. The selected project alternatives are thus all feasible from the point of view of urban finance. It should be noted that the project alternatives under consideration do not only consider economic issues, but also environmental and social aspects that are indispensable for a sustainable town development.

It should be noted here that urban sustainability planning is clearly a multi-faceted activity with many actors and many policy-makers. It is thus no surprise that the number of policy judgement criteria is high. The 35 selected criteria/indicators comprise the economic, social and environmental aspects of the urban territory considered. To ensure a manageable MCA, classification into three main groups has been made on the basis of a stratification of the three different welfare constituents (economic, social and environmental) of urban and territorial sustainability, using standard international classifications discussed above (IIUE, Dobrìs Report). As Table 13.3 shows, social and environmental indicators play a particular role according to international urban sustainability principles expressed in the above studies.

It is noteworthy that the impact assessment matrix (Table 13.3) is based on a bottom-up consultation procedure. This means that, apart from a screening on financial compatibility with the available budget, this table includes a great many choice options of local actors, which have to be judged from different perspectives. This table includes, of course, also traditional economic indicators (such as GDP) which are related to the overall value of goods and services produced within the urban territory, and related socio-economic indicators (such as the urban employment rate). Environmental criteria in the checklist are, *inter alia*, the proportion of waste recycled and the recovery of agricultural land situated in suburban areas; the former is used as a proxy of the carrying capacity of an open urban system, while the latter refers to the previously mentioned notion of the ecological footprint (Rees, 1992; Segale, 1993).

Table 13.3: MCA impact matrix of the Cremona project

Alternatives	1	2	3	4	5	6	7	8	9	10	11	12
									Derelict area			
Criteria	Rural restructuring	Park project	System of waste recycling (e.g. incinerator)	Street reconstruction-restructuring	Suburban commercial system	Parking system	Road system	Public transportation network	Centre for elderly support	Cultural centre Conservatoire/ Auditorium	University centre	Congress centre
Economic												
Variation in GDP	6	8	7	8	10	6	8	7	9	10	9	9
Variation in employment	6	6	7	8	8	5	8	7	9	7	9	7
Integration of production sectors	7	9	6	7	8	7	9	9	6	6	8	9
Job mobility	7	7	7	7	8	6	7	7	8	8	7	7
Development SME	6	9	6	6	9	9	9	9	6	6	7	9
Social												
Commuting patterns	5	6	6	7	8	10	10	10	10	6	9	6
Administrative efficiency	6	10	9	6	7	9	9	9	9	9	9	9
Accessibility to education	6	2	6	2	2	10	9	10	6	9	10	10
Accessibility to sports facilities	7	1	10	5	1	8	8	8	7	1	7	2
Accessibility to welfare provisions	6	7	10	4	7	10	8	10	10	10	9	10
Accessibility to health service	1	10	2	2	1	7	9	7	10	4	4	4
Accessibility to post services	6	6	7	5	5	9	8	8	6	8	10	10
Availability bank services	7	5	7	6	8	8	8	5	7	8	8	9
Public security	6	9	5	2	4	8	7	8	7	5	3	5
Housing quality	8	7	10	10	1	7	7	7	2	7	5	5
Quality of life	2	1	8	9	9	10	9	10	6	8	8	9
Population density	6	10	9	6	6	3	7	2	3	2	6	5
Ageing	2	9	7	6	9	10	9	10	6	7	9	7
Information systems	1	7	8	4	9	9	9	10	10	6	6	3
Cultural public relations	6	8	9	1	2	6	2	5	10	9	8	8
Racial integration	7	2	6	8	4	7	6	6	7	10	10	10
Transport and traffic quality	6	1	5	4	8	6	3	8	5	9	9	9
Facilities for handicapped children	6	7	8	6	7	10	10	10	8	8	10	10
Civil participation	7	1	7	6	9	10	10	10	10	8	9	8
Environmental												
Percentage of waste recycled	5	5	7	4	8	5	4	5	6	10	6	6
Quality of urban nature area	10	7	10	7	6	5	1	5	1	1	4	4
Energy consumption	7	7	6	5	7	5	7	7	5	4	5	7
Urban green areas per person	7	10	10	9	9	10	7	7	7	1	1	2
Landscape	10	8	10	7	7	4	6	4	6	6	6	6
Air pollution	7	3	10	5	5	1	2	1	5	5	5	5
Water pollution	4	3	10	5	5	5	5	5	3	5	5	5
Soil pollution	4	3	10	5	4	4	3	3	5	5	5	5
Noise	7	4	10	5	4	1	1	2	6	4	5	5
Renewal rural areas	10	8	10	6	2	5	3	6	6	6	5	5
Percentage derelict areas	7	7	7	5	5	5	5	5	8	8	9	8

Assigning weights to policy criteria is a standard step in MCA and is necessary, as the weighting vector represents the relative importance ('shadow price') of each criterion. As a starting point in MCA, the various weights are often assumed to be equal. In our comprehensive study approach, three different weight configurations have been applied (Table 13. 4):

- weights specified by the Cremona Town Council representatives, employing a 5-10 scale (weight set A);
- equal weights for all criteria (weight set B); and
- weights defined by the urban experts according to a 10-point scale[1] (weight set C).

In all these cases, the highest weight score is defined as the best (the higher the better). High importance (weight of 10) was on average attached by the representatives in the Council to the economic criteria (a rise in GDP and an employment increase). This fact proves how much the Cremona area needs economic strengthening. But also environmental criteria appeared to receive a very high weight confirming an increasing interest in environmental safeguarding and in future sustainable development of the city. In the same vein, information and communication channels, network services as well as social services related to handicapped children and elderly people are regarded as important. The quality of urban nature life and land recovery in the urban suburbs are apparently conceived of as a less important problem for the city of Cremona.

The experts wished to have a more refined assessment scale and hence the expert weights were measured on a 10-point scale, with 1 as the worst and 10 as the best result. It appears that for the outcomes of the weight vector assignment by experts on this scale, the highest weight was given to social and economic issues, such as employment and environmental criteria (including green areas and availability of parks), that seem to be considered as a necessity for the community.

Once the impact assessment matrix (Table 13.3) and the weight matrix (Table 13.4) have been constructed, the actual MCA technique (VISPA) has to be executed. For this purpose, a sequential software programme based on both *weighted summation* and *concordance analysis*[2] was used in the various successive operations. In the next section, the results will be compared for each of the three different weight assignment options (weight sets A, B and C) as given in Table 13.4.

13.6 MCA results

The weighted summation technique (Table 13.5) does not present essential differences among the A, B and C options, at least not for the first priority positions. The first selected alternative appears to be the Park Project (alternative 2) for each of the weights assigned to the criteria. The University Centre and the Meeting Centre alternatives (alternatives 11 and 12), which belong to the derelict areas recovery project, assume the second and third position, respectively. The remaining positions are taken by the public transport project, the road system project and the parking systems project (alternatives 8, 7 and 6, respectively) with some slight differences among the A, B and C options of the weight sets.

---

[1] A 10-point scale also has been used for alternative scores during the matrix construction phase.

[2] The concordance matrix is a matrix for a pairwise comparison of alternatives. It is an n x n matrix, where n is the number of alternatives in which for each dominance relationship in a pairwise comparison the corresponding weights are added up.

Table 13.4: Criteria and weights system for the Cremona study

	CRITERIA	A*	B*	C*
Economic	Variation of GDP	10	1	7
	Employment rate	10	1	10
	Integration of product sector	8	1	4
	Employment mobility	9	1	2
	Development small enterprises (PMI)	8	1	8
Social	Commuting patterns	9	1	2
	Efficiency of administration	8	1	4
	Education	8	1	7
	Accessibility to sporting equipment	8	1	5
	Accessibility welfare work	7	1	6
	Accessibility sanitary service	8	1	8
	Accessibility post service	8	1	3
	Availability bank services	7	1	3
	Public security	8	1	8
	Housing quality	8	1	5
	Quality of life	7	1	10
	Population density	8	1	1
	Ageing	9	1	5
	Information systems	9	1	1
	Cultural public relations	7	1	3
	Racial integration	8	1	7
	Transport and traffic quality	7	1	8
	Handicap/child facilities	9	1	10
	Citizens participation	8	1	4
Environmental	Percentage of waste recycled	7	1	5
	Quality of urban wildlife	5	1	4
	Energy consumption	7	1	5
	Urban green areas per person	7	1	10
	Landscape	7	1	4
	Air pollution	9	1	10
	Water pollution	9	1	10
	Soil pollution	9	1	10
	Noise	9	1	4
	Recovery of agricultural land	5	1	4
	Percentage derelict areas	7	1	2

* A is a weight specified by the Cremona Town Council on scale of 5-10; B is weight of 1; and C is a weight defined by experts on a 10 point scale

From the perspective of the policy-makers' weights, the public transport option in Cremona turns out to be more urgent than the other two alternatives, given the present inadequacy in both the urban and the extra-urban context. Such interventions are clearly important not only for the community, but they are also an essential condition for a greater efficiency and dynamism of productive and commercial sectors. The project alternative on social services development (through the building of retirement homes) is likely to be a good choice as well, as it regards an alternative use of abandoned areas. Generally speaking, such areas are abundantly available in Cremona and occupy a wide surface of the urban territory (University area, Meeting Centre, Social services), so that a recovery policy can produce various significant advantages in the urban social, environmental and productive contexts.

Table 13.5: Results from the weighted summation test

										Derelict Area		
	Rural restructuring (agritourism etc.)	Park project	Systems of waste recycling (incinerator)	Street furniture/ restructuring built	Suburban commercial system	Parking system	Road system	Public transportation network	Buildings eldery people	Cultural Centre Conservatoire Auditorium	University Centre	Meeting Centre
Alternatives	1	2	3	4	5	6	7	8	9	10	11	12
Weight A	11	2	10	12	9	7	6	8	5	8	1	3
Weight B	11	1	10	12	9	6	7	4	5	8	2	3
Weight C	11	1	10	12	9	8	7	5	4	6	2	3

Table 13.6: Results from the concordance test

										Derelict Area		
	Rural restructuring (agritourism etc.)	Park project	Systems of waste recycling (incinerator)	Street furniture/ restructuring built	Suburban commercial system	Parking system	Road system	Public transportation network	Buildings eldery people	Cultural Centre Conservatoire Auditorium	University Centre	Meeting Centre
Alternatives	1	2	3	4	5	6	7	8	9	10	11	12
Weight A	11	1	10	12	9	7	5	4	6	8	2	3
Weight B	11	1	10	12	9	6	5	4	7	8	2	3
Weight C	11	1	10	12	9	8	4	5	6	7	2	3

In the concordance index ranking (Table 13.6), the position of the first four choices is identical to the previous ranking in Table 13.5 (weighted summation), with only very slight variations in relation to the different weight assignments to the criteria. The result appears to be very robust, but to offer an exploratory test on the overall robustness of the model results, a sensitivity analysis is carried out in which the weight variation is fixed in intervals within which the alternative is the same. This can be obtained with an artificial variation (both as an increase and a decrease) of a weight, while maintaining the others constant. The weight investigated has no sensitivity when the resulting ranking does not change whatever the weight value is. But it has a very high sensitivity when even a small variation causes alterations in the ranking of the alternatives; clearly, if a weight has a low or no sensitivity, the corresponding policy criterion cannot significantly influence the alternative ranking whatever its importance. Table 13.7 shows the results of the sensitivity test. The gradient (+, 0, -) represents a strong, zero or weak sensitivity of the relevant criteria.

Table 13.7: The sensitivity of the criteria

	CRITERIA	Weight set C	Weight set A
Economic	Variation in GDP	+	+
	Variation in employment	+	+
	Integration of production sectors	+	+
	Job mobility	-	-
	Development SME	+	+
	Commuting patterns	+	+
Social	Administrative efficiency	-	-
	Accessibility to education	+	-
	Accessibility to sports facilities	+	0
	Accessibility to welfare provisions	0	0
	Accessibility health service	+	+
	Accessibility post service	+	+
	Availability bank services	+	+
	Public security	+	-
	Housing quality	+	0
	Quality of life	-	-
	Population density	-	-
	Ageing	+	-
	Information systems	-	-
	Cultural public relations	-	-
	Racial integration	+	+
	Transport and traffic quality	+	+
	Facilities for handicapped children	+	+
	Civil participation	+	-
Environmental	Percentage of waste recycled	+	-
	Quality of urban nature area	+	0
	Energy consumption	+	-
	Urban green areas per person	+	0
	Landscape	+	0
	Air pollution	+	0
	Water pollution	+	0
	Soil pollution	+	0
	Noise	+	0
	Renewal rural areas	+	0
	Percentage derelict areas	+	+

+ = high sensitivity; 0 = no sensitivity; – = low sensitivity

The test points out clearly the necessity, when an MCA analysis is used, to consider each criterion weight with utmost caution. In the case of weight set A, in particular, all environmental criteria have turned out to be rather sensitive. Even slight appraisal differences may produce already significant changes in the rank orders of alternatives and consequently in the final choice. As indicated previously, it is possible to repeat the method for a subset of projects, or for a more detailed analysis. It will be also possible to change the critical concordance threshold values for each indicator. The analysis has pointed out that the most preferable projects belong preponderantly to the service sector and follow the development trends of most Italian and EU economies.

13.7 Conclusions

It goes without saying that sustainable development is a general issue that may be applied to all levels of strategic spatial planning. It refers to the natural environment as well as to the urban and rural system. It offers the opportunity to explicitly and adequately value public goods by acknowledging their complex nature and by showing their relevance in a non-market environment. The case study of Cremona prompts indeed an important question: how can the development and planning of a city meet the general principles and guidelines of sustainability?

There are three basic conditions to be fulfilled. Firstly, time series and cross-section indicators are required, that can help to identify possible development distortions and the related necessary corrections. Secondly, responsibility must be given to all stakeholders involved, who have the competence to exercise priority interventions for each urban site, or to intervene in possible 'emergencies' (economic, social or environmental). Thirdly, policies must be adopted for territorial safeguarding (e.g. zoning) that minimise environmental damage and favour a general recovery process. Such a policy scenario becomes feasible when an integrated urban development model is adopted, in which land use is treated from a sustainability perspective and the social requirements involved. Long-range strategic planning is the solution for this kind of problem: its application is also essential at an urban level, in particular as it is able to exploit the connectivity between urban and agro-environmental systems.

Is it possible to map out a successful path for the above consultation process? In this context, the policy-maker has a crucial role as he or she is the link between the local system and the macro system (e.g., the European urban system). The policy-maker should therefore gain new capacities and skills and ensure more interactivity in order to be responsible for political and land use choices. An important finding from our experience from the case study was that policy-makers have little knowledge about the general European and international policy guidelines (specifically to the Local Agenda 21 and the Aalborg Charter programmes) which might also offer financial support at a EU level. A further consideration concerns the general difficulties in the use of appraisal methods for the support of policy-making. Despite analytical complexities these methods are easily managed with specific modern software packages, that are usually flexible and user-friendly.

The choice of MCA derives from the necessity to provide new decision models to policy-makers that can take into account the multidimensional nature of the development process. Given that the planning process is becoming a sort of art of communication, trying to reduce conflicts among the various social parties, this kind of model is characterized by the possibility of optimising the choice among different alternative projects, thus promoting a territorial sustainable development, and the possibility of managing the plan through the necessary negotiations among social parties or stakeholders. Social parties might modify the way of managing public goods, as they can play an active role in the promotion of investments for the creation of the basic conditions of healthy social and economic development in the city.

There are many policy opportunities through which public and private sectors can and should collaborate for the promotion of the sustainable development of the city. It is therefore also necessary to approach the issue of territorial and urban sustainability from a meso-economic perspective, and to suggest a new urban management style, that considers environmental quality at the same level as traditional production factors, i.e. an input in the production process. The application of MCA in this chapter has tried to pursue this demand by providing a methodology for ranking different alternative projects. The basic concept is that of optimising a set of indicators/criteria/objectives that are considered to be critical factors for the idea of sustainable development and of urban welfare. The main positive aspect of this approach is the rationalisation and transparency of the urban decision process, obtained through the active participation of groups and experts involved, both in the indicator-setting phase and in the project-choice phase of urban sustainability policy. Seen from this perspective, one may plausibly argue that the use of multicriteria decision aid methods – for the evaluation of competing initiatives in the city – is a promising part of a multidimensional policy analysis seeking to achieve sustainable urban development.

Acknowledgement

The authors wish to thank Professor A. Segale for his constructive comments.

References

Atkinson, G., Dubourg, W.R., Hamilton, K., Munasinghe, M., Pearce D.W. and Young, C. (1997) *Measuring Sustainable Development: Macroeconomics and Environment*, Edward Elgar, Cheltenham.

Banister, D, Button, K. and Njikamp, P. (eds.) (1999) *Environment, Land Use and Urban Policy*, Edward Egar, Cheltenham.

Banister, D. (1999) Urban sustainability, in Bergh J.C.J.M. van den (ed.) *Handbook of Environmental and Resource Economics*, Edward Eldgar, Cheltenham: 560-569.

Banister, D., Watson, S. and Wood, C. (1997) Sustainable cities: transport energy and urban form, *Environment and Planning B – Planning and Design*, 24(1): 125-143

Beinat, E., and Nijkamp P. (eds.) (1998) *Multicriteria Analysis for Land-Use Management*, Kluwer, Dordrecht.

Bergh, J.C.J.M. Van den (1996) *Ecological Economics and Sustainable Development*, Edward Elgar, Chaltenham.

Bergh, J.C.J.M. van den, and Hofkes M. (eds.) (1998) *Theory and Implementation of Economic Models for Sustainable Development*, Kluwer, Dordrecht.

Breheny, M. (ed.) (1992) *Sustainable Development and Urban Form*, Pion, London.

Bruyn, S. de (1999) *Economic Growth and the Environment*, PhD Dissertation, Dept. of Spatial Economics, Free University Amsterdam.

Camagni, R., Capello R. and Nijkamp, P. (1998) Towards sustainable city policy: an economy-environmental-technology nexus, *Ecological Economics*, 24(1): 103-118.

Capello, R., Nijkamp P. and Pepping, G. (1999) *Urban Sustainability and Energy Policies*, Springer-Verlag, Berlin.

Finco, A., and Nijkamp P., (1999) Sustainable land use: methodology and application, in Lonergan, S.C. (ed.) *Environmental Change, Adaptation, Security*, Kluwer, Dordrecht: 205-222.

Haughton, G., and Hunter, C. (1994) *Sustainable Cities*, Jessica Kingley, London.

IIUE (1995) *The ABC Approach*, Delft.

Janssen, R. (1991)*Multiple Decision Support for Environmental Problems*, Kluwer, Dordrecht.

Jenks, M., Burton, E. and Williams K. (eds.) (1996) *The Compact City: A Sustainable Urban Form*, FN Spon, London.

Newman P., and Kenworthy, J. (1989) *Cities and Automobile Dependence*, Gower, Aldershot.

Nijkamp, P. and Opschoor, J.B. (1997) Urban environmental sustainability, in Chatterji M. and Kaizhong, Y. (eds.) *Regional Science in Developing Countries*, MacMillan, London: 52-73.

Nijkamp, P., and Perrels, A. (1994) *Sustainable Cities in Europe*, Earthscan, London.

Nijkamp, P., Rietveld P. and Voogd H., (1991) *Multicriteria Evaluation in Physical Planning*, Elsevier, Amsterdam.

OECD (1978) *Urban Environmental Indicators*, Paris.

OECD (1994) *Environmental Indicators*, Paris.

OECD (1995) *Urban Energy Handbook*, Paris.

Orishimo, (1982) *Urbanisation and Environmental Quality*, Kluwer, Dordrecht.

Pearce, D.W. (1999) *Ecological Economics: Essays in the Theory and Practice of Environmental Economics*, Edward Elgar, Cheltenham.

Pelt, M.J.F. van (1993) *Ecological Sustainability and Project Appraisal*, Avebury, Ashgate.

Rees W.E. (1992) Ecological footprint and appropriated carrying capacity: what urban Economics Leaves Out, *Environment and Urbanisation*, 4(2).

Satterthwaite, D. (1997) Sustainable cities or cities that contribute to sustainable development?, *Urban Studies*, 34(10): 1667-1691.

Segale A. (1993) L'uso degli Indicatori Ambientali, *Impatto dell'Agrotecnologie nel Bacino del Po*, C.N.R.-RAISA, Franco Angeli, Milano.

Selman, P. (1996) *Local Sustainability*, Paul Chapman, London.

Stanghellini, S., and Stellin, G. (1996) *Politiche di Riqualificazione delle Aree Metropolitane: Domanda di Valutazione e Contributo delle Discipline Economico-estimative*, Atti del Convegno XXVI CESET, Milano, 17-18 ottobre.

Stanners, D. and Bourdeau, Ph. (eds.) (1995) *Europe's Environment* (The Dobris Report), European Environmental Agency, Brussels.

Veen-Groot, D. van, and Nijkamp, P. (2001) Sustainable transport and "factor four", in Kohn, A. and Kohn, J. (eds.) *Towards Sustainable Development*, Edward Elgar, Cheltenham (forthcoming).

Wackernagel, M. and Rees, M. (1996) Perceptual and structural barriers to investing in natural capital, *Ecological Economics*, 20(1): 3-24.

Wallner, H.P., Naradoslawsky, M. and Moser, F. (1996) Islands of sustainability: a bottom-up approach towards sustainable development, *Environment and Planning A*, 28(10): 1719-1908.

World Bank (1992) *World Bank Development Report, Development and the Environment*, Washington D.C.

Adele Finco
Universita degli studi du Ancona – Facolta di Agraria
DIBIAGA-Area Economico Estimativa
Via Brecce Bainchi
60131 Ancona
Italy

Peter Nijkamp
Department of Spatial Economics
Vrije Universiteit Amsterdam
De Boelelaan 1105
1081 HV Amsterdam
The Netherlands

14 A METHODOLOGY FOR THE ANALYSIS OF SPATIAL CONFLICTS IN TRANSPORT POLICIES: OVERVIEW AND APPLICATIONS

Euro Beinat

14.1 Introduction

The transport sector in the European Union (EU) accounts for a substantial share of the European gross national product. It is a source of employment, a crucial factor for the economic growth of the EU and for its international competitiveness. The EU transport policy[1] focus is manifold. However, transport infrastructures play an important role, both in terms of management of existing infrastructures (information systems, capacity management, road pricing) and in terms of provision of new infrastructures. New infrastructures are meant to increase the competitiveness of European regions and to achieve higher levels of efficiency in the transport sector. However, infrastructures are blunt-edged instruments. They address the competitiveness-efficiency nexus only indirectly and have several side effects (such as environmental and social impacts). On the other hand, they involve large investments and therefore are valuable employment generators.

Infrastructures like the Trans-European Networks (TEN) affect large spatial areas. Their costs and benefits affect several different regions, which usually share different identities, environmental conditions and priorities. Conflicts usually occur, resulting in complex decision negotiations or interruptions in the decision process. The spatial patterns and distribution of effects are a primary cause of conflicts. The evaluation perspective is also spatially dependent, since some effects will be evaluated locally, while others will have regional or national relevance. At each level, actors[2] are likely to have different priorities and values. The distribution of effects and the different evaluation perspectives are the main causes of spatial conflicts.

Policy evaluation usually involves dealing with two different types of conflicts: intra-actor and inter-actor conflicts (Bogetoft and Pruzan, 1991). Intra-actor conflicts arise all the time in constrained decision-making situations where no single outcome is seen as best on all possible counts. Individuals have to accept an outcome that is less desirable in certain respects than the ideal in order to get a better outcome with respect to other factors. This would be the case, for instance, with a transport plan that causes environmental pressure but is nevertheless approved because of its economic benefits. Trade-offs between conflicting outcomes are a distinctive feature of intra-personal conflicts, which are common to most decision situations. All decision aid techniques (such as cost-benefit analysis or multicriteria analysis) aim at resolving this type of conflict by supporting implicit or explicit weighting schemes which serve to balance between conflicting achievements.

1 The term policy is used in a broad sense to include general indications for action (e.g. to shift transport from road to rail, or to decongest urban areas); plans and programs (e.g. upgrade schemes for road and rail networks, management of the level of service); and individual policy measures (e.g. a railway project).

2 Actors may present themselves as people, as role occupants, as groupings, as organizations, as pressure groups, and more. In general, an actor is any individual or grouping affected by or affecting the decision to which we can associate a relatively consistent point of view (Burgoyne, 1994).

J. Stillwell and H. Scholten (eds.), Land Use Simulation for Europe, 193–210.

Inter-actor, or interpersonal, conflicts are different in that they deal with the preferences of different individuals. Such conflicts occur when individuals disagree on a course of action: what is best (or good, or acceptable) for one person is not so for someone else. This would be the case of regional authorities that disagree on the location of a transport facility, either because they want to prevent the negative effects from being localized in their region, or *vice versa*, because they are interested in being as close as possible to where positive effects will take place.

This chapter deals implicitly with both types of conflicts, but focuses on intra-personal conflicts, which are referred to as simply conflicts in the rest of the text. In spite of the voluminous literature, there is no universal definition of conflict. A widely used definition, also used in this chapter, is that *"conflicts are the result of incompatible activities, that is one person's actions interfere, obstruct or in some way get in the way of another's action"* (Tjosvold, 1998). There are numerous conflict factors, which can be grouped into four main classes (Bogetoft and Pruzan, 1991):

- value system factors (actors disagree on goals and concerns; they employ different criteria and/or have different priorities);
- impact distribution factors (the spatial distribution of policy effects is, or is perceived to be, unjust);
- uncertainty factors (uncertainty on the policy effects, disagreement on the likely outcomes, insufficient understanding of synergies with other policies, hesitation about preferences and values); and
- process factors (actors have difficulties in communicating with each other, or are unsatisfied with their role and influence in the decision process).

Spatial decision problems have additional characterizations, due to the location and the scale dimensions. The location dimension affects both facts and values for the decision. The (expected) effects are spatially distributed usually over a large area. This holds for direct impacts (noise and pollution, for instance) and indirect impacts (spillover economic benefits between regions, for instance). The nature, use and vocation of land is also spatially differentiated, implying the existence of different measurement criteria for the same effects in different places. Additionally, communities affected by the decision may share different value systems in different places. The scale dimension indicates that different facts and values are appropriate at different scales. For instance, the national consequences of large transport infrastructures are addressed in terms of GNP contribution, economic growth, greenhouse gasses emissions, *et cetera*. At the local level, the same infrastructure will cause noise pollution, housing relocation, *et cetera*. The links between the macro and micro level are not always evident and usually imply that different decision actors evaluate effects at different scales. There is also an intrinsic relationship between space and (a subset of) decision actors which coincide with spatial policy areas. (e.g. nations, provinces, districts). However, it is important to point out that not all spatial actors play the same role in the process. Some actors are active players and pursue their specific objectives; others may have no stakes in the decision and be involved only to protect their interests.

The location and scale dimensions also suggest the distinction between vertical and horizontal conflicts. Vertical conflicts arise between policy units nested into each other. They are mainly due to scale effects. Examples are the conflicts between the Dutch transport authorities (representing national interests) and the provincial and local authorities on the acceptability of the Betuweroute railway project (see below). Horizontal conflicts occur between policy units at the same spatial level, such as between local authorities competing for the allocation of funds. Horizontal conflicts occur mainly due to location effects.

At present, a theory which systematically addresses spatial conflicts is not available. Rather, there is a set of interdependent theories and methodologies which deal with single or a few of the aspects of spatial conflicts (Densham, 1991; Armstrong, 1994; Douwen, 1996). Broadly speaking, there are three main approaches: spatial analysis and planning, decision analysis and organizational theory. The spatial analysis and spatial planning approach focuses on the spatial component of a decision problem (Maguire *et al.*, 1991). Geographical Information Systems (GIS) play an essential role in this context due to the amounts and complexity of the information needed for spatial problems. Decision analysis aims at an analytical and systematic exploration of the value systems of individuals involved in the decision. It provides methodologies for problem structuring, value judgment assessment, decision suggestions and sensitivity analysis (Rosenhead, 1989; Keeney, 1992; Vinke, 1992). Organization theory focuses on the relationship and interaction between actors in a decision process, including the role of procedures, information flows and behavioural effects (Simon, 1991; Hersey and Blanchard, 1988; Huigen *et al.*, 1993). The common ground of these approaches is usually referred to as collaborative spatial decision support (cf. Batty and Densham, 1999), and is at the basis of the methodology presented in this chapter.

The chapter therefore presents a methodology that can be used to articulate policy objectives, to assess and compare policy alternatives, and to support the analysis of spatial conflicts in a transparent way. The methodology, outlined in Section 14.2, is based on multicriteria analysis and spatial analyses. Spatial conflicts are described, factors that cause conflict are identified and a general framework for their analysis is provided. Four large projects in Europe are used in Section 14.4. As examples to show the spatial distribution of the decision actors, their different concerns and objectives and the way this drives the evaluation of the project. Some conclusions and discussion are presented in Section 14.5.

14.2 Methodology framework

The methodology is built around three fundamental policy questions:

- What is the attractiveness of a policy option with respect to the policy goals and concerns of the policy actors?
- What are the policy consequences at various spatial locations and scales?
- What are the differences and conflicts between actors and the cause and intensity of spatial conflicts?

MultiCriteria Analysis (MCA) can be used to address these issues. MCA has evolved from a mechanism to rank alternatives to a structured approach to organize factual information, value judgment, and to argue logically about the pros and cons of policy solutions. The essence of MCA lies in structuring management and planning decisions in terms of a number of relatively precise but generally conflicting criteria. Multicriteria methods are used in two basic contexts. In the first, a decision-maker seeks to make a decision which does not seriously impact on, or require justification to, other actors. In this case, methods can be relatively informal, and the rationale for the decision does not need substantial justification. This contrasts with contexts in which an individual or group has to make a decision on behalf of a larger group or community. In this case, the rationale for the choice must be documented and consideration of each interest or point of view must be demonstrated. This requirement necessitates the use of more formal methods of analysis, even when they may be less efficient and/or may impose structures (of preference, for example) which may not be strictly justifiable empirically (Bana e Costa *et al.*, 1995, p.264).

There are various multicriteria techniques available, which usually evolve around the following common activities: structure (establish goals and concerns for the problem, select criteria for the evaluation, and predict the performances of policy alternatives); value analysis (study the attractiveness of individual impacts and the relative importance of the criteria for the actors involved), recommendation for the decision (depending on the way the problem is structured, this requires some type of aggregation, leading to a ranking of the alternatives, or the isolation of one or more promising solutions).

The link between space and MCA is made through the concept of policy unit. A unit is an area, administrative or other, associated with a distinctive evaluation perspective. Units are found at various spatial scales, for instance: national government, regional authorities, municipalities, local communities. They can also be defined in terms of functional borders (e.g., a river basin, a transport corridor) or special area features (e.g. a protected area). Due to scale and location factors, units usually employ different evaluation perspectives, making the same policy attractive for some units and unattractive for others.

By applying MCA at the level of policy units, it is possible to analyse the causes of (spatial) conflicts and how they relate to value system, impact distribution and uncertainty factors. The origin, causes and intensity of conflicts can be investigated, and information can be supplied to the negotiation and conflict management process. The interpretation of this information, and the implications for negotiation, depend on the type of setting used for policy assessment. Most practical applications correspond to one, or a combination, of the three settings described in Table 14.1.

Table 14.1: General settings for conflict analysis

Individual structuring – Individual evaluation	Group structuring – Individual evaluation	Group structuring – Group evaluation
Applies when units employ totally different perspectives to evaluate the same policy. Differences may regard any aspect, and even fundamental issues such as the policy goals. Conflicts may emerge due to a variety of reasons such as different policy priorities or unfair impact distribution.	Applies when units agree on how to assess the policy. This implies a basic agreement on what a policy should be meant for and what goals it should pursue. However, the units may disagree on the relative importance of different aspects and may be affected by different impacts, leading to conflicts.	Applies when units share the same evaluation perspective. This implies an agreement on the policy goals and on the relative importance of different factors. Conflicts may emerge only due to the impact distribution, which determines a different balance of benefits and costs for the various units, causing conflicts.

Information on type and intensity of conflicts is provided by a set of conflict indices (Table 14.2). These identify the units between which conflicts are stronger, and the differences between units which are at the basis of conflicts. The indices can be used to detect the critical conflict factors, which can be the spatial distribution of costs and benefits, differences in the fundamental objectives and concerns of the units, or difference in the relative importance of these concerns. This evidence can be used either to inform and support the negotiation process, or to drive the search for alternative policy measures, which decrease conflicts and are better suitable to respond to the interests of actors involved.

Table 14.2: Some examples of conflict indices

Index of rank reversal	Index of unattractiveness	Index of maximum loss
Measures the number of rank reversals between the rankings of two units. There is no conflict only when units share the same ranking of the policy options.	For each unit, it measures how unattractive is the preferred option of another unit. There is no conflict when the best option for a unit is attractive for the others (though not best).	For each unit, measures the sacrifices which would be necessary to accept the best option for another unit.

The methodology is illustrated in Figures 14.1 – 14.3. This example from Beinat and Bana e Costa (1998) concerns the evaluation of three highway projects, and the assessment of their pros and cons form the perspective of the region, and three local administrations (areas 1-3). Figure 14.1 and 14.2 show the main steps of the methodology applied for the region and area 2 respectively. These two units have a different list of objectives and concerns, they are affected by different types of impacts, and attach different weights to the various consequences. The impact table in Figure 14.1 lists the expected consequences of the alternatives against the concerns of the regional actors. This information together with value functions and weights leads to the ranking of alternatives for the region. While the region would prefer the implementation of Alt2, followed by Alt3 and then by Alt1, area 2 would prefer Alt3 instead. This indicates a conflict between these two units. The whole range of conflicts between all units is shown in Figure 14.3, which includes also the evaluations of the remaining units. Area 1 and the region agree on the ranking and best solution. For Area 2 the best option is the second of Region/Area 1. The perspective of Area 3 is the opposite of that of Area 2, indicating a very strong conflict.

The application of conflict indices can be used to identify the intensity of various types of conflicts, the reason for conflicts, and suggest some measures that could be useful for reducing conflicts and improving the alternatives for the whole set of units. In this case (Beinat and Bana e Costa, 1998), the issues causing conflicts were essentially the noise levels in area 2 and area 3, the impacts on natural areas in area 1 and area 2, and the limited accessibility achieved by area 3. Conflict indices can be used to assess the type of improvements needed for these factors and thus provide useful information for the design of mitigation or compensation measures.

14.3 Inside the methodology

The tools comprising the methodology can be organized into three main classes of instruments: tools for structuring, for analysis of attractiveness and for conflict analysis.

Figure 14.1: The evaluation of three highways from the perspective of the region

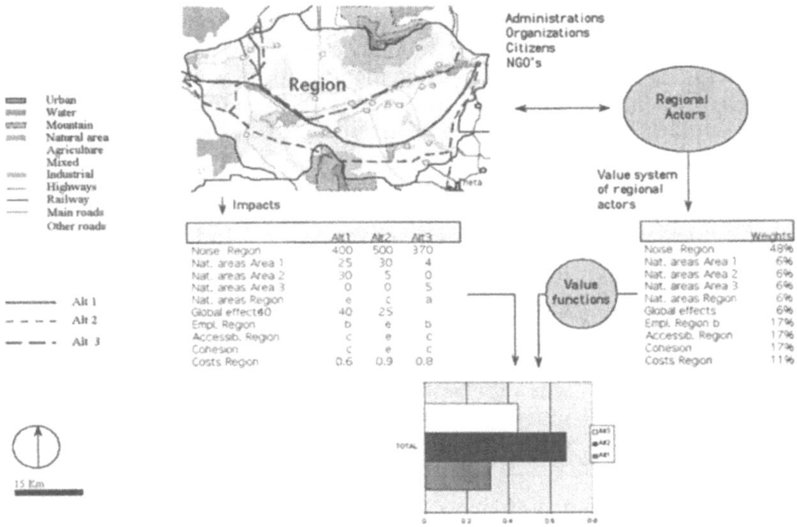

Figure 14.2: The evaluation of the three highways from the perspective of area 2

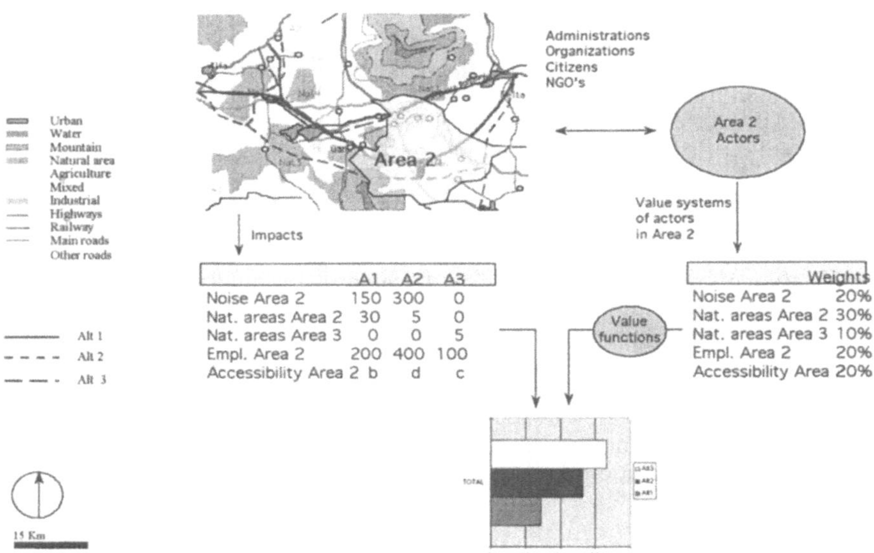

Structuring means to explore and identify the objectives/concerns of a policy unit, and to organize them into a value tree which specifies areas of concern for a unit (e.g. social effects of the policy, environmental impacts), individual concerns (e.g. accessibility and noise) and the criteria selected by the unit to measure the impacts of interest (e.g. number of houses exposed to high noise levels) (Bana e Costa *et al.*, 1998a). The value tree and the criteria may be the same for all units, or be partly or completely different for different units. The tools for structuring include various forms of interview techniques, which lead to qualitative problem representation, such as mind maps, cognitive maps, or rich-picture diagrams (Rosenhead, 1989).

The concerns of a policy unit include the objectives of the policy (the reasons why there is interest in a policy action, such as 'to remove constraints to economic growth') and the preoccupation for the adverse effects that the policy actions may have (such as 'to minimize the noise disturbance'). Concerns cannot be strictly justified empirically, as they stem from the actors' value system. However, they should respect some fundamental properties to be useful for the analysis, such as independence and absence of double counting (Keeney, 1992). Concerns are made operational by introducing the *impact descriptors*. For instance, the concern 'to minimize the disturbance to residents' could be associated with the impact descriptor 'number of houses exposed to high noise levels', which is the quantity that is measured to assess the policy performances. A policy measure is thus described by a list of impacts on the descriptors. Descriptors can be quantitative, verbal, continuous, discrete, and even pictorial. Also, there may be several suitable descriptors for an individual concern. The choice depends on the degree to which a descriptor is able to clarify and communicate as unambiguously as possible the implications of policy actions, with respect to a given concern.

Impact analysis serves to predict, as accurately as possible, the consequences of a policy for a unit (Wood, 1997). The results are summarized in an impact table. Impacts cannot be defined in isolation from the interests and expectations of the units involved. Since the value tree of the units may be different, there may be several impact tables, each one associated to one or more units. Geographical Information Systems (GIS) are used for impact assessment due to the large information demands which are necessary to perform impact calculations for multiple units at multiple spatial scales and locations (Scholten and Stillwell, 1990). A section of an impact table for two policy units is illustrated in Figure 14.4. The impacts of the railway have been computed with a GIS. Noise contours, for instance, have been estimated taking into account the noise produced by the trains and the effects of mitigation measures, such as noise screens. The number of houses exposed to levels higher than 57 dB(A) was used as a descriptor for the concern 'noise disturbance to residential areas'. Impacts were then allocated to different policy units, based on their administrative contours.

Impact analysis results in an impact profile associated with each policy option for each policy unit. However, this information does not state if the policy consequences are attractive or unattractive. Questions such as 'Does an increase in accessibility of 10 per cent justify the costs of the policy?', or 'Is a land take of 500 hectares acceptable?', require additional inputs from the policy actors. These inputs are embedded in value functions. They apply to single criteria and translate impacts into attractiveness scores (Beinat, 1997). Value functions are usually the result of interviews with policy actors or sector experts. Specific interview schemes have been designed to simplify the task of the respondents and allow for qualitative and imprecise estimates. Examples use the Macbeth technique (Bana e Costa and Vansnick, 1994) which extracts value functions from qualitative, verbal judgments. The technique is based on a interview protocol: with the support of software, the user is shown a series of impact levels and is asked to judge their attractiveness in plain language. By requiring more judgments than strictly necessary, and by judging the same impacts from different angles, it is possible to assess value functions based on qualitative, imprecise responses, which are tested for internal coherency. This approach proved useful and relatively easy to apply. An example of value function is shown in Figure 14.5. Accessibility of social facilities (such as hospitals) needs to be guaranteed by the public bus service of Lombardy. One of the provinces involved in the evaluation of regional policies for public transport (Province of Mantova) considers 13 kms as an acceptable distance to the nearest facility, while more than 19 kms become unacceptable. Intermediate distances receive an intermediate rating.

Figure 14.3: Comparison of the rankings by the region, area 2 and the other two units

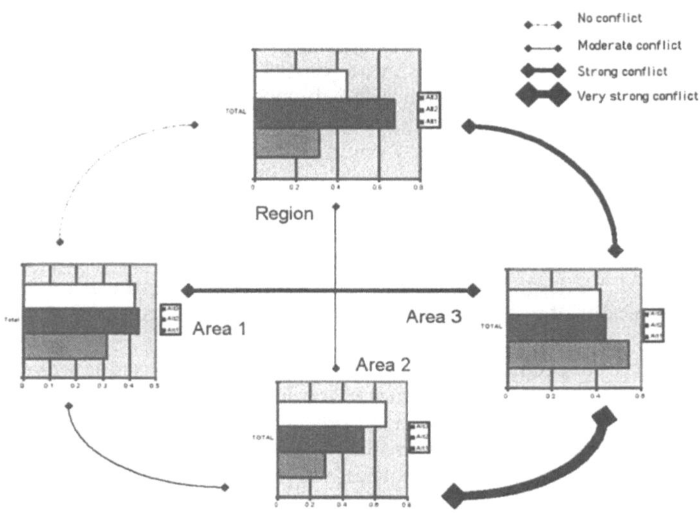

The overall attractiveness of a policy depends on these values and on the weight associated with individual objectives/concerns. Weights are assessed and used, implicitly or explicitly, every time the assessment of a transport policy depends on the combination of several impacts. However, the practice of weight assessment is often very poor and the concept of weights is frequently misused and misunderstood. The most critical mistake is the use of weights in isolation of the range of impacts of a policy and of the method used to aggregate impacts (Keeney, 1992). This may result in weights which do not represent policy priorities and which cannot be used to aggregate impacts.

Weights are intrinsically subjective and are meant to represent the decision strategy of a policy unit. The assessment of weights requires interviews with policy actors, for which specific interview protocols have been developed. Beinat (1998) lists several techniques for weight assessment. The EValue and the Macbeth techniques are suitable in cases in which only verbal and qualitative weight assessments can be made. Similar to the procedure for value functions, the weight assessment is based on qualitative verbal responses. These are checked for internal coherency and used to extract a set of weights representing the decision strategy of the respondents.

Figure 14.4: Impact analysis for the Betuweroute in the Netherlands

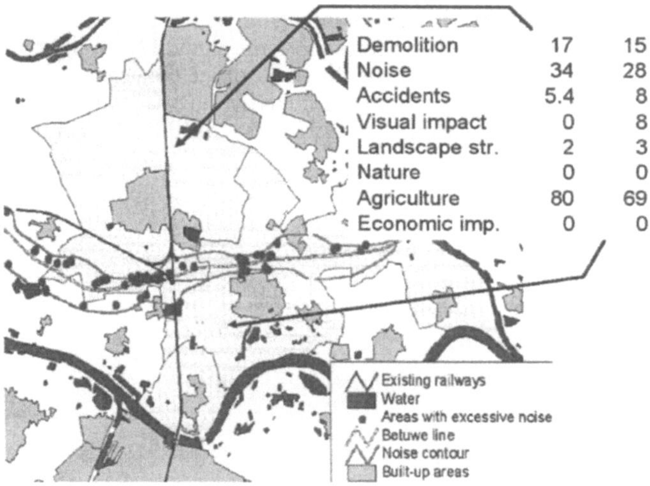

Figure 14.5: Acceptability of accessibility to social facilities

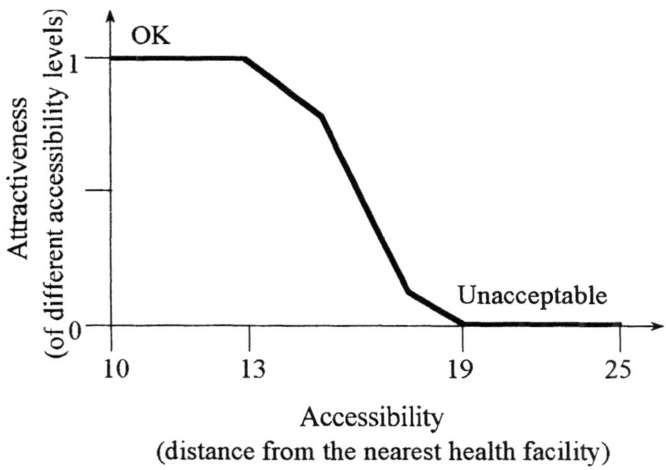

Through a weighted scheme, an overall value score can be associated with a policy option, reflecting its overall attractiveness (for a policy unit). These results are sensitive to various types of uncertainty, such as errors on impact predictions, uncertain value functions and uncertain weights. Sensitivity and robustness analyses test the credibility of results against imprecise impact estimates; uncertainty on the attractiveness of a given impact (value functions); and uncertainty and hesitation about the relative importance of different concerns (weights). An example of sensitivity analysis is shown in Figure 14.6.

Figure 14.6 shows the attractiveness of five policy options assessed in the Lisbon case (see below), and how this changes depending on the weight of noise. For a weight of about 25 per cent (the original estimate of the regional authorities), the quality of the road network would be substantially improved by implementing policy proposal A_3. However, this proposal would generate a high noise pollution in some areas. If the weight of noise remains below 50 per cent (less than the total weight of all other concerns), then A_3 is the best balance between transport improvements and environmental concerns. If the concern for noise increases above that level, then the proposal called A_1 should be preferred.

Figure 14.6: The attractiveness of five policy options

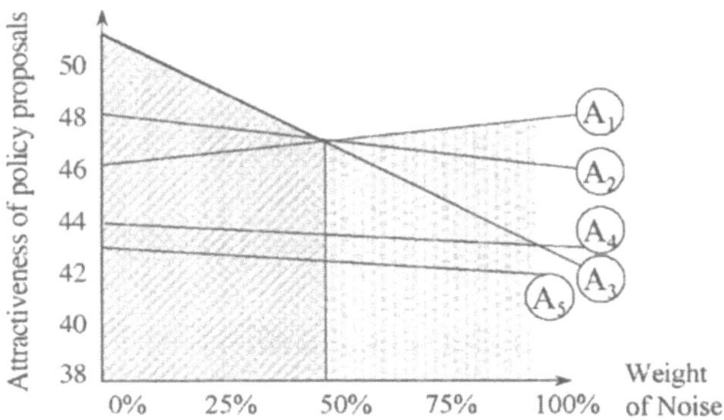

Conflicts between units correspond to disagreements on the preferred course of action. They may depend on the distribution of costs and benefits of the plan, but they may also depend on different objectives, specific evaluation criteria and individual weights employed by the units. Conflict analysis is based on conflict indices. Each index applies to a different situation and highlights a particular dimension of conflicts. For instance, ordinal conflict analysis applies to ordinal rankings, and measures the disagreement between units as concerns their rankings of policy options. Cardinal conflict analysis may be used to measure the intensity of conflicts between units. Also, indices have been developed to assess conflicts between pairs of units, or between one unit against all others.

As an example, conflict indices have been used to support the evaluation of road improvement schemes in the Lisbon region (Bana e Costa *et al.*, 1998b; see also Section 14.4). Four packages of roads were considered for improving the road system in the region. After a first analysis, one package seemed very promising. It combined good accessibility improvements in the region with limited environmental effects and few planning constraints. However, this package was very unattractive for two municipalities (out of eighteen). Five other municipalities considered the plan positive, but little better than the status-quo. This implied a conflict between a minority of the municipalities, the region and the remaining municipalities. Conflict indices pointed out the weak points of the proposal and measured the extent of the changes needed to improve the plan and reduce conflicts. This led to the identification of a new road package, in which some road links were replaced while respecting engineering and transportation constraints. The new package retained most benefits of the previous one, but significantly decreased the local costs. It represented a suitable solution to the conflict. The whole methodology, its phases and support tools are summarised in Table 14.3.

Table 14.3: Policy question, methodology and tools

POLICY QUESTION	PHASES	SUPPORT TOOLS
STRUCTURING and IMPACT ASSESSMENT		
Who are the policy actors and what are the spatial levels for evaluation?	Identification of policy units and their spatial relation	Institutional analysis, stakeholder analysis and territorial analysis
What are the goals and concerns of the actors and what criteria do they use to assess policy options?	Problem structuring and construction of a value tree	Structuring tools, such as cognitive maps, mind maps, rich-picture diagrams
What impacts are relevant for the units to substantiate their position?	Impact analysis; isolation of impacts for each policy unit	Impact prediction models. Impact scales for ordinal, cardinal, verbal, descriptive, pictorial impacts
ANALYSIS of ATTRACTIVENESS		
Are the individual impacts attractive or unattractive?	Analysis of attractiveness based on value functions	Interview technique to assess value functions
What is the relative importance of different concerns?	Analysis of trade-offs and of weights of different concerns	Interview technique to assess weights
What option is more attractive in global terms?	Aggregation of impacts based on attractiveness and importance	Additive aggregation rule
Are the results credible?	Sensitivity analysis to test robustness of results	Methods for sensitivity of impacts, value judgments and policy priorities
ANALYSIS of CONFLICTS		
Are there conflicts between the different units?	Identification of differences between the results of different units	Conflict indices to identify causes and intensity of conflicts, and the policy options which generate conflicts
What are the causes of conflicts and which measures could be nvestigated to reduce conflicts?	Feedback analysis	Methods to identify critical conflict factors (impacts, priorities, etc.) and changes in the policy measures to reduce conflicts

14.4 Applications

The methodology has been tested in four case studies:

- an assessment of alternative station locations and route options for the London Channel Tunnel high-speed railway (Norman and Vickerman, 1998; Vickerman and Norman, 1998);

- an assessment of the pros-and-cons of the freight railway at the national, regional and local levels for the Betuweroute in the Netherlands (Beinat 1998; Beinat *et al.*, 1998);
- an analysis of attractiveness of road investment packages for the Lisbon region and its municipalities (Bana e Costa *et al.*, 1998b); and
- an evaluation of the suitability of spatial zoning for increasing public-transport efficiency for the region and the provinces of Lombardy (Foresti *et al.*, 1998).

The Channel Tunnel rail link in England and the Betuweroute in the Netherlands were analysed *ex-post*, while the reorganization of the public bus network in Lombardy (Italy) and the re-organization of the road system in the Lisbon metropolitan area (Portugal) used the methodology as a decision support tool during the actual decision process.

The Channel Tunnel rail link case focuses on the location of the station in Ashford, the route through the Medway section (a natural area) and the number and location of intermediate stations (Figure 14.7). The balance between business and economic concerns (e.g. economic generation capacity, penalty to international travelers), quality of life (e.g. traffic, noise) and ecological impacts proved difficult to achieve resulting in conflicts in the vertical sense (e.g. the national versus the local level for the location of the Ashford station) and horizontal sense (e.g. between regions for the location of intermediate stations).

Figure 14.7: The Channel Tunnel rail link

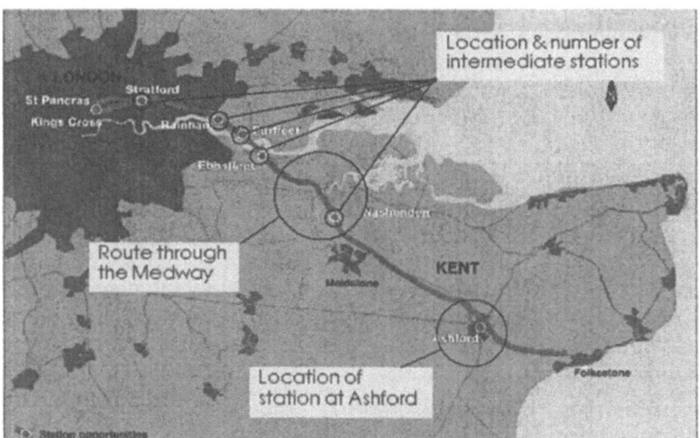

For the crossing of the Medway section, the application of the methodology *ex-post* justifies the choice of the short-tunnel option, which represents a compromise between the national and regional authorities on economic and environmental concerns. For the Ashford route, the conflicts between the national unit and the regional/local units could be distilled down to a trade-off between economic regeneration capacity and costs. The methodology also justifies the choice of the least conflicting solution that was selected in the decision process.

For the location and number of intermediate stations, the methodology was applied *ex-ante*. It highlighted the difficulty in finding a solution suitable for the national level (concerned with the quality of domestic and international connections) and the regional levels (interested in the economic regeneration effect of a station). The only way to find a station option, which is agreeable to the national and regional units, is by increasing environmental mitigation or economic compensation.

The Betuweroute case focuses on the position of national, regional and local actors as concerns the final route of the railway (Figure 14.8). The spatial distribution of benefits (which peak in Rotterdam) and costs (distributed across the whole corridor) was a main cause of conflicts. Beinat and Drunen (1998) analysed the reactions of provinces, municipalities, citizens and stakeholders presented to the national authorities when the project was completed (more than 5,000 documents).

Figure 14.8: The Betuweroute corridor and the regional and local policy units involved in the case study

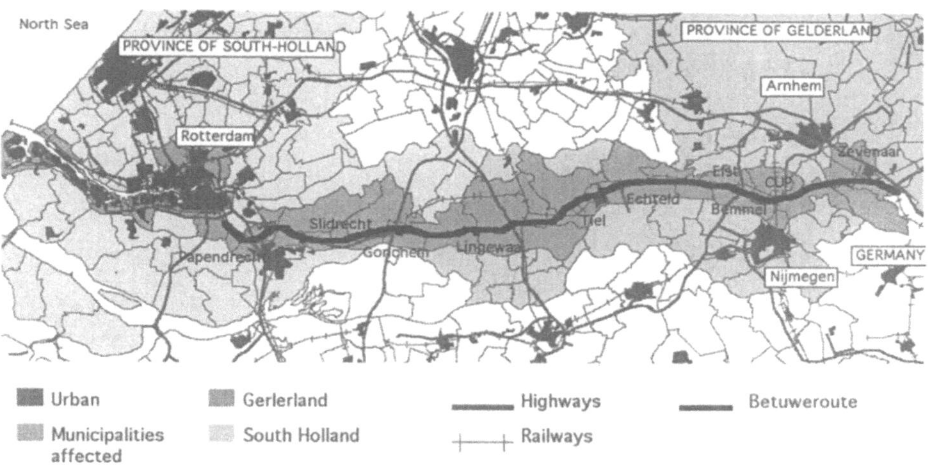

Aside from the predictable correlation between complaints and impacts, the large share of procedural remarks (25 per cent) indicates that local levels are clearly concerned about their participation in the strategic decisions. The analysis was also able to demonstrate that top-down decisions complemented by public participation may be inefficient, ineffective and trigger strategic behaviours that undermine the usefulness of participation. The methodology was then applied *ex-post*, to analyse the perspective of the regional and local units from a different angle, and requiring their direct involvement in the assessment of the plan. For the units analysed (province of Gelderland and its municipalities), the plan was unattractive, underlying a conflict with the national authorities. However, systematic differences in the positions of units could be identified. They were the consequence of impact distribution, but also of different concerns and weights employed by the units. Mitigation of noise and landscape effects emerged as the most effective strategy to reduce conflicts. However, only radical re-designs, such as the tunneling of the whole railway, could change the position of regional and local units from opposition to neutrality.

An alternative approach, based on the combination of the Betuweroute and other regional plans which produce regional benefits, emerged as a possible strategy to remove the deadlock produced by the conflict between national, regional and local authorities.

The zoning of Lombardy in Italy requires the division of the region into areas which are homogeneous in terms of transport demand (Figure 14.9). This process was part of the whole reorganization of the bus service in the region. Within each area of a zoning solution, licenses to operate the public bus network are allocated through public tenders. This system is meant to increase the competition between bus operators, to increase the efficiency of the Lombardy bus network and to decrease the amount of regional subsidies to bus companies. The most suitable zoning solution was negotiated between regions and provinces, taking into account the national guidelines, the interests of large urban areas, bus companies and trade unions. Conflicts emerged between the region (which prefers small areas to maximize competition and minimize subsidies), the provinces (which prefer large areas to minimize changes in the service level), and bus companies (large companies prefer large areas to keep their leadership; small companies prefer small areas to prevent being excluded from the tenders).

Figure 14.9: The final zoning solution (51 areas) adopted by regional and provincial authorities in Lombardy

The methodology was applied *ex-ante*, to support all stages of assessment, evaluation and negotiation. Several alternative zoning options were generated based on spatial transport demands. Each option was evaluated taking into account multiple concerns (such as accessibility levels in each area and suitability of the zoning to generate competition between bus companies). The methodology was also applied to compare different zonings and to generate new solutions that could decrease conflicts. The approach guaranteed transparency and the possibility of handling large amounts of information in a clear and rational way. It also provided information to support all stages of the negotiation between provinces and the region, which eventually agreed upon a final zoning (51 areas).

The reorganization of the road network in the Lisbon Metropolitan Region involves the allocation of investments to packages of road links, which are meant to increase the quality of the regional road network and to decrease congestion (Figure 14.10). The whole set of road links proposed would require a budget twice as high as the one currently available. The choice of which sub-set to implement requires the balancing of budget constraints, improvements in the regional transport system, economic benefits and environmental impacts. Since the costs and benefits of each package are not uniformly distributed across the 18 municipalities which compose the region, the spatial distribution of impacts has also to be taken into account. The methodology was applied *ex-ante*, to structure the perspective of the regional and municipal administrations.

Figure 14.10: The road links considered for improving the quality of the road network in the Lisbon Metropolitan area

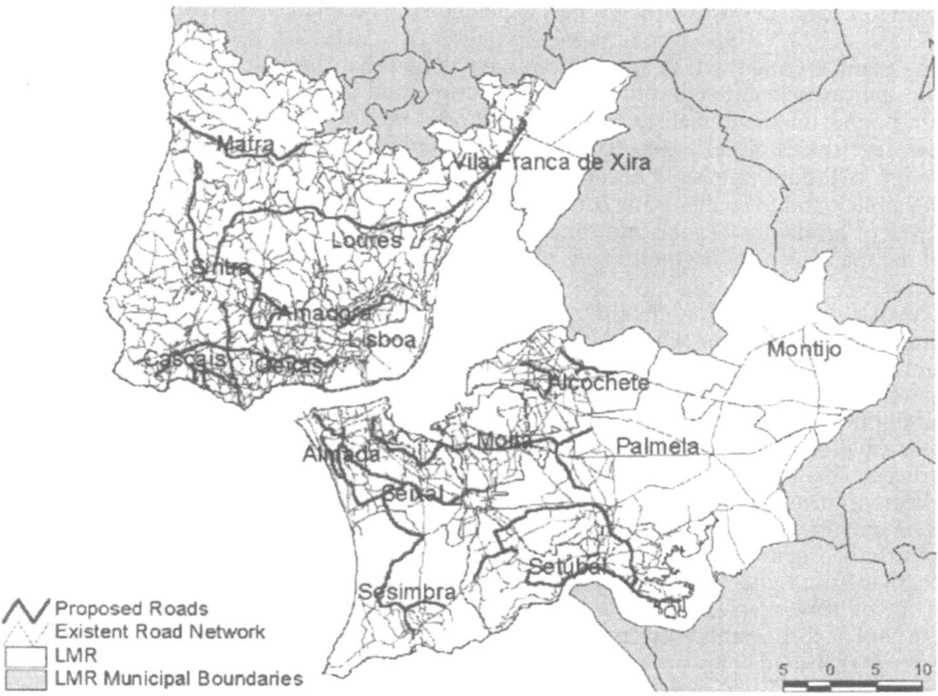

Impacts were estimated on a GIS platform. One of the four packages which was first evaluated emerged as a promising solution, attractive for the region and for ten municipalities. However, this package raised conflicts with the other municipalities, for instance, due to high environmental impacts and limited economic effects. Conflict indices suggested the generation of two additional options that could potentially diminish conflicts. One of the new packages was then selected. This package was able to satisfy regional and local actors to a large extent, and to diminish the conflicts between regional and local authorities.

14.5 Conclusions

The methodology requires a close interaction with the actors in the units. This interaction can take on different forms, such as meetings, workshops, interviews and questionnaires. It is an implicit assumption that actors can participate in the decision process, that their interests are considered, and that there is a positive engagement of actors in the search for an acceptable solution.

There is a wide set of instruments that can be used to support policy assessment and conflict analysis, from simple qualitative guidelines to sophisticated quantitative models. The complexity and diversity of real situations is such that there would be no point in specifying a single best instrument for decision aid. Flexibility and adaptability to the specific contexts are required to be able to provide the type of support that is most appropriate in each situation. The tools developed for structuring, impact analysis, policy assessment, and conflict analysis are components of a toolbox of instruments, which need to be combined in ways which strongly depend on the necessity of individual applications.

The examples and the trial applications have shown that the articulation of policy goals and concerns is either implicit, poorly documented or simply neglected in practice. Although it may be unrealistic to expect policy actors to be candid about their preferences, it is nevertheless necessary to articulate at least the fundamental goals and concerns of a policy. Without this basis, it becomes extremely difficult to investigate the pros-and-cons of a policy, and to facilitate any constructive debate between actors. However, thinking in terms of fundamental goals and concerns did not seem to be the normal way of thinking of the majority of policymakers.

The case studies explored the role of multiple actors involved in each case on the basis of different participation schemes. A wide spectrum of types of organized participation emerged from the analysis of the cases. The two extremes are the Portuguese and the Dutch case studies. In the Portuguese case, a collective body including representatives of all policy units was involved in all stages of the assessment. This is an example of open participation based on a permanent negotiation forum, which affects the whole decision process. This arrangement was suitable for structuring, assessment and evaluation of policy effects at various scales and locations. In the Dutch case, local administration and citizens were only involved after the strategic decisions were made. The Dutch case is an example of strictly organized participation, in which the main actors (the national level) allocate participation rights to other actors (provinces, municipalities, citizens), who have a given scope of influence, which can be used at specific stages of the decision process. Although the pool of participants was very broad (every citizen could express its view), the procedural restrictions made this process ineffective and limited its usefulness to the choice of mitigations and compensations. This procedural arrangement was inadequate to include the views and opinions of a wide spectrum of actors, and increased conflicts and opposition instead of favoring cooperation. Together with the experiences in the other case studies, this seems to suggest that co-decision processes can deploy their benefits only when systematically adopted, while only a partial implementation may even be counterproductive.

The methodology developed in the project introduces structure and organizes subjective judgments. The tools that support this analysis offer techniques to gain insight in this domain and to increase clarity. However, they are necessarily a simplified representation of people's values and tastes. Therefore, it is of fundamental importance to interpret these instruments as a means to organize discussion, debate and argument, rather than a means of engineering efficient policy solutions.

Finally, it is clear that no analytical approach can substitute political negotiations and debate. Special attention was necessary to dispel any such doubt during the applications. Also, the degree to which the use of this methodology can trigger strategic behaviours is still to be verified (for instance, the exploitation of the approach to achieve over-compensation of damage). Such behaviour is present in any situation which mixes cooperative and competitive motives, but the effect of using a structured and transparent approach to conflict analysis on the negotiation behaviour of policy actors is still to be fully understood.

References

Armstrong, M.P. (1994) Requirements for the development of GIS-based group decision support systems, *Journal of the American Society for Information Science*, 45(9): 669-677.

Bana e Costa, C.A., Ensslin, L. and Costa, A.P. (1998a) Structuring the process of choosing rice varieties at the South of Brazil, in Beinat, E. and Nijkamp P. (eds.) *Multicriteria Analysis for Land-use Management*, Kluwer, Dordrecht, pp. 33-48.

Bana e Costa, C.A., Joaquina Ramos, I.A., Nunes da Silva, F. and Gomes Veiga, R.J. (1998b) Restructuring the Lisbon metropolitan region network, in Beinat E. (ed.) *A Methodology for Policy Analysis and Spatial Conflicts in Transport Policies*, Report R98/08, Institute for Environmental Studies, Vrije Universiteit Amsterdam, pp. 87-110.

Bana e Costa, C.A., Stewart, T.J. and Vansnick, J.C. (1995) Multicriteria decision analysis: some thoughts based on the tutorials and discussion sessions of the SIGMA meeting, in R. Slowinski (ed.) *Proceedings of the 14th European Conference On Operational Research*, pp. 261-272.

Bana e Costa, C.A. and Vansnick, J.C. (1994) MACBETH: an interactive path towards the construction of cardinal value functions, *Int. Trans. Opl. Res.*, 1(4): 489-500.

Batty, M., and Densham, P.J. (1999) Decision Support, GIS, And Urban Planning, (accessed March 2000) *http://www.geog.ucl.ac.uk/~pdensham/s_t_paper.html.*

Beinat, E. and Bana e Costa, C. (1998) An example for the DTCS methodology for spatial conflicts in transport policies, in Beinat E. (ed.) *A Methodology for Policy Analysis and Spatial Conflicts in Transport Policies*, Report R98/08, Institute for Environmental Studies, Vrije Universiteit Amsterdam, pp. 87-110.

Beinat, E. (1997) *Value Functions for Environmental Management*, Kluwer, Dordrecht.

Beinat, E. (1998) Spatial conflicts in transport policies: structuring the analysis for large infrastructure projects, in Stewart, T. and Honert, R. van den (eds.) *Trends in Multicriteria Decision Making*, Springer, Berlin, pp. 385-394.

Beinat, E. and Drunen, M. van (1998) Spatial conflicts in transport policies: an exploration of the perspectives of regional and local actors, in Beinat, E. and Nijkamp, P. (eds.) *Multicriteria Analysis for Land-use Management*, Kluwer, Dordrecht, pp. 139-162.

Beinat, E., Drunen, M. van, Jones, M. and Janssen, R. (1998) Analysis of conflicts for the Betuweroute, in E. Beinat (ed.) *A Methodology for Policy Analysis and Spatial Conflicts in Transport Policies*, Report R98/08, Institute for Environmental Studies, Vrije Universiteit Amsterdam, pp. 147-169.

Bogetoft, P. and Pruzan, P. (1991) *Planning with Multiple Criteria*, North-Holland, Amsterdam.

Burgoyne, J.G. (1994) Stakeholder analysis, in Cassell, C. and Symon, G. (eds.) *Qualitative Methods in Organisational Research*, Sage, London.

Densham, P. J. (1991) Spatial decision support systems, in Maguire, D., Goodchild, M.F. and Rhind, D. (eds.), *Geographical Information Systems: Principles and Applications*, Wiley, New York.

Douwen, W. (1996) *Improving the accessibility of spatial information for environmental management*, PhD thesis, Free University of Amsterdam, Amsterdam, The Netherlands.

Foresti, E., Laniado, E., Muratori, S. and Tagliavini, P. (1998) Analysis of conflicts for the restructuring of the Lombardy public transport network, in E. Beinat (ed.) *A Methodology for Policy Analysis and Spatial Conflicts in Transport Policies*, Report R98/08, Institute for Environmental Studies, Vrije Universiteit Amsterdam, pp. 171- 190.

Hersey, P. and Blanchard K.H. (1988) *Management of Organisational Behaviour*, Prentice-Hall, N.J.

Huigen, J., Frissen, P.H.A. and Tops, P.W. (1993) *Het project Betuwelijn: spoorlijn of bestuurlijke co-productie?* PhD thesis, KUB, Tilburg (in Dutch).

Keeney, R.L. (1992) *Value-Focused Thinking*, Harvard, Cambridge.

Maguire, D., Goodchild, M.F. and Rhind D. (eds.) (1991) *Geographical Information Systems: Principles and Applications*, Wiley, New York.

Norman, C. and Vickerman, R. (1998) Local and regional implications of Trans-European Transport Networks: The Channel Tunnel Rail Link, *Environment and Planning A*, forthcoming.

Rosenhead, J. (1989) (ed.) *Rational Analysis for a Problematic World*, Wiley, Chichester.

Scholten, H.J. and Stillwell, J.C.H. (1990)(eds.) *Geographical Information Systems for Urban and Regional Planning*, Kluwer, Dordrecht.

Simon, H.A. (1991) Bounded rationality and organisational learning, *Organisation Science* 2: 125-139.

Tjosvold, D. (1998) The cooperative and competitive goal approach to conflict: accomplishments and challenges, *Applied Psychology: An International Review*, 47: 285-313.

Vickerman, R. and Norman, C. (1998) Analysis of conflicts for the Channel tunnel rail link, in E. Beinat (ed.) *A Methodology for Policy Analysis and Spatial Conflicts in Transport Policies*, Report R98/08, Institute for Environmental Studies, Vrije Universiteit Amsterdam, pp 119-146.

Vincke, Ph. (1992) *Multicriteria Decision-Aid*, Wiley, Chichester.

Wood, C. (1997) *Environmental Impact Assessment: A Comparative Review*, Longman, Harlow.

Euro Beinat
SPINLab: Spatial Information Laboratory
Vrije Universiteit
De Boelelaan 1115
1081 HV Amsterdam
The Netherlands

15 VIRTUAL REALITY AND THE SIMULATION OF EUROPE'S LAND USE IN THE TWENTY FIRST CENTURY

António Câmara

15.1 Introduction

There are several trends in spatial visualisation and interaction that go beyond the use of traditional maps and remote sensing images. Methods that may be applied to visualise and interact with simulation models of Europe's land uses include:

- the association of linked views including 3D models, graphs and databases to maps, such as in Cook *et al.* (1997) and Anselin (1999), linking mapping and exploratory data analysis software, and augmenting geographical information with digital video and sound (Shiffer, 1993);

- the superimposition of transparent representations of thematic information on maps, aerial photographs or satellite images as discussed in Boice (1992) and Chakraborty and Armstrong (1996). Monmonier (1999) presents related visualisation examples from weather forecasting;

- the use of animation in dynamic mapping, as proposed by DiBiase *et al.* (1992) and Mitas *et al.* (1997);

- the visualisation of uncertainty of spatial information as discussed by Pang *et al.* (1997) and Fisher (1999); and

- the exploration of three-dimensional representations of the terrain and associated data and simulation models in virtual environments.

Virtual reality or VR-based systems will be increasingly used in the visualisation of planning activities by younger generations of professionals trained in three-dimensional video games. Such virtual systems will integrate other techniques mentioned above and will offer unmatched browsing capabilities. Currently hampered by their lack of quality, virtual environments will become highly realistic by 2005 with the fast growing computer power, broadband networks and availability of high resolution satellite imagery.

This chapter will introduce virtual reality technologies and their fundamental concepts in Section 15.2, paying particular attention to the visualisation of terrains. The coupling of simulation models and virtual environments is considered in Section 15.3 whilst in Section 15.4, some suggestions for the application of new techniques for visualising and interacting with European spatial information are presented. Conclusions are drawn in Section 15.5.

J. Stillwell and H. Scholten (eds.), Land Use Simulation for Europe, 211–231.

15.2 Virtual reality methods and the visualisation of terrains

15.2.1 *Virtual reality methods*

The development of virtual environments consists of modelling and visualising real or imaginary worlds in three dimensions and in real time. The emergence of Virtual Reality Modelling Language (VRML) offers now the possibility of developing virtual environments at reduced cost and making them available to a wide audience via the Internet. Câmara *et al.* (1998) argued that site investigations are reality-based and provide sensory immersion but allow only limited exploration in space and time dimensions. Traditional interactive computer systems, despite their graphic interfaces, rely on symbolic computations, do not provide sensory immersion, and allow analysis along the space and time dimension in a compressed period of time.

Virtual reality (VR) technologies, providing the real-time generation of quasi-realistic three-dimensional graphics and sound, combine the advantages of realism of site investigation with the what-if capabilities of interactive systems. In addition, they provide sensory immersion. Most VR applications to spatial information are related to terrain visualisation. They enable one to fly at different altitudes and walk on photo-textures draped over digital terrain models. Virtual reality has been also applied to the immersive visualisation of spatial statistical data as presented by Cook *et al.* (1998). Finally, virtual representations have been used as front-ends for users of spatial databases and simulation models. Virtual Cheaseapeake (Wheless *et al.*, 1996) and Virtual Tejo (Câmara *et al.*, 1998) are illustrative examples of virtual decision support systems.

Virtual environments (VEs) provide the real-time generation of quasi-realistic three-dimensional graphics and sound and enable object manipulation in real time (Burdea and Coiffet, 1994). In traditional VR, the user is interacting in three-dimensional worlds generated by a computer using appropriate displays and position/orientation sensors. The three dimensional perspective results from the slightly different versions of a scene that each of our eyes may have. This binocular disparity provides the mechanism allowing one to see three-dimensional information. There are now several computer displays and head mounted displays that show three-dimensional images (Jones, 1996; Brice, 1997). They fall under non-immersive and immersive categories. Non-immersive solutions include the use of glasses whose lenses consist of fast shutters synchronised to the computer display. The shutters, using liquid crystal display (LCD) technology, alternately display different images for each eye (Jones, 1996). The head-mounted displays (HMDs) are popular immersive approaches. HMDs are usually helmets containing two displays, one for each eye, with the images filling the user's field of view. Coupled with devices to track the position and orientation of the user's head (Meyer *et al.*, 1991; McKenna and Zelter, 1992), they give the computer the parameters to render the images and sounds in real time.

The Cave Automatic Virtual Environment (CAVE) (Cruz-Neira *et al.*, 1992) is based on a projection system (the user is inside a system defined by projected images). It is not fully immersive but enables the sharing of virtual environment by groups. The Immersadesk (Reed *et al.*, 1997) supporting interaction with a large graphical surface viewable in stereo 3D using LCD eye pieces is another display used to visualise virtual environments.

Objects in VEs have a hierarchical structure: nodes group several objects, objects group several polygons, and polygons include several vertices. Vertices are the atomic elements described by their (x, y, z) co-ordinates. The objects that populate a virtual environment can be produced by defining the vertices and polygons in a text editor, using a three-dimensional modelling package or working directly from real objects applying a 3D digitising system. They can also be produced in real time in the virtual environment as a result of user modelling or as an output of a simulation process (Stuart, 1996). Objects may be defined by attributes such as: position and orientation; scale; colour, texture and shape; visibility; attached sound; behavioural properties; and interaction relationships with other objects.

A key characteristic of virtual reality systems is latency. Latency is the accumulation of lag introduced by each system component required to generate the system response as a result of the following (Stuart, 1996): the data rate and update rate of position trackers; the time spent by the computer in processing the tracker position, running the simulation and rendering the graphical and sonic images; and the refresh rate of display devices and the cumulative transmission time.

Virtual Reality enables one to fly, walk and dive into representations of natural systems, select any point of view, manipulate objects and scale them. The visualisation of terrains has been the most frequent application of virtual reality to spatial problems. The generation of terrain images in real time, both in standalone and networked environments, has been based on the Level-Of-Detail (LOD) management technique. Here we introduce the technique and methods that have been developed to improve LOD management for the interactive visualisation of large terrains. It also presents non-traditional methods to interact with virtual representations of terrains.

15.2.2 *Level-Of-Detail (LOD) management*

The Muchaxo *et al.* (1999) approach illustrates the issues involved in the visualisation of large terrains. They developed a method that starts by considering a digital terrain model (DTM) draped with aerial photographs or satellite images. The digital terrain model is represented using a triangulated irregular network (TIN). The rendering requirements of large terrains implied the adoption of an LOD management approach. An LOD implies that the rendering detail in the terrain should be inversely proportional to the distance of the viewpoint. LODs are based on partitioning the terrain data following methods such as the ones proposed by Muchaxo *et al.* (1999). These implementations enable the generation of images according to pre-established error bounds for geometry (related to the digital terrain model) and texture detail (related to the photo-textures draping the terrain model).

The approach followed by Muchaxo *et al.* (1999) was to design a tool that subdivides a terrain hierarchically into four tiles (or quad cells) and that generates TINs for each quad cell. The process is repeated for each of the sub-terrains until a maximum recursion depth is achieved. A quadtree is assembled in the process, resulting in several LODs for each level of the tree. The TINs are created using a modified version of DeFloriani's triangulation (DeFloriani *et al.*, 1985). DeFloriani's triangulation consists of starting with a coarse regular triangulation and breaking each polygon in tree polygons using an 'inner' point. That point is the surface point that deviates the furthest from the triangle. The subdivision is repeated until the triangulation is within a given tolerance. That tolerance is a function of the quadtree depth. The selected point can also be over the edge, making the triangle to be broken into two triangles.

The Muchaxo *et al.* (1999) algorithm also takes into account terrain features by simply giving more relevance to the deviations over the edges. This method intends to capture the terrain's critical lines such as valleys, ridges or channels. Along with the creation of the meshes, the large texture that one wants to map on the terrain is also broken into several tiles according to the quadtree subdivision. One can then create textures with different levels of detail for each level: four pixels at the quadtree depth N+1 (fine level) are averaged into one pixel in the quadtree depth N (coarser level). Quadtrees have to be created from leaves to root. One has to start at the finer level (the leaves of the quadtree depth M), create triangulations and break the texture into tiles and only then go to depth M-1 (where one quadcell corresponds to four quadcells on level M). In real time, according to the distance to the viewer, one simply has to go down the quadtree (root to leaves) and select the proper quadcell to be rendered. As a result, the scene generated is such that it provides decay in polygon size and in texture detail from viewpoint position to infinity.

This LOD technique presents a critical problem if one wants the land use cover to be changed over time; any change in a level has to be immediately updated in all other levels. This is difficult to achieve in a real time generated environment, even with powerful machines. Muchaxo and Neves (1999) developed a method to handle the different resolution levels in the visualisation in real time of large terrains using wavelet functions. In simple terms, Muchaxo and Neves proposed that instead of changing geometry, one only has to change the coefficients that represent the different resolution (or detail) levels. Any change at a given level can then be updated instantly at all other levels. Gross *et al.* (1996) developed a related work by using wavelets and quadtrees to model terrains. However, their technique does not check the generation of images according to pre-established error bounds.

Digital Portugal was a successful implementation of this algorithm. In Digital Portugal (Neves *et al.*, 1998), a mosaic with aerial photos, covering Lisbon and Porto with one metre resolution, and LANDSAT images covering the rest of the country at 25metre resolution, was used. This mosaic draped a digital terrain model provided by Portugal's Instituto Geográfico do Exército. The users could fly seamlessly over the whole country by manipulating a joystick. In neighouring screens, forty thematic databases on the areas being flown over could be browsed. Figure 15.1 shows the Digital Portugal installation. Figure 15.2 displays a view obtained from a virtual flight. This project was shown at Expo98 and used by hundreds of thousand visitors. Digital Portugal's virtual reality component is installed in a Silicon Graphics Onyx2 Infinite Reality. The real time rendering was done at a rate of 20 frames per second.

Figure 15.1: Digital Portugal installation

Figure 15.2: View from a virtual flight over Digital Portugal

15.2.3 *VRML representations of terrains*

Vince and Earnshaw (1999) provide a recent review of virtual environments on the World Wide Web where VRML is the key word. VRML is a file format for describing 3D objects and worlds on the World Wide Web. It grew out of Silicon Graphics' object-oriented Open Inventor 3D toolkit. VRML worlds can use formats such as JPEG, MPEG, and MIDI, by way of the Multipurpose Internet Mail Extensions (MIME). These media formats can be scripted within VRML using Java and Java Script and externally by using a Java application. VRML worlds can be developed using authoring tools or programmed directly.

Examples of where VRML 2.0 (Ames *et al.*, 1997) has been used in the visualisation of terrains on the World Wide Web include: the use of VRML in cartography by Fairbairn and Parsley (1997); work in the interface of GIS and VRML (Rhyne, 1997); interactive visualisations in VRML of meteorological and environmental phenomena (Djurcilov and Pang, 1998); the Virtual Field Course project led by three UK university groups (Figure 15.3 shows a screen of LandSerf, a related Java based project); and TerraVision II, a browser for massive terrains which uses Java scripting to extend VRML base capabilities and an External Authoring Interface for managing specific applications in the virtual geographic environment. TerraVision II has been designed to enable the browsing of the Earth (Reddy *et al.*, 1999).

VRML 2.0 has been now updated into VRML 97. A VRML 97 world is essentially a hierarchic graph with nodes that are event generators and others that are event receivers. Most event generator nodes are sensor nodes that enable the interactivity in VRML (by detecting touching, proximity, collisions, and visibility). Other event generator nodes may be script nodes that can be defined by Java or Java Script to trigger events in the VRML world.

Figure 15.3: LandSerf Java based visualise for raster images of terrains overlaid with vector representations

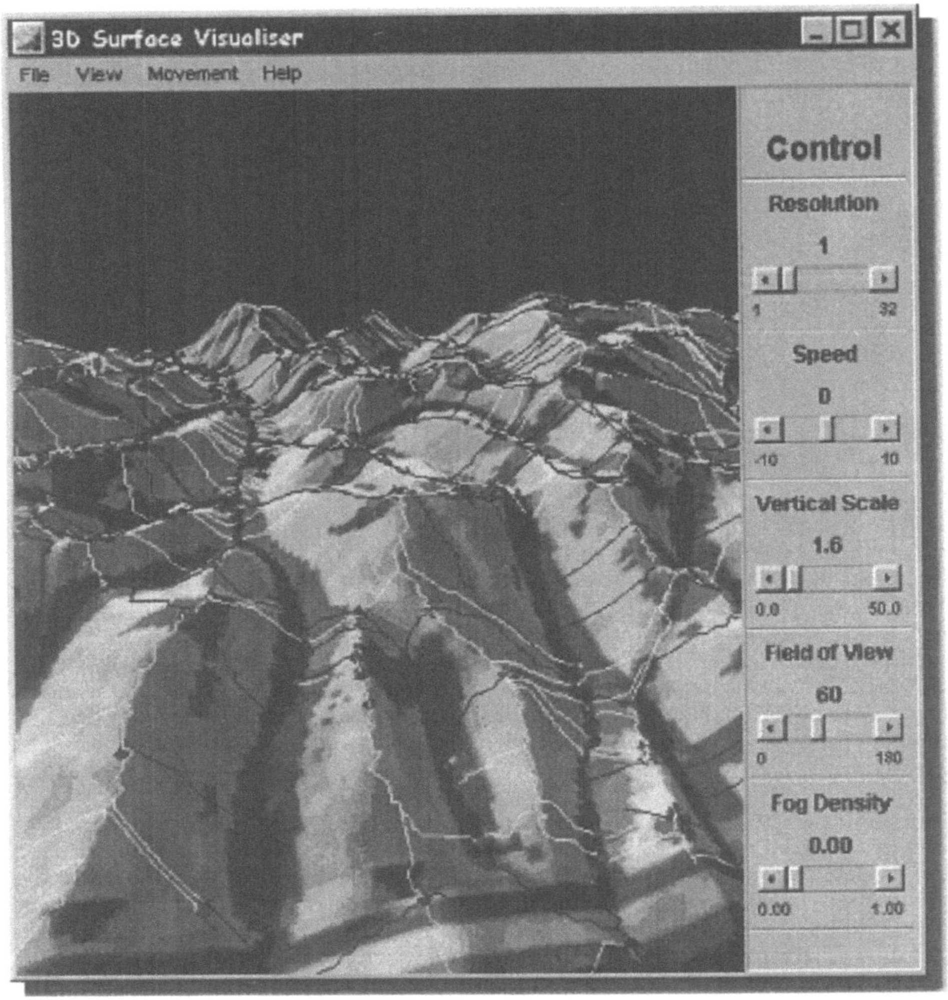

Source: http://www.geog.le.ac.uk/vfc. Image courtesy of Jo Wood

15.2.4 *Interacting with terrains*

Methods to interact with virtual terrains may fall under three categories: methods that are associated to virtual environments; methods that result from the interoperability of virtual front ends and geographical information systems; and those that were developed for other applications but that may be ported to virtual environments (i.e. filters, sketching).

When designing a virtual reality interface for a real space, an estuarine basin system, Câmara *et al.* (1998) identified the following tasks:

- fundamental tasks include problem perception, such as navigation (fly-over, walking and diving), object manipulation (object selection, information and retrieval), and visualisation (land uses, environmental quality levels);

- intermediate tasks involve navigation for problem investigation (terrain/object following) and visualisation (emission sources, compliance with standards); and

- advanced tasks are management actions, including the creation, deletion, scaling and movement of natural and artificial entities.

Figure 15.4 shows an interface developed for this system that was named Virtual Tejo. To perform the VR tasks, a virtual pointer was selected. A wireless mouse may control this pointer. To track the mouse position, a Polhemus Isotrack sensor was attached.

Figure 15.4: Virtual Tejo interface

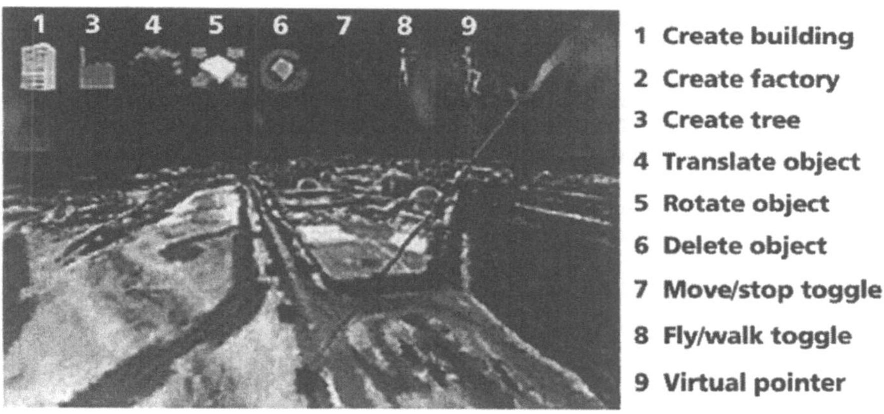

To complement this interface, Dias *et al.* (1995) and Almada *et al.* (1996) developed a tool, named BITS (after Browsing In Time and Space), for a user to make annotations (logging) while exploring a virtual ecosystem. Earlier work in real time annotation for video (Weber and Poon, 1994) and audio (Wittaker *et al.*, 1994) inspired the development of this tool. BITS was designed following a virtual pad and pen metaphor. It allows for travelling in space and time, navigation, and object selection and manipulation besides logging.

A related interface was proposed by Stoakley *et al.* (1995). These authors developed the World In Miniature (WIM), where a user has in his/her hands a miniature of the world he/she is exploring in a virtual environment. In WIM, an object can be manipulated either in miniature or in the original virtual environment, because its corresponding counterpart suffers the same modifications.

The Virtual GIS Room (Neves *et al.*,1997) is an attempt to merge virtual reality and geographical information systems technologies. It is inspired by Mark's concept of spatial data handling in three levels or spaces (Mark, 1992): haptic space, that is understood by haptic perception and sensorimotor experiences; pictorial space, which is mainly visually sensed; and transceptual space is a mental production resulting from haptic and pictorial experiences over time. Wayfind metaphors in virtual environments are structuring paradigms for this space.

Neves *et al.* (1997; 1999) suggested that by combining virtual environments and geographical information systems, these three spaces could be combined in one single computation environment called Virtual GIS Room, where users would:

- work in the haptic space by using a digitising tablet, where a pen could be used to select parts of a map, paint the DTM, select objects and travel in virtual space;
- see different perspectives of data in the computer screen and invoke commands in the desktop environment. This pictorial space is the space typically used by GIS systems; and
- immerse themselves in the transceptual space, materialised by a HMD, and explore the virtual space.

The Virtual GIS Room is based on a workbench, where the DTMs are displayed in two and a half dimensions (2.5 D) (2D plus altimetry), with texture mapping of the images and several layers of information represented as 2D posters on the wall (Figure 15.5). The users have direct access to the information in the posters and hear a 3D sound with a description of a layer after selecting it. The layers on the wall are two-dimensional layers that change to 2.5 D after draping the DTMs on the table *(Neves et* al., 1997).

Figure 15.5: The Virtual GIS Room

Source: Neves et al. (1999)

An extension of the Virtual GIS Room was the connection between a geographical information system (GIS) and a virtual environment also presented in Neves *et al.* (1997).

Actions on a map (a symbolic representation) can be readily seen in a realistic virtual representation of the terrain. There have been several other efforts to bridge geographical information systems and virtual environments, namely by Koller *et al.* (1995) and Rhyne *et al.* (1997) (a review is presented in Neves and Câmara, 1999). Koller *et al.* have developed an integrated real-time three-dimensional geographical information system. Rhyne *et al.* have converted information from ARC/INFO (GIS) to AVS (visualisation system) to VRML. They have also incorporated animation functions into digital elevation models available in VRML.

Two other interaction strategies for virtual representations of terrains are the use of 'magic lenses' and sketching. Magic lens filters were originally proposed by Fishkin and Stone (1995). Magic lenses use a lens emulator that changes the view of the objects viewed through that operator. By spatially bounding the filter, one can gain from the user's experiences with physical lenses. Neves (1999) inspired by Fishkin and Stone's work, has proposed that filters inspired by animal vision may be used to browse spatial information in virtual environments. In Figure 15.6, a filter is used to focus on a certain area while flying in a virtual world.

Figure 15.6: Browsing virtual terrains using lenses

Fernandes *et al.* (1997) proposed sketching tools to interact with mosaics of orthophotos. These tools enable the interactive digitising of points, lines and polygons over a background of digital orthophotos. Vector data created with this procedure can then be used to define thematic layers (i.e. roads, streams, and administrative boundaries), associate these layers with information, and store them in a database. This tool, implemented in Java, can also be used for spatial measurements such as area calculation, perimeters, line length or distance between two points. A sketching approach was also used to trigger cellular automata simulation models in the Virtual Tejo system. This method is based on the work presented by Nobre and Câmara (1999).

The application of sketching tools can be greatly facilitated by the use of computer vision (Freeman *et al.*, 1998). Rather than sketching with a mouse or a pen in appropriate tablets (or expensive touch screens), users could use pointing commands and gestures to manipulate objects in the screen. This alternative is appropriate in shared screen applications such as the ones that may be used for planning. Current limitations are related to response times and reliability (Freeman *et al.*, 1998).

15.2.5 *Spatial statistics*
Cook *et al.* (1998) used the CAVE2 environment to explore spatial data. The system enables the exploration of water chemistry data from sampling sites. The viewing box shows the scatterplot of measured variables. Interaction with data in the three-dimensional environment is shown in Figure 15.7. In this environment, one may paintbrush data with different colours and geometric representations. Any viewpoint can be also assumed.

Figure 15.7: Exploring spatial statistics in a CAVE2 environment

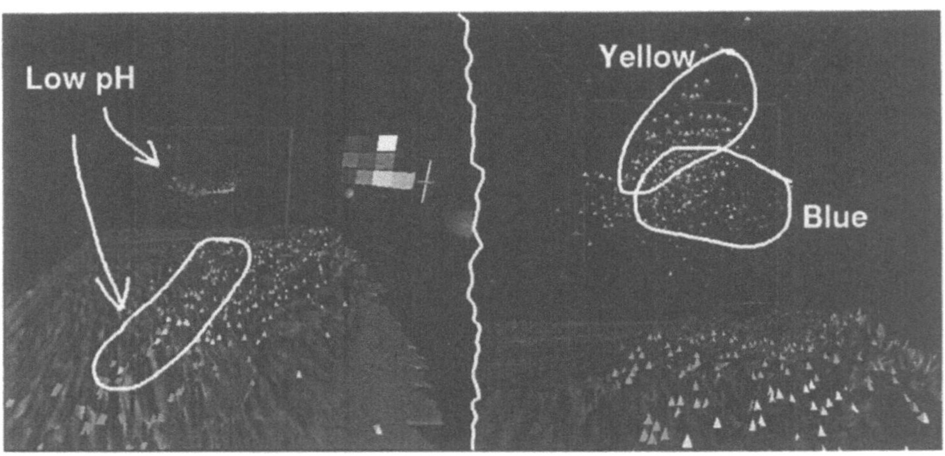

Source: Image courtesy of D. Cook

15.3 Coupling simulation models to virtual environments

Wheless *et al.* (1996) have shown how to link water quality models to virtual representations of the Cheasapeake Bay basin. Gonçalves *et al.* (1997) and Câmara *et al.* (1998) developed links between spatial models based on cellular automata formulations and virtual representations of terrains. Gonçalves *et al.* (1997) showed how cellular automata models can be visualised in Excel spreadsheets and simultaneously in virtual environments using the interoperability provided by Microsoft's Component Object Modelling (COM) and Virtual GIS Room. In this case, the cellular automata model and the virtual environment are running independently but exchanging objects with each other. The Virtual GIS Room application updates the digital terrain model, where a correspondence between cell sizes and the virtual environment and the simulation environment is established when a forest fire model starts running. The user can, at the same time, change model inputs in an Excel spreadsheet. Figure 15.8 displays the multiple views offered by this arrangement.

Figure 15.8: Visualising on a virtual environment (a) the results of a forest fire model (b)

a

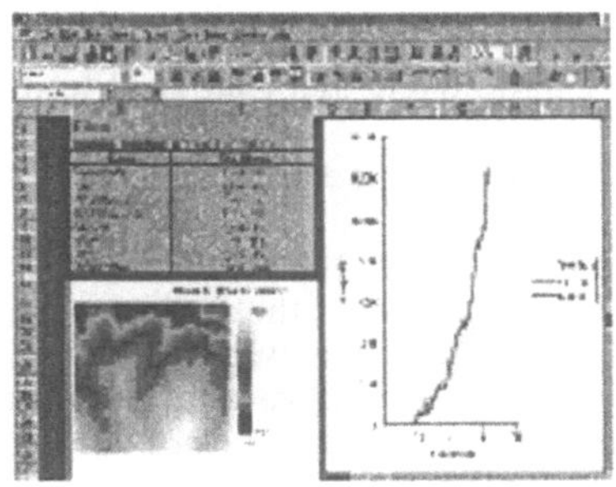

b

Câmara *et al.* (1998) have shown how to link cellular automata based models for non-point pollution and water quality to virtual representations of an estuarine basin using Microsoft's COM. An illustrative example is provided in Figure 15.9.

Figure 15.9: Visualisation of the output of a cellular automata model on a virtual environment

Vision depends on the reflection of light from surfaces. Vibrating materials causes sound. Gaver (1989) pointed out that, as a result, sound can provide information that vision can

not such as information about occluded materials or the internal mechanisms of complex objects. The use of sound in visualisation implies (Kramer, 1994; Barrass and Kramer, 1999): the association of sounds to data points; the control of the sound attributes by data values; and the triggering of the sound on some event.

For sonification exercises, the National Center for Supercomputing Applications has made available the NCSA Vanilla Sound Server. The use of sound in environmental visualisation (or auralisation) complements the role of images (Shepherd, 1994) as:

- multidimensional data can be interpreted because there are several sound parameters and each one can be used for each dimension;
- parameters that change in time can be represented; and
- outliers and extreme values can be highlighted.

In an animation of the evolution of Yellowstone National Park, Scaletti and Craig (1993) provided examples of the application of sound in the visualisation of spatial phenomena. Shiffer (1993) has also used sound to show the impact of airplays in taking off from airports in a multimedia application. In another example, Shiffer (1993) has linked a video on urban traffic with the corresponding noise levels in a planning support system for a central area in Washington DC. Other spatially related applications of sound are presented by Krygier (1994).

Wenzel *et al.* (1990) described the techniques used to generate three-dimensional sounds. Those techniques involve the mapping of the qualities of sounds as they reach the ears from many directions. In pioneer experiments, Wenzel and colleagues moved sound sources to various points in the room and made measurements. From these data, head-related transfer functions (HRTFs) were created to modulate the sounds as the sound source moved relative to the user's head. If HRTFs could also account for changes in orientation and location of the user's head, measured by position sensors, they proved that it was possible to add three-dimensional sound to virtual environments. Sound can then help the user locate sources of information that are outside of the field of view in virtual environments. In these environments, it is also recommended to vary the location of the sound sources horizontally, as the vertical judgement is less accurate (Begault, 1994; Hereford and Winn, 1994).

Ellis (1998) discusses the use of sound in VRML environments. VRML2 deals with sound through the Audio Clip node and the Sound node. The Audio Clip node may be used to retrieve and play a sound file from any Universe Resource Locator. Using the Audio Clip node as a source, the Sound Node enables the spatialisation of sound. To determine the extent and shape of the sound, this node allows the specification of the direction and of how far in front and behind the sound extends.

The use of sound in visualisation efforts still has unsolved problems such as the lack of models (unlike colour) and the choice of perceptual auditory parameters (i.e. pitch, intensity, timbre). The modelling of different sound sources is another hard problem common to noise assessment studies.

Robinett (1992), Holloway and Lastra (1993) discussed the alternative methods of providing synthetic experiences involving senses other than vision and hearing. For spatial systems, the most relevant are haptic senses (kinaesthetic, tactile senses).

In recent years, the game industry enabled the development of inexpensive force-feedback devices. These follow the pioneer work of Brooks *et al.* (1990). Current systems such as Phantom from Sensible Technologies have already capabilities such as enabling one to feel shapes and textures and modulate and deform objects. In addition, these capabilities can be programmed.

15.4 Visualisation of Europe's land use simulation model

Human society in the twenty first century will rely on three types of screens: those of information appliances such as cellular phones and personal digital assistants; those of desktop computers; and the large screens in meeting rooms and home theatre rooms. Visualisation of simulation models in information appliances will be mostly connected to games. There can be, of course, games on the evolution of Europe (a SimEurope) but they are beyond the scope of this discussion which will focus on the desktop for professional users and on large screens for senior managers and politicians.

15.4.1 *Desktop visualisations*

The previous section demonstrated that VR can provide realistic views and facilitates browsing of land use systems. VR systems may also be applied as a front-end to professional spatial analysis tools. Among these tools, data mining will become increasingly important to retrieve information from data infrastructures. One can envisage a system where the results of queries to data and models on European land use scenarios are visualised in a VR front end based on satellite images draping a digital terrain model. For this purpose, Spotfire and SGI's Mineset software may be used. Spotfire, the result of research by Ahlberg and Shneiderman (1994), proposes multi-dimensional scattergrams with each additional dimension controlled by a slider to dynamically query spatial databases. This principle is illustrated in Figure 5.10.

Figure 15.10: An illustrative screen of Spotfire software for visual data mining

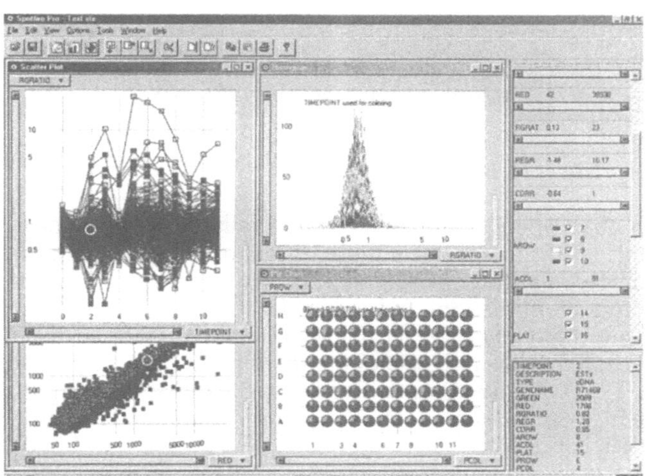

Source: Image courtesy of C. Ahlberg

Mineset's Scatter Visualiser is similar. Mineset also includes the Map Visualiser, where strong spatial relationships are explored. Data is displayed as graphical elements on a map, with user defined variables indicated by the height and colour of each element. Users can drill down for more detailed information about specific areas.

A VR system coupled with data mining tools exploring existing data and models could be used over the Internet by professionals and researchers interested in simulating Europe's land uses. To facilitate these collaborative efforts, a Collaboratory (Kouzes *et al.*, 1996) sould be established by the European Community. To support this Collaboratory, three systems should be investigated: Ishii and Kobayashi's Clear Board; Rhyne's metacomputer concept; and Thiebaux's Virtual Director. ClearBoard, authored by Ishii and Kobayashi (1992), enables the interaction, on the same screen, of users remotely located over a shared drawing as illustrated in Figure 15.11. Both users can share a common orientation. Both users can read all the text and graphics in their correct orientation. In addition, ClearBoard provides the capability that Ishii and co-workers call 'gaze awareness', the ability to monitor the direction of a partner's gaze and thus his or her focus of attention, something that can not be achieved with a whiteboard. With increasing availability of broadband communications and flat screens, this proposal may be become feasible for professional applications.

Figure 15.11: Illustrative examples of ClearBoard

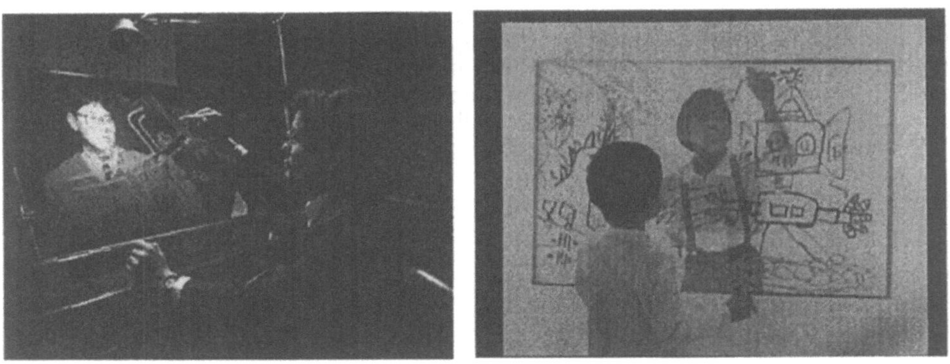

Source: Image courtesy of H. Ishii

Rhyne's idea is that when resources become available to workers in a networked environment, this results in a metacomputer that should be used as easily as a personal computer. In a metacomputing environment, WWW browsers are a tool of choice as they enable the running of animations and the exploration of 3D worlds in VRML. Using Multicast Backbone (Mbone) further facilitates collaborations through the Internet. Mbone provides video-conferencing and enables the sharing of visual information (Macedonia and Brutzman, 1994). Rhyne also mentions that collaborative computing means that research problems may be spread across several computers (Rhyne, 1998), making computers work in concert. A visualisation example is provided by LinkWinds.

Reed *et al.* (1997), referring to previous studies, assert that recordings of results and histories are key components in collaborative data analysis. This means that collaboration environments should enable users to record interactions for future review when navigating in large data sets, something that can be accomplished with Virtual Director (Thiebaux, 1997). Another proposal by Reed *et al.* (1997) relates to the possibility of project team members manipulating software components and their behaviour while immersed in shared virtual environments.

15.4.2 *Large screen visualisations*

For high level managers and politicians, a more ambitious large screen based vision of Europe's land uses in the twenty first century could be developed. This vision could extend the Digital Portugal concept to the whole European Union. Users would be able to fly across Europe and visualise interactively the results of macro policies. The visual and sound standards expected for such interactive visualisations on large screens, in early twenty first century, will be those set by the movie industry. Actually, in many ways, interactive visualisations of simulation models will tend towards the interactive movie concept.

Interactive movies, in Altman and Nakatsu's (1997) terms, are viewed as being interactive stories that may use one or more of the following models:

- *non-linear story trees*: story telling may use maze, character and alternate reality models. In the maze model, the story may be represented as a complex graph, where nodes are events. The character model means that the story is based on following a character. Finally, in the alternate reality model, there are multiple concurrent realities. Movies based on non-linear story trees may become interactive by introducing decision points. They can be implemented digitally using scripting languages;

- *gradual discourse*: many movies reveal their story gradually. Computerised interactive cinema inspired by this technique may divide the movie into levels of complexity. Then, the viewer may move along the different levels by pressing instruction icons;

- *parallel action*: in this technique, the movie jumps back and forth between two concurrent events. In a digital implementation, behaviours of characters have to be co-ordinated. The viewer may then command when to move from one event to the other; and

- *camera movement*: techniques of camera movement may include the selection of view point, focus and level of detail. They can become interactive by using virtual reality technologies. A virtual reality implementation would also enable viewing movie scenes from different viewpoints.

This movement towards the interactive cinema approach will require a significant participation from artists. Historically, this participation has enhanced visualisation efforts. A paradigmatic example is the Breathing Earth developed by the Sensorium Team for Japan's Pavillion in the Internet's World Expo (Malamud, 1997). Figure 15.12 shows how the Japanese team represented earthquakes happening around the Earth during two weeks.

Virtual scenarios for Europe's land uses could be displayed in large shared screens such as the one shown in Figure 15.13. Shared screens do present problems when large numbers of people want to interactively control the system, which for many functions, such as zooming or panning, is a technical impossibility. Robinett (1994) indicates that group interactions directly with the screen, in such a setting, are limited to voting and elimination of entities (i.e. elimination meant shooting in Robinett's words as he was discussing the use of shared screens for entertainment).

Figure 15.12: The Breathing Earth

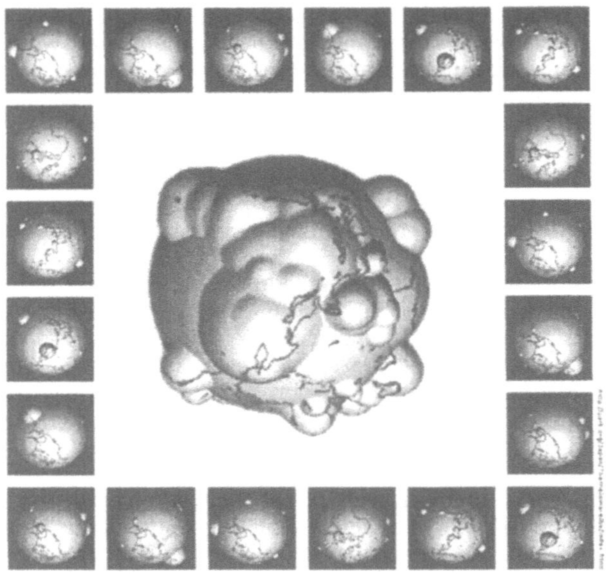

Source: Image courtesy of C. Malamud

Figure 15.13: Example of a shared screen

A large screen can be divided into a number of shared screens if the system can be divided into as many sub-systems. Each of the screens may have an associated projector such as Winograd's Interactive Works Spaces (Winograd, 1998). Each screen may also be a Liveboard (Elrod *et al.*, 1992) as in GMD's I-Land system. Winograd and GMD's projects relate to the development of working murals using multiple projectors or Liveboards. Power Wall, developed at the University of Minnesota, in co-operation with Silicon Graphics, is a high-end manifestation of the concept. Power Wall shows the potential of murals for high-resolution co-operative visualisations. A high-end shared screen system is the Virtual Great Barrier Reef installation in the DOME immersive virtual reality system (Refsland *et al.*, 1999). Fig. 15.14 displays the installation that was designed to hold 70-100 visitors interacting over 360 degrees with three-dimensional sights and sounds.

Figure 15.14: Virtual Great Barrier installation

Source: Image courtesy of S. Refsland

The system was designed to make users feel immersed in a story, not looking to a computer screen. An immersive 360x270 degree high-definition television camera system, located under the water in the reef, transmits live pictures to the DOME. These images provide a background that is complemented by artificial life graphic fish. These use pattern recognition to identify the real fish of their own species and learn to emulate behaviours, collision detection and depth perception (Refsland *et al.*, 1999). The environment includes also biobots, which are part human and part artificial life. Biobots enable users to play different marine life roles. The DOME is made of an inflatable fabric that is highly reflective for projection viewing. From a computational standpoint, it is composed of a large distributed network, which uses four WALLs (see above), and three CAVE systems. Interaction was done using computer vision techniques and input devices such as the Dive Panel and the FishGlove. The Dive Panel displays on a small LCD monitor information about the species under investigation and other information. The FishGlove was designed to facilitate control of the biobots.

The Virtual Great Barrier is an experimental installation with a scope beyond the project under discussion. However, it provides a vision of what may become common for high-end visualisation efforts including those that will be developed for Europe's land use simulation models.

15.5 Conclusions

Younger generations of professionals used to the three-dimensional video games will demand VR-based visualisations for land use planning. The building block of such efforts is the interactive visualisation of large terrains in virtual environments. This chapter has discussed interaction methods with virtual terrains. Among the approaches presented are: those dependent on the generation of images in real time, that are common in the exploration of virtual worlds; those resulting from interoperability with geographical information systems; and applications of filters and sketching that may be ported to virtual environments. Virtual terrains may be also seen as front ends to spatial data and models as shown herein.

The visualisation of Europe's land uses in early twenty first century may be approached from two perspectives, one for professional uses relying on the desktop, and another targeted for high level managers and politicians based on large shared screens. The development of an interface between virtual reality, visual data mining tools and European spatial data infrastructures was recommended for the professional application. This application could be developed across Europe using a Collaboratory set up on the Internet. It was proposed that the large screen application should be developed following an interactive cinema model. Some visions for a high end system have also been offered.

References

Ahlberg, C. and Shneiderman, B. (1994) *Visual Information Seeking: Tight Coupling of Dynamic Query Filters with Starfield Displays*, Proceedings CHI94, ACM Press, New York, pp.313-321.

Almada, A., Dias, A., Silva, J.P., Santos, E., Pedrosa, P. and Câmara, A.S. (1996) Exploring virtual ecosystems, in Proceedings AVI96 Advanced Visual Interfaces, ACM Press, pp. 166-174.

Altman, E.J. and Nakatsu, R. (1997) Interactive Movies: Techniques, Technologies, and Content. *Course 16, SIGGRAPH 97*, ACM, New York.

Ames, A.L., Nadeau, D.R. and Moreland, J.L. (1997) *VRML 2.0 Sourcebook*, John Wiley, New York.

Anselin, L. (1999) Interactive techniques and exploratory spatial data analysis, in Longley, P.A. Goodchild, M.F., Maguire, D.J. and Rhind, D.W. (eds.) *Geographical Information Systems*, John Wiley, New York, pp. 253-266.

Barrass, S. and Kramer, G. (1999) Using sonification, *Multimedia Systems*, 7: 23-31.

Begault, D.R. (1994) *3D Sound for Virtual Reality and Multimedia*, Academic Press Cambridge, MA.

Boice, M. (1992) How To Create a Toxic Plume Map, *CEC Fact Sheet No. 2*, Citizens Environmental Coalition Albany, NY.

Brice, R (1997) *Multimedia and Virtual Reality Engineering*, Newnes, Oxford.

Brooks, F.P., Ouh-Young, M., Batter, J.J. and Kilpatrick, P.J. (1990) Project GROPE- Haptic Displays for Scientific Visualisation, *Computer Graphics*, 24(4): 177-185.

Burdea, G. and Coiffet, P. (1994) *Virtual Reality Technology*, John Wiley, New York.

Câmara, A.S., Neves, J.N., Gonçalves, P., Gomes, J.M., Muchaxo, J., Silva, J.P. and Pinheiro, A. (1999) Towards Virtual Ecosystems.Unpublished report, Environmental Systems Analysis Group, New University of Lisbon, Monte de Caparica, Portugal.

Câmara, A.S., Neves, J.N., Muchaxo, J., Fernandes, J.P., Sousa, I., Nobre, E., Costa, M., Mil-Homens, J. and Rodrigues, A.C. (1998) Virtual environments and water quality management, *Journal of Infrastructure Systems*, ASCE, 4(1): 28-36.

Chakraborty, J. and Armstrong, M.P. (1996) Using geographic plume analysis to assess community vulnerability to hazardous accidents, *Computers, Environment and Urban Systems*, 19: 341-356.

Cook, D., Cruz-Neira, C. and Kohlmeyer, B.D. (1998) Exploring environmental eata in a highly immersive virtual reality environment, *Environmental Monitoring and Assessment*, 51(1-2): 441-450.

Cook, D., Symanzik, J., Majure, J.J. and Cressie, N. (1997) Dynamic graphics in a GIS: more examples using linked software, *Computers and Geosciences*, 23(4): 371-385.

Cruz-Neira, C., Sandin, D.J., DeFanti, T., Kenyon, R.V. and Hart, J.C. (1992) The CAVE: Audio Visual Experience Automatic Virtual Environment, *Communications of the ACM*, 35: 65-72.

DeFloriani, L., Falcidieno, B., Nagy, C. and Pienovi, C. (1985) Delaunay-based representation of surfaces defined over arbitrarily shaped domains, *Computer Vision, Graphics and Image Processing*, 32(1): 127-140.

Dias, A.E., Silva, J.P. and Câmara, A.S. (1995) BITS: Browsing in Time and Space, in *Comp. ACM CHI'95, 248-249*, ACM Press, Denver.

DiBiase, D., MacEachren, A.M. and Krygier, J.B. (1992) Animation and the role of map design in scientific visualisation, *Cartography and Geographical Information Systems*, 19(4): 201-214.

Djurcilov, S. and Pang, A. (1998)Web visualization of environmental data, SPIE's Symposium on Electronic Imaging, San Jose, CA (oral presentation).

Ellis, S. (1998) Sound and VRML, *Proceedings of VRML98, 95-100*, ACM Press, Monterey Bay.

Elrod, S., Bruce, R., Gold, R., Goldberg, D., Halasz, F., Janssen, W., Lee, D., McCall, K., Pedersen, E., Pier, K., Tang, J. and Welch, B. (1992) LIVEBOARD: A Large Interactive Display Supporting Group Meetings, Presentations, and Remote Collaboration, *Proceedings of ACM CHI '92, 599-607*, ACM Press, Monterey.

Fairbairn, D. and Parsley, S. (1997) The Use of VRML for cartographic visualisation, *Computers and Geosciences*, 23(4): 475-482.

Fernandes, J.P., Fonseca, A., Pereira, L., Faria, A., Figueira, H., Henriques, I., Garção, R. and Câmara, A.S. (1997) Visualisation and interaction tools for aerial photograph mosaics, *Computers & Geosciences*, 23(4): 465-474.

Fisher, P. (1999) Models of uncertainty in spatial data, in Longley, P.A., Goodchild, M.F., Maguire, D.J. and Rhind, D.W. (eds.) *Geographical Information Systems*, John Wiley, New York, pp. 191-205.

Fishkin, K. and Stone, M.C. (1995) Enhanced dynamic queries via movable filters, *Proceedings of ACM CHI' 95, 8-11*, ACM Press, Denver.

Freeman, W.T., Anderson, D.B., Beardsley, P.A., Dodge, C.N., Roth, M., Weissman, C.D., Yearzunis, W.S., Kage, H., Kyuma, K., Miyake, Y., Tanaka, K. (1998) Computer vision for interactive computer graphics, *IEEE Computer Graphics and Applications*, 18(3): 42-53.

Gaver, W.W. (1989) The SonicFinder: an interface that uses aditory icons, *Human-Computer Interaction*, 4(1): 67-94.

Gonçalves, P.P, Neves, J.N., Silva, J.P, Muchaxo, J. and Camara, A.S. (1997) Interoperability of Geographic Information: From the Spreadsheet to Virtual Environments, *International Conference and Workshop on Interoperating Geographical Information Systems*, NCGIA, Santa Barbara.

Gross, M.H, Staadt, C.G. and Gatti, R. (1996) Efficient triangular surface approximations using wavelets and quadtree data structures, *IEEE Transactions on Visualisation and Computer Graphics*, 2(2): 130-143.

Hereford, J. and Winn, W. (1994) Non-speech sound in human-computer interaction: a review and design guidelines, *Journal of Educational Computing Research*, 11(3): 211-233.

Holloway, R. and Lastra, A. (1993) Virtual Environments, *Technical Report*. University of North Carolina, Chapel Hill.

Ishii, H. and Kobayashi, M. (1992) ClearBoard: A Seamless Media for Shared Drawing and Conversation with Eye Contact, *Proceedings of ACM CHI'92, 525-532,* ACM Press, Monterey.

Jones, C. (1996) *Visualisation in Optimization.*, Kluwer, Norwell.

Koller, D., Lindstrom, P., Ribarsky, W., Hodges, L.F., Faust, N. and Turner, G. (1995) Virtual GIS: a Real-time 3D Interface for Geographical Information Systems. *Technical Report,* Georgia Tech., Atlanta. Available at *ftp://ftp.gvu.gatech.edu/pub/gvu/tech-reports/95-14.ps.Z.*

Kouzes, R.T., Myers, J.D. and Wulf, W.A. (1996) Collaboratories: doing science on the Internet, *IEEE Computer,* 29(8): 40-46.

Kramer, G., (ed.) (1994) *Auditory Display: Sonification, Audification and Auditory Interfaces*, Addison-Wesley, Reading.

Krygier, J.B. (1994) Sound and geographic visualisation, in MacEachren, A. and Taylor, D.R.F. (eds.) *Visualisation in Modern Cartography,* Pergamon Press, Oxford.

Macedonia, M.R. and Brutzman, D.P. (1994) Mbone provides audio and video across the Internet, *IEEE Computer,* 27(4): 30-36.

Malamud, C. (1997) *A World's Fair for the Global Village,* MIT Press, Cambridge, MA.

Mark, D. (1992) Spatial metaphors for human-computer interaction, *Proceedings 5th International Symposium on Spatial Data Handling,* Charleston.

McKenna, M. and Zelter, D. (1992) Three dimensional visual display systems for virtual environments, *Presence,* 1(4): 421-458.

Meyer, K., Applewhite, H.L. and Biocca, F.A. (1991) The ultimate tracker: a survey of position trackers, *Presence,* 1(2): 173-200.

Mitas, L., Brown, W.M. and Mitasova, H. (1997) Role of dynamic cartography in simulations of landscape processes based on multivariate fields, *Computers and Geosciences,* 23(4): 437-446.

Monmonier, M. (1999) *Air Apparent,* University of Chicago Press, Chicago.

Muchaxo, J., Neves, J.N. and Câmara, A.S. (1999) A real-time, level-of-detail editable representation for phototextured terrains with cartographic coherence, in Câmara, A.S. and Raper, J. (eds.) *Spatial Multimedia and Virtual Reality,* Taylor & Francis, London, pp. 137-146.

Muchaxo, J. and Neves, J.N, (1999). Visualisation of Large Terrains Using Wavelet Functions, Unpublished report, Imersiva, Monte de Caparica, Portugal.

Neves, J.N. (1999) Filters for Browsing Virtual Environments, Unpublished report, Imersiva, Monte de Caparica, Portugal.

Neves, J.N. and Câmara, A.S. (1999) Virtual environments and GIS, in Longley, P.A., Goodchild, M.F., Maguire, D.J. and Rhind, D.W. (eds.) *Geographical Information Systems,* John Wiley, New York, pp. 557-565.

Neves, J.N., Bento, J. and Gouveia, C. (1998) Digital Portugal, Paper presented at GIS Planet, Lisbon, Portugal.

Neves, J.N., Silva, J.P., Gonçalves, P., Muchaxo, J., Silva, J.M. and Câmara, A.S. (1997) Cognitive spaces and metaphors: a solution for interacting with spatial data, *Computers and Geosciences,* 23(4): 483-488.

Neves, J.N., Gonçalves, P., Muchaxo, J. and Silva, J.P. (1999) A Virtual GIS Room: interfacing spatial information in virtual Environments, in Câmara, A.S. and Raper, J. (eds.) *Spatial Multimedia and Virtual Reality,* Taylor & Francis, London, pp. 147-156.

Nobre, E. and Câmara, A.S. (1999) Spatial simulation by sketching, in Camara, A.S. and Raper, J. (eds.) *Spatial Multimedia and Virtual Reality,* Taylor & Francis, London, pp. 103-110.

Pang, A.T. Wittendrink, C.M., Lodha, S.K. (1997) Approaches to uncertainty visualisation, *The Visual Computer,* 13(8): 370-390.

Reddy, M., Leclerc, Y. and Iverson, L. (1999) Terra Vision II: visualising massive terrain databases in VRML, *IEEE Computer Graphics and Applications*, 19(2): 30-38.

Reed, D.A., R.C. Giles, R.C. and C.E. Catlett, C.E. (1997) Distributed data and immersive collaboration, *Communications of the ACM*, 40(11): 39-48.

Refsland, S., Ojika, T., Loeffler, C. and DeFanti, T. (1999) Virtual Great Barrier Reef, in Heudin, J.C. (ed.) *Virtual Worlds*, Perseus, Reading, MA, pp.153-179.

Rhyne, T. (1997) Going virtual with geographic information and scientific visualisation, *Computers and Geosciences*, 23(4): 489-492.

Rhyne, T. (1998) Interactive Visualisation & Web-Based Exploration in the Physical and Natural Sciences, *SIGGRAPH Tutorial Notes*, Los Angeles.

Rhyne, T., Brutzman, D. and Macedonia, M. (1997) Internetworked graphics and the Web, *IEEE Computer*, 30(8): 99-101.

Robinett, W. (1992) Synthetic experience: proposed taxonomy, *Presence*, 1(2): 229-247.

Robinett, W. (1994) Interactivity and individual viewpoint in shared virtual worlds: the big screen vs. networked personal displays, *Computer Graphics*, 28(2): 127-130.

Scaletti, C. and Craig, A.D. (1993) Using sound to extract meaning from complex data, *Technical Report*, Urbana-Champaign: CERL, University of Illinois.

Sheperd, I. (1994) Multi-sensory GIS: mapping out the research frontier, *Proceedings Spatial Data Handling 94*, 356-370, Edinburgh, Scotland.

Shiffer, M. (1993) Augmenting geographic information with collaborative multimedia technologies, *Proceedings of AUTOCARTO 11*, Minneapolis, pp. 367-376.

Stoakley, R., Conway, M. and Pausch, R. (1995) Virtual Reality on a WIM: Interactive Worlds in Miniature, *Proceedings of ACM CHI' 95*, ACM Press. Denver, CO, pp. 265-272.

Stuart, R. (1996) *The Design of Virtual Environments*, McGraw Hill, New York.

Thiebaux, M. (1997) Virtual Director, Master's Thesis, Department of Electrical Engineering and Computer Science, University of Illinois at Chicago.

Weber, K. and Poon, A. (1994) Marquee: a tool for real-time video logging, *Proceedings of CHI' 94, 58-64*, ACM Press, Boston.

Wenzel, E.M., Stone, P.K., Fisher, S.S. and Foster, S.H. (1990) A System for Three-Dimensional Accoustic Viewing in a Virtual Environment Workstation, *Proceedings of the First IEEE Conference on Visualisation*, IEEE Computer Society Press, Los Alamitos.

Wheless, G.H., Lascara, C.M., Valle-Levinson, A., Brutzman, D.P., Sherman, W., Hibbard, W.L. and Paul, B.E. (1996) Virtual Cheasapeake Bay: Interacting with a Coupled Physical/Biological Model, *IEEE Computer Graphics and Applications*, 16(4): 52-57.

Winograd, T. (1998) Interactive Work Spaces – A Human Centered Architecture, Unpublished paper, Stanford University, Palo Alto, Available at *http://graphics.stanford.edu/projects/iwork*.

Wittaker, S., Hyland, P, and Wiley, M. (1994) Filochat: Handwritten Notes Provide Access to Recorded Conversations, *Proceedings of CHI'94*, ACM Press, Boston.

António Câmara
ILAB
The Idea Laboratory
New University of Lisbon
2825 Monte de Caparica
Portugal

PART IV:
THE DEVELOPMENT OF A EUROPEAN LAND USE SIMULATION SYSTEM

16 EUROSCANNER: A SIMULATION MODEL FOR LAND USE CHANGE IN EUROPE

Piet Rietveld, Henk Scholten and John Stillwell

16.1 Introduction

The present generation of proprietary geographical information systems (GIS) can be used to support strategic planning processes in several ways. GIS are able to store, manage and analyse the enormous amount of data needed and they are able to provide facilities to visualise the data from a wide range of perspectives including those of different interest groups and stakeholders involved in particular planning applications. However, proprietary systems do not usually provide any applied modelling functionality and for this we require more intelligent GIS that have modelling capability available either through direct or indirect coupling (Birkin *et al.*, 1996). In order to simulate land use change, it is necessary to maximise the GIS capabilities for data storage, manipulation and display whilst simultaneously using a statistical or mathematical modelling software 'engine' to undertake the prediction functions. Such a model, that can be used to simulate the effects of autonomous or planned developments, is outlined in this chapter, The model is called *EuroScanner* because it has been designed to be applicable across different parts of Europe. This chapter explains how it works whilst examples of national and regional applications are contained in the two chapters that follow. We begin by reviewing alternative approaches to modelling land use in Section 16.2 and then explain the structure of the *EuroScanner* system and identify some of its characteristic features in Section 16.3. Section 16.4 introduces the categories of land use data used in the model, the constraints that are used to ensure consistency at regional level and the concept of suitability mapping. The detailed mathematical formulation of the basic 'spatial interaction' model is presented in Section 16.5 and some extensions to the doubly constrained model are presented in Section 16.6. Finally, Section 16.7 contains some conclusions.

16.2 Approaches to modelling land use

Various dynamic modelling approaches can be distinguished in the field of land use change. These have been identified by Grothe (1998) as follows:

- *Planning models* that compute an optimal allocation of land in order to arrive at a maximum value for some optimisation criterion (for example, maximum profit of a farm), or to maximise some social welfare objective sought after by a public sector planner (see, for example, Schipper, 1996)

- *Individual choice models* that describe the locational preferences of individual actors on the basis of individual decision process (Timmermans, 1984).

- *AI models* derived from the field of Artificial Intelligence. Examples of these allocation algorithms are neural networks, genetic algorithms and cellular automata

J. Stillwell and H. Scholten (eds.), Land Use Simulation for Europe, 235–244.

(CA). Applications of AI models in the field of allocation planning include, for example, those described by Raju *et al.* (1998), Pereira *et al.* (1994), Wu, (1996) and White *et al.* (1997). Engelen *et al.* (1999) describe the structure of a dynamic CA model linked to GIS and applied to simulate future land use change in the IJmond region of the Netherlands.

- *Equilibrium models* explaining land use in terms of supply, demand and equilibrium prices. These models are part of the tradition of the classical approaches of Von Thünen, Lösch and Alonso. In most of these models, transport and accessibility play an important role as determinants of the equilibrium outcomes of land use. Examples of operational models used in planning contexts are those of Anas (1982) and Landis (1994).

In this chapter, we will make use of a model of the latter category. An interesting development is that these land use models are linked to environmental models in order to be able to assess the environmental consequences of land use change (Hayashi and Roy, 1996). For example, changes in land use will have consequences for spatial interaction patterns and these, in turn, will have an impact on emission and noise caused by transport (Young and Bowyer, 1996).

16.3 System structure and model features

The basic process of simulation involves 'calibrating' a model based on observed data and 'testing' this model in a historical or pseudo-predictive context, before using the model to generate future projections of land use. *EuroScanner* is a prototype spatial information system which has, at its core, a simulation model that is provided with information about historic and current land use, projected information relating to a particular (long term) scenario, and information in the form of what have been called land attractiveness maps (Figure 16.1). The model then produces projections of future land use for the grid system that is being used.

Figure 16.1: EuroScanner system structure

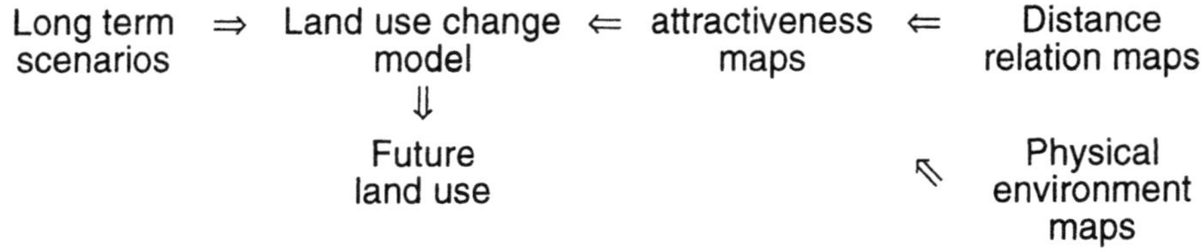

The concept of scenarios is used (Schoute *et al.*, 1994) in order to cope with expected increases or declines in demand for space that result from the dynamics of economic and social change. Alternative economic scenarios may involve different rates of national growth, for example, whereas social scenarios may relate to different demographic structures arising from alternative assumptions about fertility, mortality or migration. Under any one scenario, the future demands for land for each category of land use have to be known or estimated. In practice, these land use demands or 'claims' are mostly derived from models for particular sectors that quantify the demands for each land use type. Thus, in broad terms, demographic models are required to calculate the demand for residential land, agricultural models determine the demand for agricultural land and economic models provide the projections for industrial land. Typically, these projections are provided as exogenous inputs from independent models.

In order to determine the geographical locations of the land use demands of different sectors, the *EuroScanner* system makes use of the concept of 'attractiveness maps' in which all the driving forces or explanatory factors that determine the spatial allocation of each land use category can be incorporated. An attractiveness map can be based on a regression equation derived from statistical analyses of historic land use developments. It can also be based on a spatial planning concept translated by an expert into the driving forces.

Some of the characteristic features of the *EuroScanner* model itself are as follows. Firstly, the model has a raster structure or is *grid based*. The data is collected and the simulations run to generate values for matrices of grid cells. This means that the model describes the relative proportions of land to be used for a number of land use types in each grid cell in a system. The specification of the model and the software allow large numbers of grid cells. Typically the size of each grid cell would be 500 metres by 500 metres. Secondly, the model is *integrated* in that it provides an integration framework for sectoral databases, as implied by the grid cell framework, and for sectoral policy proposals by combining these inputs in a context that allows spatial analysis to be undertaken. Thirdly, the model is *exhaustive* in the sense that all grid cells in a spatial unit are considered. All types of land use are explicitly considered; thus there are no residual or 'rest' categories left untreated. The model can be formulated in such a way that the transfer of water cells (sea, lakes) into land cells is allowed. Fourthly, the model is *dynamic* because it deals with changes in land use taking into account present land use patterns. The attractiveness or suitability of the grid cells for certain types of land use is not assumed constant, but may change as the result of changes in land use in the course of time. Fifthly, the model has a *satellite structure*, as it is driven by forecasts at a national or regional level of variables such as population, agricultural production, infrastructure, *et cetera*, that constitute the scenarios. Sixthly, the model is *stochastic* in that its outcomes are interpreted as expected proportions of land to be used for various categories of land use.

The main purpose of the model is not to predict land use in particular small grid cells in the future very precisely, since this task is virtually impossible. The aim of *EuroScanner* is to show the broader impacts of processes like population growth, increasing production (manufacturing, agriculture, *et cetera*) and natural conservation on spatial patterns of land use. Given this aim, it is critically important that the model is policy-orientated and can be of value to physical and strategic planners operating at different spatial scales and with different area jurisdictions. *EuroScanner* attempts to make explicit the future patterns of land use change under alternative scenarios that may or may not be influenced by policy decisions. In particular, the model helps to solve questions referring to those grid cells in which major policy conflicts are expected to emerge under a likely or foreseeable scenario. The model can also be used to investigate the implications of sectoral and macro policies for human settlement and land use patterns. Scholten *et al.* (1999) use this modelling approach to assess the projections of land use in ten categories and illustrate the impact of locating a new airport near Breda in the south of the Netherlands. In addition to these aims, the property of integration means that *EuroScanner* can function as a tool to improve communication between analysts working in various (and frequently separate) sectors of land use (for example, urban development versus agriculture versus natural areas). The model also helps to improve consistency between projections made in these sectors.

16.4 Land use data, regional constraints, suitability maps and government policy interventions

As with all types of modelling, the accuracy and value of the final simulations will depend to a large extent on the quality and level of disaggregation of the data on land use that is used to calibrate the historical model. The following types of land use are typically distinguished in the model:

- urban: residential, industrial, roads, railways and airports;
- agriculture: pasture, corn, arable land (potatoes, beets, cereals), flower bulbs, orchards, cultivation under glass, and other agriculture;
- natural areas: wood, nature; and
- water.

This classification gives 15 different land use types although the number can be extended if more disaggregated information is available.

As indicated above, the *EuroScanner* model is driven by the outcomes of other sector-specific models that generate results at a much lower spatial detail. In the Netherlands, for example, projections are made of agricultural land under most types of crops for 14 agricultural regions, and projections of residential and industrial land are made for 40 COROP regions that correspond with labour market areas. The exogenous estimates and projections for these sets of regions relate to some planning horizon (say 2020). However, they can be used as constraints to ensure consistency in the observed data sets and the simulated projections for agricultural, residential or industrial land use that are made at a more spatially disaggregated level. For natural areas, regional constraints can be included to ensure that the total amount of land used for nature is at least the minimum amount that is set for each region in government plans.

The demand side of the land market is represented by attractiveness or suitability maps. A suitability map is generated for each land use type to indicate the suitability of each grid cell for that type of land use. These suitability maps can be interpreted as bid prices. The suitabilities depend on factors such as physical characteristics (soil quality and groundwater conditions), accessibility of facilities and infrastructure, land use transition costs (given previous land use) and the amount of similar land use in the neighbourhood. The physical characteristics constitute the 'physical usability map' whilst the neighbourhood characteristics provide the 'gravity map' (Figure 16.1). In addition to suitabilities expressing bid prices of actors in the land market, governmental planning regulations also have an impact on land use developments. These include, for example, policies towards building permits for residential and industrial land use, or policies regarding the preservation and development of natural areas. 'Policy maps' can therefore be produced. The policy interventions can be interpreted as subsidies and taxes for certain types of land use so that the final market outcome does not merely reflect the willingness to pay of actors at the market. Also the intentions of the public sector to stimulate or discourage land use of particular types are to some extent reflected by the resulting land use patterns. Further details of how suitability maps are computed can be found in Hilferink and Rietveld (1999). In the next section, we turn our attention to the methodology for estimating the suitability of land in each grid cell for different types of use.

16.5 Mathematical formulation of the model

In this section of the chapter, we specify the equations that comprise the model that is at the heart of the *EuroScanner* system. We begin by considering a set of grid cells and a number of land use categories. A core variable of the model is the suitability, s_{cj}, for land use of type j in grid cell c. This suitability represents the net benefits (benefits minus costs) of land use type j in cell c. The higher the suitability for land use type j, the higher the probability, x_{cj}, that the cell will be used for this land use. In the simplest version of the model, a logit type approach is used to determine this probability:

$$x_{cj} = \frac{\exp\left(\beta \cdot s_{cj}\right)}{\sum_{j'} \exp\left(\beta \cdot s_{cj'}\right)} \qquad \text{for all } c \text{ and } j. \qquad (16.1)$$

Thus, when the β parameter in equation 16.1 is zero, all types of land use have the same probability; i.e. the suitability or attractiveness factors, s_{cj}, do not play any role in determining these shares. On the other hand, when β goes to infinity, the limit of the probability that the category with the highest suitability gets the cell is equal to 1. In terms of expected values, the expected volume of land use, L_{cj}, for category j in cell c equals:

$$L_{cj} = x_{cj} \cdot L_c \qquad \text{for all } c \text{ and } j, \qquad (16.2)$$

where L_c denotes the total volume of land in cell c. With equal-sized cells, L_c would of course be equal for all c. Cells of unequal size may occur in the case of areas located near a national border, or cells partly containing areas of water (if a transfer from water to dry land use is not allowed), or cells that may contain a preset land use based on exogenous data, such as infrastructure developments.

The model as formulated here does not guarantee that the allocation of space across possible land uses is in accordance with overall demand conditions. Therefore, demand side constraints have to be imposed in order to ensure that, at the relevant levels of aggregation, total demand is met. This leads to a reformulation of the model to include the constraints on land uses where appropriate. Let D_j be a restriction on total demand for land use category j. In addition, let M_{cj} denote the expected amount of land in cell c that will be used for category j taking into account the supply side constraints. This results in a doubly constrained model:

$$M_{cj} = a_j \cdot b_c \cdot \exp \beta \cdot s_{cj} \qquad \text{for the constrained } j \text{ and all } c, \qquad (16.3)$$

where a_j and b_c are balancing factors such that the following constraints are satisfied:

$$\sum_c M_{cj} = D_j \qquad \text{for the constrained } j \qquad (16.4)$$

and

$$\sum_j M_{cj} = L_c \qquad \text{for all } c. \qquad (16.5)$$

Equation 16.4 guarantees that the predicted amount of land allocated for land use type j equals the imposed amount, D_j. In addition, equation 16.5 implies that the sum of the expected volumes of the various land use types per cell is equal to the total area of each cell. We use the expression 'for the constrained j' when an aggregate constraint has been formulated for the particular land use type j. It is clear that the constraints may imply that no feasible solution exists. This can be checked by searching for a starting solution for the system in a linear programming context. When no feasible solution is found, the aggregate constraints have to be reconsidered before the model can be used. For those land use types j for which no aggregate constraint applies, we arrive at a singly constrained solution:

$$M_{cj} = b_c \cdot \exp \beta \cdot s_{cj} \qquad \text{for unconstrained } j \text{ and all } c \qquad (16.6)$$

under constraint equation 16.5. This assumes that a_j has been set equal to 1. Note that in the extreme case where none of the land use types has any *aggregate constraints*, we have:

$$M_{cj} = L_{cj} \qquad \text{for all } c \text{ and } j \qquad (16.7)$$

and

$$b_c = \frac{L_c}{\sum_{j'} \exp\left(\beta \cdot s_{cj'}\right)} \qquad \text{for all } c. \qquad (16.8)$$

The above reformulation as a doubly constrained land use model (equation 16.3) is helpful for an understanding of the structure of the model. The structure of the model is quite similar to doubly constrained spatial interaction models used in transportation research (see, for example, Fotheringham and O'Kelly, 1989) or migration analysis (Stillwell, 1978).

We now turn to the interpretation of the balancing factors. From equations 16.3 to 16.5, it follows that:

$$b_c = \frac{L_c}{\sum_j a_j \cdot \exp\left(\beta \cdot s_{cj}\right)} \qquad \text{for all } c \qquad (16.9)$$

$$a_j = \frac{D_j}{\sum_c b_c \cdot \exp\left(\beta \cdot s_{cj}\right)} \qquad \text{for the constrained } j \qquad (16.10)$$

and

$$a_j = 1 \qquad \text{for the other } j \qquad (16.11)$$

$\sum_c b_c \cdot \exp(\beta \cdot s_{cj})$ in equation 16.10 can be interpreted as the aggregate suitability of land for land use type *j*; when the suitability of land use type *j* would be low in terms of s_{cj}, the value of the denominator in equation 16.10 is low as well. The balancing factor is high when a high constraint D_j is combined with a low aggregate suitability. In a similar way, $\sum_j a_j \cdot \exp(\beta \cdot s_{cj})$ in equation 16.9 can be interpreted as a measure of demand for land use in cell *c*. A high value of this expression means that the demand for the land in cell c for the various land use types (taking into account the urgency of the land use types as represented by a_j) is relatively high. It leads to a low value of the balancing factor b_c, and thus ensures that in equation 16.5, the total amount of land finally allocated in cell *c* does not exceed the supply of available land, L_c. Thus, the solution of the doubly constrained model yields, as a side-product, the shadow prices of land in the cells. Another way to interpret the balancing factors is to rewrite equation 16.3 as:

$$M_{cj} = \exp\left(\beta \cdot \left[s_{cj} + \beta^{-1} \cdot \log(a_j) + \beta^{-1} \cdot \log(b_c)\right]\right) \text{ for the constrained } j \text{ and all } c. \quad (16.12)$$

A large value of a_j implies a strong pressure on land use type *j*. It can be interpreted as a subsidy to this type of land use; the subsidy is given to ensure that the aggregate target for land use type *j* is achieved. The reverse case is a small value for a_j this can be interpreted as a tax on this land use type to prevent that excess of the related target. Note that the case in between occurs when a_j equals 1, implying $\log(a_j)=0$. In order to clarify the role of the balancing factors, we perform the following transformation on equation 16.12. Define land use price p_c in cell *c* as $-(1/\beta).\log(b_c)$ and price λ_j for constraint *j* as $+(1/\beta).\log(a_j)$. Then M_{cj} can be considered as a demand function of land use price p_c.

$$M_{cj}(p_c) = exp\ (\beta\ .\ s_{cj} + \lambda_j - p_c) \quad (16.13)$$

This formulation also sheds light on the b_c factor. A high value of b_c means that use of cell *c* is discouraged. It can therefore be interpreted as an indicator of demand/supply conditions in each cell. An increase in the aggregate demand of category *j* will lead to a shift in land use in the following way. The higher value of D_j will lead to a higher balancing factor a_j which will lead to a corresponding increase in the expected land use in the various grids depending on the relative suitability of the grids for the various types of land use.

The solution of the model is achieved through iteration to establish the values of the balancing factors, a_j and b_c, Starting with arbitrary values for a_j (say 1), the values for b_c can be computed using equation 16.9. Then the computed b_c values are fed into equation 16.10 to calculate the a_j values that are subsequently reused in equation 16.9. After successive iterations, each of the balancing factor values will have converged to a constant value and once this convergence has been achieved, the implied land use pattern can be found by generating the expected amount of land in cell *c* that will be used for category *j* taking into account the constraints using equation 16.3. In order to compute a solution for the model, equation 16.9 is used instead of b_c in the model equation 16.3 in order to allow for the land supply restrictions given by equation 16.5. In order to find the values of a_i that satisfy the land use restrictions given by equation 16.4, a_j is iteratively adjusted until the restrictions are met. The balancing factor iteration continues until convergence has been achieved. This is when there is a close fit between projected land use D_j and observed land use A_j and the β parameter has been optimised.

16.6 Extensions to the doubly constrained land use model

In reality, the model is more complex than presented here. One complication is that the constraints are not always in terms of equalities, but in terms of inequalities. Consider the constraint that:

$$\sum_{c} M_{cj} \geq D_j \tag{16.14}$$

Then, in the case when the constraint is not binding, we have $a_j = 1$, and when the constraint is binding we have a_j as defined in equation 16.10. Thus, in this case, we arrive at:

$$a_j = \max\left\{1, \frac{D_j}{\sum_c b_c \cdot \exp\left(\beta \cdot s_{cj}\right)}\right\} \quad \text{for all lower bounded } j \tag{16.15}$$

In the case of a ≤ constraint we arrive at:

$$a_j = \min\left\{1, \frac{D_j}{\sum_c b_c \cdot \exp\left(\beta \cdot s_{cj}\right)}\right\} \quad \text{for all upper bounded } j \tag{16.16}$$

Another complication is that the aggregate constraints not only function at the level of the country, but also at various regional levels. This means, for example, that population predictions have been made for labour market regions, or that agricultural production has been predicted at the level of agricultural areas. This leads to an extended version of the model where the balancing factor a_j becomes specific for each region for which a constraint has been formulated. The solution of the model occurs in an iterative way. In the case of equality constraints, it comes down to finding the values of a_j and b_c in equations 16.9 to 16.11. Beginning with arbitrary values for the $a_{j,}$ one can compute the resulting values for b_c by means of equation 16.9. Then these b_c values are fed into equation 16.8 leading to revised values for $a_{j.}$ In the case where some of the constraints are in terms of inequalities, one should use equations 16.15 and 16.16 instead of 16.10. Once these factors have been determined, the implied land use pattern can easily be found by means of equation 16.3.

In Section 16.2, we indicated that the *EuroScanner* model belongs to the class of equilibrium models. Equilibrium takes place at the macro level: aggregate supply for land is given (it is assumed to be inelastic) and aggregate demand for land for various land use types is also given at the level of regions. Inequality constraints (equation 16.14) ensure that aggregate demand does not exceed aggregate supply. Prices do not play a role at the aggregate level in the present version of the model since both demand and supply are price inelastic at this level. Future development of the model may lead to the specification of elastic demand functions at the aggregate level. Also at the level of individual cells, demand and supply of land are in equilibrium (equation 16.5). Balancing factors are used to achieve this. These balancing factors can be interpreted in terms of shadow prices (see equation 16.13). Cells with high suitability for various types of land use will have a high shadow price to ensure that demand does not exceed supply. Note, for example, that in equation 16.13, the shadow price, p_c, functions as a correction factor for a high level of suitability, s_{cj}.

16.7 Conclusions

The modelling 'engine' of the *EuroScanner* simulation system is a doubly constrained spatial interaction model in which the expected amount of land in any single grid cell that will be used for one particular type of land use is predicted as a function of the relative suitability of that cell to have that land use. The first of the two constraints incorporated is that the total amount of land allocated to one land use type in each cell adds up to the given total amount of land in that category across the whole system of grid cells. The second constraint is that the total amount of land use of different types in any one cell sums to the total amount of land available in that cell. As in traditional constrained spatial interaction models (Wilson, 1974), balancing factors are utilised to ensure that the constraints are maintained. They also may be interpreted in the land use change context. A high value of the first balancing factor for one particular land use suggests strong pressure on cells to adopt this land use whereas a high value of the second balancing factor for any one cell indicates that the use of the cell is discouraged. The balancing factors are computed through an iterative routine and the best fit model parameter is achieved when the predicted and observed land use totals are equivalent.

The system of equations that comprise the spatial interaction model in the *EuroScanner* system is an innovative solution very different in its conception from the other types of approach that were introduced earlier in the chapter. Examples of the results of model data sets, components and simulations have not been included in this chapter since these are the focus of the last two chapters of the book.

References

Anas, A. (1982) *Residential Location Models and Urban Transportation*, Academic Press, New York.

Birkin, M., Clarke, G., Clarke, M. and Wilson, A.G. (1996) *Intelligent GIS Location Decisions and Strategic Planning*, GeoInformation International, Cambridge.

Grothe, M. (1998) Keuze voor ruimte, ruimte voor keuzen. De ontwikkeling van GIS applicaties voor locatieplanning; een objectgeoriënteerde analyse, Unpublished Phd dissertation, Vrije Universiteit, Amsterdam.

Engelen, G., Geertman, S., Smits, P. And Wessels, C. (1999) Dynamic GIS and strategic physical planning support: a practical application, Chapter 5 in Stillwell, J.C.H., geertman, S. And Openshaw, S. (eds.) *Geographical Information and Planning*, Springer, Heidelberg, pp. 87-111.

Fotheringham, A.S. and O'Kelly, M.E. (1989) *Spatial Interaction Models: Formulations and Applications*, Kluwer, Dordrecht.

Hayashi, Y. and Roy, J. (1996) Transport, Land Use and the Environment, Kluwer, Dordrecht.

Hilferink, M. and Rietveld, P. (1999) Land Use Scanner: an integrated model for long term projections of land use in urban and rural areas, *Journal of Geographical Information Systems*, 1: 155-177.

Landis, J. (1994) The California Urban Futures model: a new generation of metropolitan simulation models, *Environment and Planning B*, 21: 399-420.

Pereira, A.G., Munda, G. and Paruccini, M. (1994) Generating alternatives for siting retail and service facilities using genetic algorithms and multiple criteria decision techniques, *Journal of Retailing and Consumer Services*, 1: 40-47.

Raju, K.A., Sikdar, P.K. and Dhingra, S.L. (1998) Micro-simulation of residential location choice and its variation, Computers, *Environment and Urban Systems*, 22(3): 203-218.

Schipper, R.A. (1996) *Farming in a Fragile Future*, Wageningen Agricultural University, Wageningen.

Scholten, H.J., Velde, R. van de, Rietveld P. and Hilferink, M. (1999) Spatial information infrastructure for scenario planning: the development of a Land Use Planner for Holland, in Stillwell, J., Geertman, S. and Openshaw, S. (eds.) *Geographical Information and Planning*, Springer, Berlin, pp. 112-134.

Schoute, F.Th., Finke, P.A. Veeneklaas, F.R. and Wolfert, H.P. (eds.) (1997) Scenario Studies for the Rural Environment, *Proceedings of the Symposium Scenario Studies for the Rural Environment*, Wageningen, Kluwer Academic Publishers, Dordrecht.

Stillwell, J.C.H. (1978) Interzonal migration: some historical tests of spatial interaction models, *Environment and Planning A*, 10, 1187-1200.

Timmermans, H.J.P. (1994) Decision models for predicting preferences among multiattribute choice alternatives, in Bahrenberg, G., Fisher, M.M. and Nijkamp, P. (eds.) *Recent Developments in Spatial Data Analyses: Methodology, Measurement and Models*, Gower, Aldershot, pp. 337-355.

Wilson, A.G. (1974) *Urban and Regional Models in Geography and Planning*, Wiley, Chichester.

Wu, F. (1996) A linguistic cellular automata simulation approach for sustainable land development in a fast growing region, *Computers, Environment and Urban Systems*, 20(6): 367-387.

White, R., Engelen, G. and Uljee, I. (1997) The use of constrained cellular automata for high-resolution modelling of urban dynamics, *Environment and Planning B*, 24: 323-343.

Young, W. and Bowyer, D. (1996) Modelling the environmental impact of changes in urban structure, *Computers, Environment and Urban Systems*, 20(4/5): 313-326.

Piet Rietveld
Department of Spatial Economics
Vrije Universiteit Amsterdam
De Boelelaan 1105
1081 HV Amsterdam
The Netherlands

Henk Scholten
Department of Spatial Economics
Vrije Universiteit Amsterdam
De Boelelaan 1105
1081 HV Amsterdam
The Netherlands

John Stillwell
School of Geography
University of Leeds
Leeds LS2 9JT
United Kingdom

17 A NATIONAL PLANNING APPLICATION OF EUROSCANNER IN THE NETHERLANDS

Kees Schotten and Camiel Heunks

17.1 Introduction

Physical planning is a strategic policy-making activity aimed at finding the optimum arrangement of space and society. The planning process can be regarded as a set of governance practices for the development and implementation of spatial strategies, plans, policies and projects by regulating the location, timing and form of development (Healey, 1997). These practices are shaped by the dynamics of economic and social change which give rise to the demands for space. Frequently, the parties involved in development initiatives have different interests, objectives and preferences which are difficult to reconcile, Consequently, it is often necessary for governments to be involved in the planning process.

Spatial planning at various scales has returned to the forefront of public and political attention. At a European level, this is illustrated by the recent adoption of the European Spatial Development Perspective (ESDP). At a national level, the spatial planning key decision regarding the Fifth National Physical Planning Report has been produced in the Netherlands, whilst at a regional level, the (first) Regional Plan for the Lisbon Metropolitan Area (PROTALM) in Portugal has been scheduled for release and in the United Kingdom, regional physical planning strategies are being produced as part of Regional Planning Guidance (RPG) for local authorities.

This chapter is the first of two that demonstrate applications of *EuroScanner*, the GIS-based land use simulation system described in detail in the previous chapter. In this case, *EuroScanner* is applied in the context of the Fifth National Physical Planning Report in the Netherlands. The simulation model generates future land use in the Netherlands under one of a number of alternative growth scenarios and a spatial strategy that comprises both forecast strategic and autonomous developments. The planning framework and the scenarios are discussed in Section 17.2. In this application, we focus on the use of the simulation model to illustrate the location of residential areas throughout the Netherlands from 1995 to 2020 in Section 17.3, and the simulation of an integrated land use map for 2020 in Section 17.4. These maps are used to assess the effects of land use changes on a wide range of environmental indicators in order to facilitate planning decisions (Young and Bowyer, 1996) in Section 17.5 and some conclusions are drawn in Section 17.6.

17.2 Planning background and scenario development

The Fifth National Physical Planning Report in the Netherlands has recently been published. In the preparation phase, the National Institute of Public Health and Environment (RIVM) investigated the impact of various land use change scenarios and the effects of these on the environment for the period 1995 to 2020 (the time horizon of the Fifth Report). In this study, the results acquired until October 1999 are described. The

J. Stillwell and H. Scholten (eds.), Land Use Simulation for Europe, 245–256.

aim of the RIVM study was not to deliver a land use blueprint for 2020, nor to simply quantify the impacts of land use change on the environment, but it was to establish a methodology that made use of models and would enable discussions to take place about future land use developments. Of course it was expected that discussions would embrace the necessary operational assumptions made in order to obtain results but should also be used to gain understanding of land use dynamics and (indirect) effects on the environment. However, the main goal of the study has been to facilitate the evaluation of spatial policies and to assess the environmental effects of alternative spatial policies.

An important element that shapes spatial planning practices is the expected rise in the demand for space that results from the dynamics associated with economic and social change. In order to cope with uncertainties about the future demands for space, the concept of scenario evaluation is often used (Schoute *et al.*, 1997). Here, we adopt three internally consistent and coherent scenarios, developed by The Dutch Central Planning Bureau (CPB, 1997), to describe demographic, economic, social and policy developments until 2020. These scenarios can be characterised as follows:

- *Divided Europe (DE)*: where protectionist tendencies in the different European countries are strong; where economic growth is low with an annual rise in GDP of approximately 1.7 per cent; where the number of inhabitants of the Netherlands grows to 16.3 million while the number of dwellings rises to 7.5 million (currently 15 and 6.2 million respectively); and where the Common Agricultural Policy (CAP) is continued as at present.

- *European Co-ordination (EC)*: where the world economy is dominated by trade organisations resulting in an accelerated integration of the European countries; where the annual economic growth is moderate (3 per cent) but the growth of inhabitants in the Netherlands increases to 17.7 million which is the highest of the scenarios; where the number of dwellings grows to 7.8 million; and where a unified European policy is implemented implying a certain level of protection of agriculture in the common market.

- *Global Competition (GC)*: where market forces dominate the world economy and the economic growth is high (3.3 per cent); where the number of inhabitants rises to 16.9 million but, because of a more individualistic lifestyle, the number of dwellings grows to 8.1 million, faster then in the EC scenario; and where the CAP is abandoned, forcing the agricultural sector to compete on the world market.

Taking the developments described in the CPB scenarios, different sector-specific models are used to calculate the amount of land needed for agriculture, housing, industrial and commercial areas in each of the three scenarios (Figure 17.1). The demands for land for nature and land for (rail and motorway) infrastructure in the Netherlands are determined by policies already implemented, so the claims of these land use types are identical in each of the scenarios. Nature (including forest) is also the land use type that claims the highest amount of land until 2020 in each of the scenarios. While the claims on land for housing, industry and commerce are less than that for nature, they also vary due to the difference in forecast dwellings and economic growth within the scenarios. In all the scenarios, the area used for agriculture declines, and it is only the rate that differs in each scenario: relatively high in the Divided Europe and Global Competition scenarios and relatively low in the European Co-ordination scenario. This results in surplus land for the Netherlands in the GC scenario, a more or less balanced supply in the DE scenario and a situation in which the demand exceeds the supply of land by 46,000 hectares in the EC scenario.

Figure 17.1: Sectoral land use demands in the three scenarios

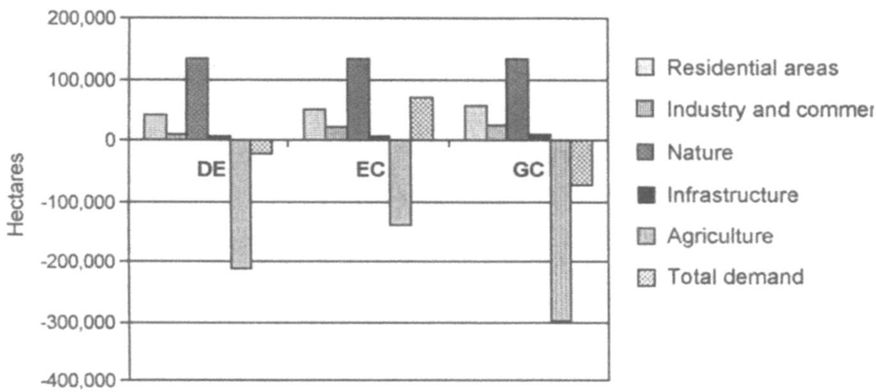

The aim of the *EuroScanner* model is to identify the implications of demands from different sectors for space and thus the spatial policies that determine the location of sectoral land uses for the whole of the Netherlands. The model is grid-based and simulates land use in relative proportions at a spatial resolution of 500 by 500 metre squares. The outcome of the model can be interpreted as expected proportions of land in each grid cell. The starting point for every simulation is is a map with the current location of land uses. The sectoral demands for space are derived from sectoral models which forecast land use claims at a much broader level of spatial resolution (e.g. the whole country or regions). Spatial strategies for each distinct land use type can be made explicit with suitability maps which are constructed within the system by combining maps that show political regulations (e.g. housing planning regulations and nature conservation plans), physical properties (e.g. seismic hazard, yield reduction) and characteristics in relation to neighbouring cells (e.g. the presence of infrastructure and the deduced accessibility). For the simulation of future land use, (i) the current land use, (ii) claims and (iii) suitability maps are fed into the model engine. The model engine then allocates land use based on a doubly constrained logit approach as described in the previous chapter.

In this study, *EuroScanner* is used in two successive approaches ways in order to allocate future land use claims. Firstly, the location of residential areas between 1995 and 2020 is determined using a suitability map that is based on a regression equation that describes the housing developments in the period 1980-1993. Secondly, an integrated land use map for the year 2020 is simulated using suitability maps based on expert knowledge.

17.3 Simulation of residential areas

The period of simulation is divided into two parts due to differences in strategic policies. In the period 1995-2010, the current housing policy formulated in the Fourth Physical

Planning Report Extra (VINEX) is applied, whilst for the period 2010-2020, no spatial policies have been defined as yet, except for nature conservation where the time horizon is 2018. Thus, current land use, residential land use demands and suitability maps have to be fed into *EuroScanner* for simulation, each of which is now discussed in more detail.

Current land use is derived from two databases, the land use statistics in 1993 (CBS, 1997) and the land use Netherlands database or LGN (Thunnissen and Noordman, 1997). The 1993 land use statistics are derived from aerial photographs and focus on urban and natural land use types, whilst LGN is derived from Landsat TM satellite images and differentiates various agricultural land use types. Of both databases the versions with a 25 metres resolution have been combined into a current land use map that gives relative proportions within the 500 metre grid cells for a total of 45 land use types. For the simulation of residential areas, these 45 land use types are aggregated to two types:

- land use types that are not allowed to be converted into new residential land but will keep their present use, such as urban, infrastructure, water and nature, as outlined in the present policies; and
- other, mainly agricultural land use types that can be converted into residential areas.

Figure 17.2: Regression based probabilities for the allocation of single and multiple family dwellings for 1995-2020

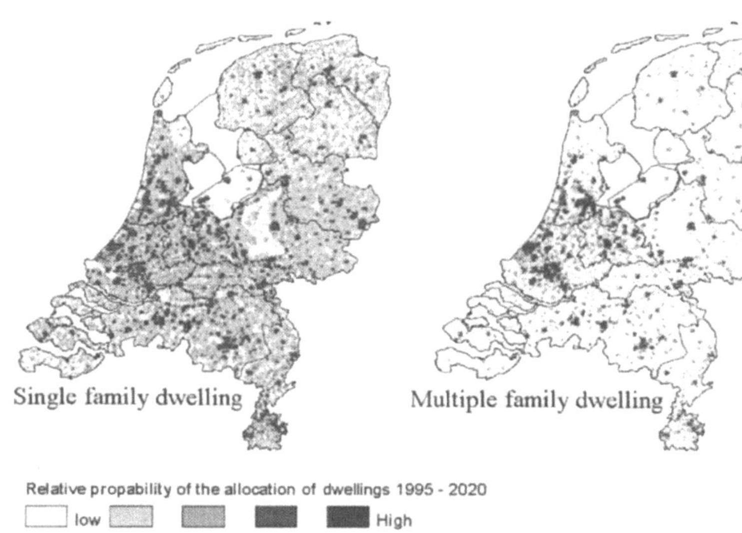

The forecasts of the number of single and multiple family dwellings, shown in Figure 17.2, are derived from the SONAR, TRENDRAPPORT and PRIMOS models at the COROP level (Heida, 1993). The 40 COROP regions are purely statistical units that correspond approximately with labour market regions and aggregate to provinces. In order to determine *demand for residential land* in terms of hectares for the years 2010 and 2020, the forecasts are combined with standard figures representing the current policies concerning location features, quotas and building densities (Kolpron Consultants, 1996). Up to 2010, an average of 26 per cent of the total number of new dwellings will be created in existing built-up areas, 44 per cent in planned building locations and the remaining 30 per cent in other locations. In the period 2010-2020, an average of 26 per cent will be built within urban areas and the remaining dwellings will be built on other locations. The average densities are assumed to remain constant over the whole period: 57.3, 26.2 and 18.5 dwellings per hectare respectively in urban areas, planned residential areas and other locations. It is assumed that current built-up areas do not change. Therefore, the 26 per cent of dwellings that are built in urban areas are not taken into account when simulating new residential areas.

In order to make the locational preferences of dwellings explicit, *suitability maps* for both single and multiple family dwellings have to be established. This is done by means of stepwise regression of housing developments over the period 1980-1993. On the basis of a review of the literature, a set of 17 variables were considered, of which the following location features are found to have a significant effect (Wagtendonk and Rietveld, 1999):

- proximity and extent of existing residential areas;
- proximity and extent of industrial and commercial areas;
- proximity and extent of natural areas, forest and leisure areas;
- distance to points of highway access and railway stations;
- presence of motorways and railways; and
- presence of spatial policy plans.

Because the locational preferences differ between the most urbanised area (Randstad Holland) and the rest of the Netherlands, two sets of regression equations are derived for single and multiple dwellings, whose parameters are summarised in Table 17.1.

In general, the chance that a grid cell will be converted into residential land is very small, as can be observed from the negative intercept in each case. This can be explained by the small increase in residential areas over the time period compared to the size of the Netherlands. The intercept is less negative for both types of dwellings in the Randstad because the number of houses built there is higher than in the rest of the Netherlands. The chance that a grid cell will be converted increases if neighbouring cells already have a residential function. Also, the proximity of infrastructure entrance points (access points to highways and railway stations), areas where people work and natural areas increase this change. However, the most important factor is spatial policy. Grid cells within municipalities designated as growth centres, catering for overspill within a city region in the Third Physical Planning Report, have the highest probability of change. The presence of motorways and railways decreases the likelihood of conversion, probably due to environmental effects such as noise nuisance.

Table 17.1: Parameters of the four regression equations that explain the location of new residential areas over the period 1980-1993

	One dwelling house				Multiple dwelling house			
	Randstad Holland		Other regions		Randstad Holland		Other regions	
	B	*T-value*	*B*	*T-value*	*B*	*T-value*	*B*	*T-value*
Intercept	-7.02	-389.6	-20.35	-255.6	-11.81	-462.2	-15.02	-329.8
Industry	Not applicable		Not applicable		0.28	121.4	Not applicable	
Knowledge services	Not applicable		1.00	160.0	Not applicable		0.34	79.3
Forest for leisure purposes	-0.02	-27.5	0.06	36.0	0.04	39.9	Not applicable	
Wetlands	Not applicable		Not applicable		0.07	69.1	0.11	39.6
Motorways	-0.34	-94.7	-0.36	-69.6	-0.11	-42.5	-0.15	-27.0
Railways	-0.58	-91.3	0.05	7.0	-0.37	-85.1	-0.30	-35.8
Highway access points	0.04	33.0	0.12	61.3	-0.12	-117.2	Not applicable	
Railway stations	-0.14	-103.5	-0.12	-87.4	0.05	56.6	-0.18	-109.7
Growth centres	1.13	264.0	0.74	101.0	1.63	366.0	1.08	121.6
One family dwellings	0.73	548.0	0.54	242.1	0.83	934.1	0.88	300.8
Multiple family dwellings	0.06	75.7	0.39	331.7	0.10	126.0	0.50	328.7

Source: Wagtendonk and Rietveld (1999)

To simulate future land use, it is necessary to update the maps used for the historical analyses as indicated in Table 17.2. Some operational problems had to be overcome; the most recent land use statistics that are used to get the location of different land use categories in 1981 as well as in 1993 do not include forest used for recreational purposes. It was therefore necessary to replace this category by planned green recreational areas in Randstad Holland.

The regression-based probabilities for single and multiple family dwellings for the period 1995-2020 are presented in Figure 17.2. Other approved spatial policies that intervene with the allocation of residential areas have to be given geographical locations. These locations are derived from a digital GIS database that includes all plans from the last planning report (VINEX) to be implemented before 2010. Once locations of residential areas are identified, industrial and commercial areas, leisure areas and new water areas are outlined. Other databases provide data on the areas planned for conversion to natural areas by 2018 and the motorways and railways that will be constructed before 2020. On all these planned locations, the assumption is made that they will not be converted into residential land but into the planned land use.

By using the derived regression equations for the simulation of future land use (Figure 17.3), the assumption is made implicitly that the factors determining the chance of converting land into residential areas are still valid. This accounts for human behaviour as well as for spatial politics. However, this assumption can be contested because, with the publication of the Fourth Physical Planning Report and the Fourth Physical Planning Report Extra (VINEX), the new planning concept of concentrating development was adopted (Faludi and van der Valk, 1994). But more importantly, planned residential areas are delimited in much more detail than the designated municipality growth centres that were the major policy in the period 1980-1993.

Table 17.2: Maps used in the historical analyses and the simulation of future residential areas

Maps used in the historical analyses	Maps used for future simulation
Industry 1981	Industry 1993
Knowledge services 1981	Knowledge services 1993
Forest used for leisure 1981	Planned green recreational areas
Wetlands 1981	Wetlands 1993
(Main) Motorways 1981	(Main) Motorways 1998 and extensions in 2010 and 2020
Railways 1980	Railways 1998 and extensions in 2010 and 2020
Railway stations 1980	Railway stations 1993
Highway access points 1980	Highway access points 1995
Municipalities designated as growth centres	Vinex and other planned residential areas
Single family dwellings 1980	Single family dwellings 1995
Multiple family dwellings 1980	Multiple family dwellings 1995

Figure 17.3: Location of new residential areas simulated for 1995-2010 and 1995-2020

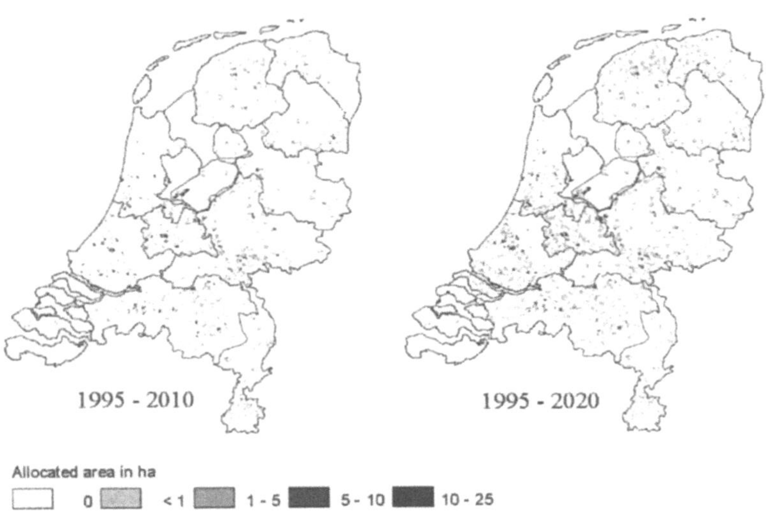

17.4 Integrated land use map

Other land use types have dynamics that will have impacts on the environment as well. To take these impacts into account, an integrated land use map is made for the European Co-ordination scenario 2020. Unlike the new residential areas, this will be done in one period and suitability maps are derived on the basis of expert judgement. For the simulation of the future land use map, a mixed approach is chosen which comprises a hierarchic land use scheme and the simulation of the agricultural land use types. For the first set of calculations, the spatial analysis tools of *EuroScanner* are used; for the latter, the simulation facilities are used.

Firstly, the 45 land use types distinguished in the present land use map are reclassified into 14 types that are relevant to assess their effects on the environment, landscape and recreation:

- built-up areas including residential areas, industrial and commercial areas and infrastructure (roads, railways and airports);
- agricultural land classified as pasture, arable land (potatoes, beet, cereals and other arable land types), maize, flower bulbs, orchards and greenhouses;
- natural areas classified as brushwood, heath(land) and forest; and
- land uses like water, leisure areas, foreign country and other land use type (dump sites, mining sites *et cetera*).

On the present land use map, the following land use types are overlaid: the earlier simulated residential areas (time horizon 2020), planned industrial and commercial areas (2010), infrastructure (2020), leisure areas (2010), water (2010), natural areas of which the cover is given (2018), orchards and other agricultural land. In case the amount of land use exceeds the 25 hectares, the area of the last mentioned type (in the row) will be reduced until the cell contains the 25 hectares that can be allocated in a 500 by 500 metre grid cell. It is assumed that the demands for industry, leisure and nature up to 2020 will not change. No demands for industrial and commercial areas are known on a sufficiently accurate regional level after 2010. The emergence of leisure areas, water and natural areas is mainly policy driven and no spatial policies for these categories up to 2020 have been formulated yet.

Following this, an overlay procedure is implemented for the agricultural land use types so that greenhouses, flower bulbs, maize, arable land and pasture remain. These are simulated. The regional claims of these categories are supplied by the regionalized Dutch Agricultural Model (DRAM) on the level of municipalities (Bethe *et al.*, 1999). The suitability maps of each type are made by combining yield reduction maps and both the location and proximity of the concerned land use type in the present situation. The integrated land use map for 2020 that is the result of the above procedure is presented in Figure 17.4.

17.5 Environmental impact assessment

The results derived above have been used to determine the effects of land use change on the environment my means of a set of indicators. The indicators have been selected on the basis of political relevance, sensitivity for land use change, the broad coverage of environmental themes and the availability of operational models. Some of the indicators can be directly computed by means of the *EuroScanner* model. However, in most cases, land use as predicted for 2020 (generated with the *EuroScanner model*) is just one of the essential inputs to compute the relevant indicators. Due to the broad coverage of the environmental themes, these indicators are calculated using knowledge and a wide range of models available at the RIVM and other institutes including Alterra and AVV. We give some details for a selection of indicators.

Figure 17.4: Integrated land use map 2020 for the European Co-ordination scenario, existing spatial planning policies and autonomous developments

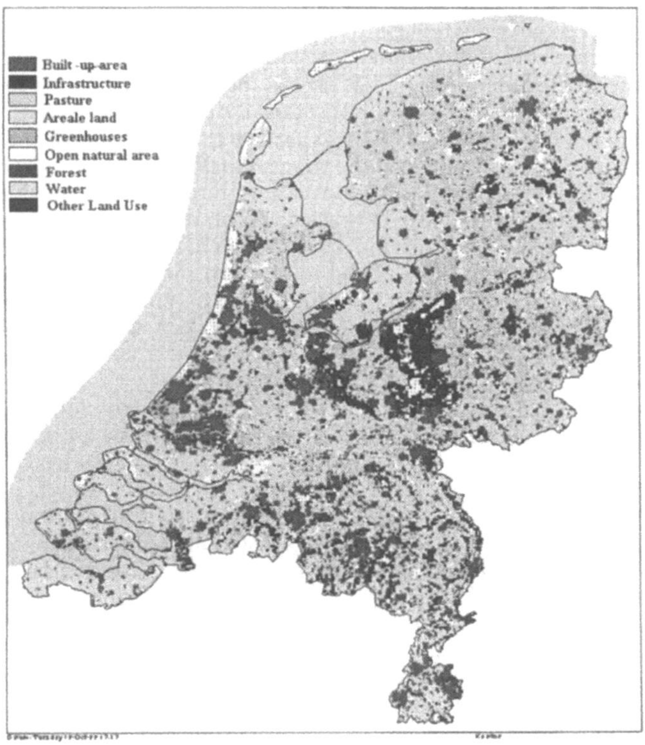

The expansion of nature areas as delimited by the provinces will increase the area of nature in the Netherlands significantly. However, spatial analysis shows that the locations of new natural areas are so scattered that the cohesion of the natural areas develops in a less favourable way than the increase in the natural area. The quality of the flora in the Netherlands is calculated with the Nature planner model. The value of the indicator is a product of nature quantity and the nature quality expressed as the occurrence of selected plant species (van de Hoek *et al.*, 2000). The increase in nature quality can be preliminary contributed to the restoration of the ground water levels. Reduction in acidification and fertilisation are too small to have a significant influence.

Spatial analyses regarding safety show that new built up areas are located in areas in danger of flooding and water excess, taking into account the water drainage in 2020. While other land use types like natural areas are also not planned at locations were they can have a reducing effect on flooding and water excess.

The model FACTS (AGV, 1999) is used to calculate the ownership, fuel cost and energy use of passenger cars in 2020. Car ownership and fuel cost are, in conjunction with the *EuroScanner* results, used to calculate the use of cars on an average working day making use of the LMS-Model (HCG, 1997). The outputs of the LMS model are on their part again used to calculate the noise nuisance (making use of the model described in VROM, 1997) and the accessibility of jobs by the labour force with a commuting time lower than 45 minutes (Geurts and Ritsema van Eck, 2000).

The overall conclusion of the comparison between the present situation and the simulation for 2020 is that the environmental quality is improving, although one has to keep in mind that currently implemented policies without spatial aspects also have positive impacts on the environmental quality in the year 2020. Especially the indicators dealing with nature are positive owing to the realisation of more than 150,000 hectares of new nature. This is about 4 per cent of the total area of the country and implies a considerable shift in land use. The background of this increase is that agricultural areas are replaced by natural areas. Not only the total size of natural areas is improving, also the cohesion of natural areas is getting better. Cohesion is measured in terms of the degree to which the cutting up of natural areas has been avoided. It is an important qualitative aspect of nature since in addition to total area also the existence of barriers between natural areas has an impact on ecosystems.

The environmental indicators show that an increase in CO_2 emissions is expected. The improvement of energy efficiency of cars is more than off-set by an increase in car use and an increase in the weight of cars. The amount of nitrate that leaches to groundwater, and threatens its use for drinking water, decreases.

The new housing developments have a negative impact on the landscape indicators. Even though new residential areas tend to be built near to existing residential areas there is a tendency that the openness of landscapes (measured via the extent to which built areas interfere with open space) deteriorates. Another aspect of the expected residential construction activities is that in vulnerable regions in the Netherlands the flooding hazard and the chance on water excess increases moderately.

17.6 Conclusions

The spatial analytical possibilities of GIS are improved significantly with the linkage of allocation models. This will help the planner to assess the pros and cons of various spatial principles or to value the effects of spatial strategies on policy targets or societal aims. Moreover, it opens the way to a much more quantitative approach to planning that can complement the present more qualitative planning practice. This application shows that it is possible to anticipate complex land use dynamics and their environmental impacts. Due to the developments in the IT world in general, and the world of GIS in particular, the problems encountered are no longer technological. In an application whose main objective is to support decision makers, the main concerns are to ensure that the numerous different models involved make use of the same validated data and that the quality and limitations of intermediate results are know to those researchers that use them for further processing.

The simulated land use dynamics, which form the final result of a complex allocation methodology, give an explanatory overview of the locational preferences of the selected land use types and their mutual interactions. However, it is useful that auxiliary information regarding land use demands and suitability maps is also made available to the decision maker(s) involved in strategic planning. This not only helps these individuals to gain insight into the driving forces that influence land use change but also to help with the formulation of alternative spatial strategies.

Acknowledgements

Thanks to Wideke Boersma (Geodan) and Roland Goetgeluk for their help running the *EuroScanner* model, to John Helming (LEI) for the calculations of the agricultural demands with the DRAM model, to Alfred Wagtendonk and Piet Rietveld (VU) for their work on the historical regression, and to Karst Geurs and Rob Alkemade (RIVM) for their work on the impact assessments .

References

AGV (1999) FACTS3.0, *Forecasting Airpollution by Car Traffic Simulation*, AGV Adviesgroep Verkeer en Vervoer, Nieuwegein.

Bethe, F.H., Helming J.F.M., Heusden M. and Leopold R. (1999) *Referentiebeeld landbouw Vijno-toets.* Alterra, Wageningen.

CBS (1997) *Statistick van het Bodemgebruik* (1993) CBS, Voorburg/Heerlen.

Central Planning Bureau (1997) *Economie en Fysieke Omgeving*, Sdu Publishers, The Hague.

Faludi, A. and Valk, A. van der (1994) *Rule and Order: Dutch Planning Doctrine in the Twentieth Century*, Kluwer, Dordrecht.

Geurts K.T. and Ritsema van Eck J.R.(2000), *Effecten van een compacte verstedelijkingsvariant op mobiliteit, bereikbaarheid en geluid; Analyses voor de Vijfde Nota Ruimtelijke Ordening.* RIVM report 711931 003. RIVM. Bilthoven.

Healey, P. (1997) The revival of strategic spatial planning in Europe, in Healey, P., Khakee, A., Motte, A. and Needham, B. (eds) *Making Strategic Spatial Plans: Innovations in Europe*, UCL Press, London, pp. 3-19.

HCG (1997), LMS 5.0: *Modelbeschrijving. Documentatie LMS 5.0 – Deel D.* Hague Consulting Group, Den Haag.

Heida, H.R. (1993) *Primos Huishoudenmodel: Dynamische Ssimulatie van Houshoudensontwikkeling*, Ministerie van VROM, Den Haag.

Kolpron Consultants (1996) *Inventarisatie van VINEX-bouwlocaties*, Kolpron Consultants, Rotterdam.

Schoute F. Th., Finke, P.A., Veeneklaas, F.R. and Wolfert, H.P. (eds) (1997) Scenario studies for the rural environment, Selected and edited Proceedings of the Symposium Scenario studies for the rural environment, Wageningen, Kluwer Academic Publishers, Dordrecht.

Thunnissen, H.A.M and Noordman, E. (1997) National land cover database of The Netherlands: classification methodology and operational implementation, *BCRS report 96-20*, BCRS, Delft.

Wagtendonk, A.J. and Rietveld, P. (1999) Ruimtelijke ontwikkelingen woningbouw Nederland, 1980 – 1995; Een historisch-kwantitatieve analyse van de ruimtelijke ontwikkelingen in de woningbouw in de periode 1980 – 1995, ter ondersteuning van de Omgevingseffectrapportage Vijfde Nota Ruimtelijke Ordening. Vrije Universiteit Amsterdam.

Van der Hoek, D.C.J., Bakkenes M. and Alkemade, J.R.M. (2000), *Natuurwaardering in de Natuurplanner; Toepassing binnen de Vijno.* RIVM report 408657004, RIVM. Bilthoven, 2000.

VROM (1997), *Naar een landelijk beeld verstoring.* Publicatiereeks Verstoring nr. 12/1997, Ministerie van Volkshuisvesting, Ruimtelijke Ordening en Milieubeheer, Den Haag.

Young, W. and Bowyer, D. (1996) Modelling the environmental impact of changes in urban structure, *Computers, Environment and Urban Systems,* 20(4/5): 313-326.

Kees Schotten
Institute of Public Health and the Environment (RIVM)
P.O. Box 1
NL-3720 BA Bilthoven
The Netherlands

Camiel Heunks
Institute of Public Health and the Environment (RIVM)
P.O. Box 1
NL-3720 BA Bilthoven
The Netherlands

18 A REGIONAL PLANNING APPLICATION OF EUROSCANNER IN PORTUGAL

Alfred Wagtendonk, Rui Pedro Julião and Kees Schotten

18.1 Introduction

Socio-economic change is becoming more complex and appears to be having greater environmental and land use impacts throughout Europe. Land use morphology reflects changes in the size, shape and function of buildings and is linked with changes in transport infrastructure. It is also related to a general increase in urban and built-up areas (partly a result of urban sprawl) and is closely related to the changing role and function of rural areas. It can be argued that analysing medium to long term spatial planning scenarios is important in order to incorporate appropriate mitigation action in plans to alleviate the potentially detrimental effects of development. Scenario-based land use planning can also be extremely useful in helping to develop an understanding of the driving forces behind land use change (ESDP, 1997).

EuroScanner aims to be a multi-scale, spatial decision support tool that policy-makers at European, national and regional levels can use in an integrated fashion to help plan development in different areas. This chapter presents a regional example of the application of *EuroScanner* by focusing on the Lisbon and Tagus Valley area of Portugal as a case study. It is one of three case studies being undertaken, the other two involving the Paris and Randstad regions of France and the Netherlands respectively. The study aims to demonstrate how *EuroScanner* can assist decision-makers in the spatial evaluation of different planning policies. It also aims to describe in detail the interactions between rural and urban land use throughout the history of the area and to investigate how historical trends can be used to develop theoretical understanding of the spatial factors that affect land use and land use change. The case study offers a way in which to calibrate the *EuroScanner* model to evaluate change and policy plans in other areas where less detailed information is available.

The chapter begins with a description of the main physical and socio-economic characteristics of the Lisbon and Tagus Valley area. It attempts to distinguish rural from urban land use, to identify the boundary of the Lisbon area and to suggest what spatial and temporal scales are appropriate to use. Future plans affecting land use in the region are outlined in Section 18.3. Section 18.4 discusses the results of the land use classification based on the Landsat images of 1984 and 1998. In Section 18.5, an exact definition is given of the land use changes studied, followed by a detailed analysis of the driving forces behind these changes, based on existing literature and the actual situation in the Lisbon region as described in Section 18.2. Section 18.6 explains how the chosen driving forces are operationalised in spatial map variables, whilst in Section 18.7, the results of the correlation and regression analyses of land use changes between 1984 and 1998 are given. The chapter considers which land use simulation scenarios should be investigated to best meet the objectives of the case study in Section 18.8 and reports on the results of the simulation exercise undertaken in Section 18.9. Finally, there are some conclusions and suggestions for further research in Section 18.10.

J. Stillwell and H. Scholten (eds.), Land Use Simulation for Europe, 257–291.

18.2 Lisbon and the Tagus Valley

Lisbon was selected as a case study area as it provides an example of a capital city that is fast becoming a global city in its own right but also serving a special function of being a gateway to the European Union. Lisbon is aiming to be the Atlantic capital of Europe. The Lisbon and Tagus Valley region (LTVR) is the most dynamic area of Portugal in terms of land use change and economic development. Its area, depicted in Figure 18.1, corresponds to 13.5 per cent of the mainland area of Portugal (11,927 sq km) and contains 35.1 per cent of the mainland population (3,292,108 inhabitants). This gives an average population density 2.5 times higher than the national average.

Figure 18.1: The Lisbon and Tagus Valley area

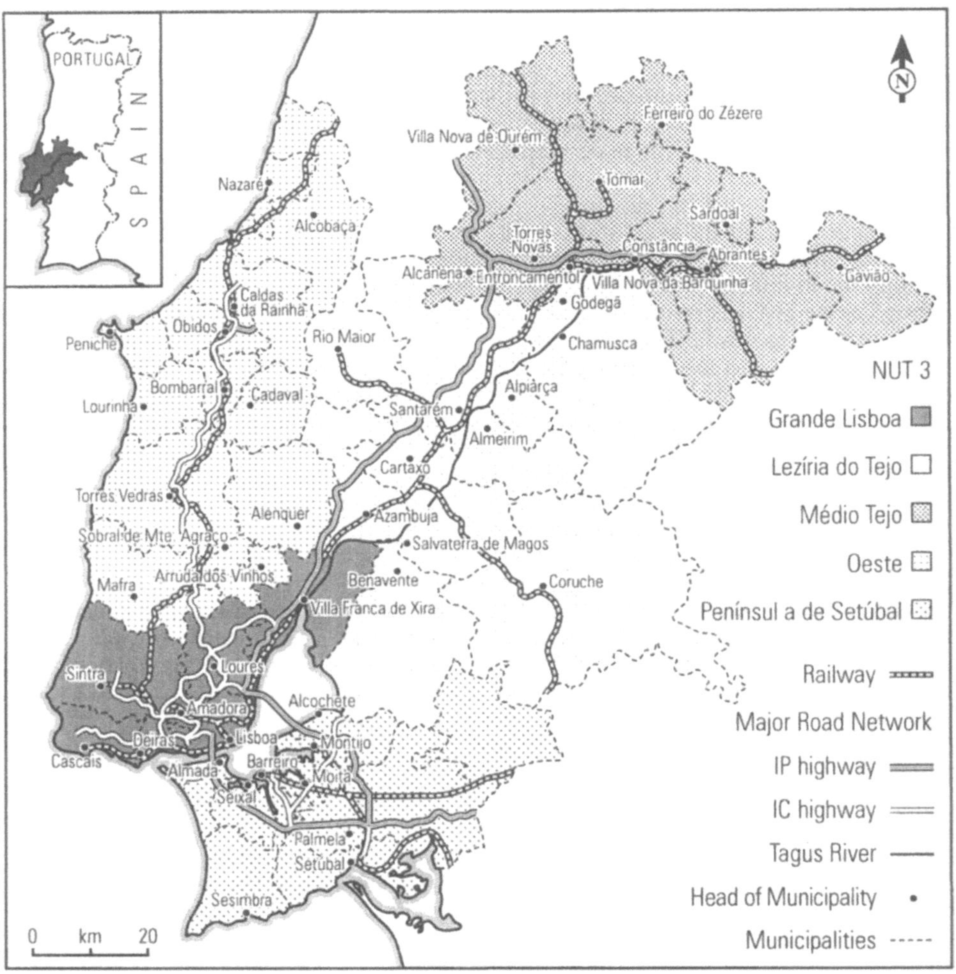

The whole region is divided by the River Tagus into two very distinct northern and southern parts. The area on the northern bank accommodates a variety of active industrial units. Agricultural land use shown in Figure 18.2 is mainly vineyards, fruit and vegetables. On the southern side of the river, little industrial activity can be found, except on the Setúbal Peninsula (just south of Lisbon). Agricultural activity is also fairly limited, except for the Ribatejo region (north east of Lisbon). Forest is the most expansive agricultural activity south of the river.

Figure 18.2: Agricultural and forest land cover

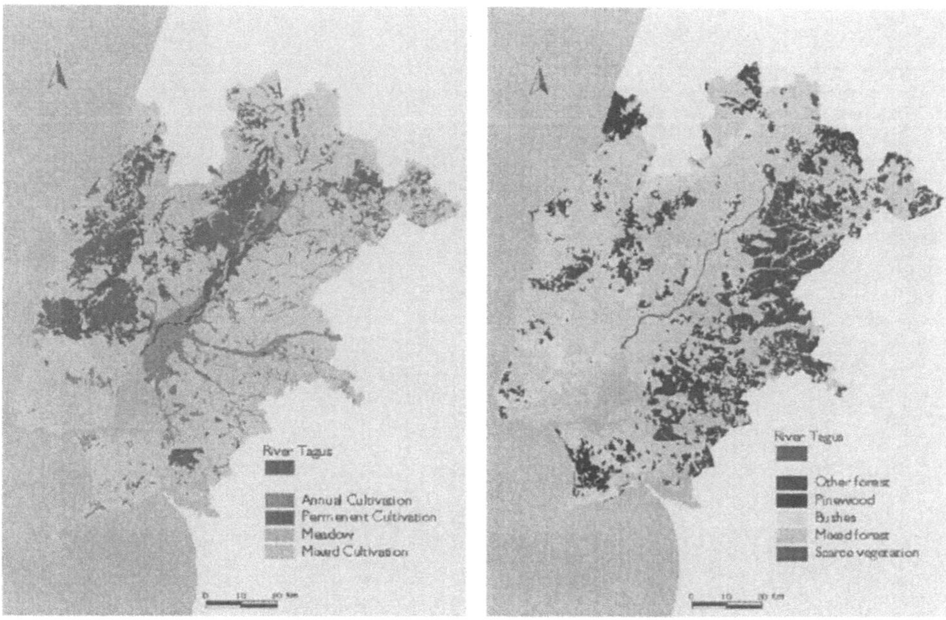

The map of population density by *freguesias* (neighbourhoods) indicates that the majority of the urban areas are concentrated around Lisbon and on the north side of the river (Figure 18.3). Both the differences in land use and population density shown in Figures 18.2 and 18.3 indicate that the Tagus forms a natural barrier separating two very different areas within the LTVR and the Lisbon Metropolitan Area.

Changes in population density and economic activity in the Lisbon area between 1940 and 1991 indicate strong land use dynamics in this area. Between 1989 and 1993, 35.4 per cent of total European Union funds and 70.7 per cent of EU industrial support funds received by Portugal were granted to the LTVR. Some selected economic indicators for 1994 are shown in Table 18.1. Another indication of the economic dynamism of this region is the recent exclusion of this area from the Objective 1 regions for the next European Framework Programme.

During the last half century, and more intensively since 1986, when Portugal became an EU member, major investments were made to improve transport infrastructure and accessibility which have transformed land use in the LTVR. In fact, even by the end of the 1970s, the Portuguese road network was very underdeveloped. In the Lisbon Metropolitan Area (LMA), by the end of the 1970s, there were still only a few kilometres of highway running from Lisbon to the west (10 km), the north (25 km) and to the south (15 km). Until 1999, when the Vasco da Gama Bridge was finished, the river could only be crossed in Lisbon or 25 kilometres to the north at Vila Franca de Xira. Since the bridge was constructed, north-south accessibility has improved considerably, as can be seen in Figure 18.4. Accessibility to Lisbon is good from both north and south of the river now as a dense highway infrastructure covers the LMA. Despite this, the highway density is considerably higher north of the river in general.

Figure 18.3: Population density by *freguesia*

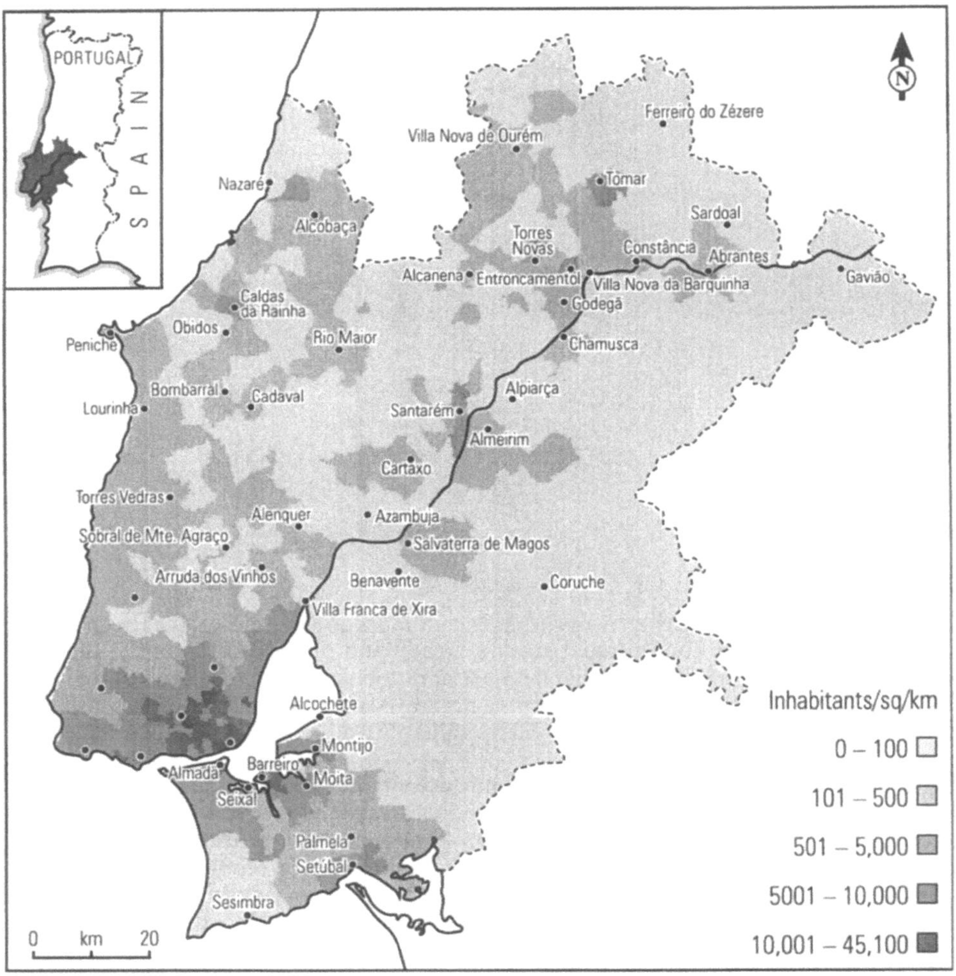

The choice of case study area boundary was influenced by the following factors: not all the LTVR is characterised by the dynamism presented above; rural-urban interaction is much more evident within an area that integrates the actual LMA and some municipalities near its border; the proposed localisation of the new Lisbon Airport on the north side of Tagus, near Alenquer (Ota); the extension of the PROTAML (Regional Plan for the LMA) which is currently under development; and the extension of the available satellite images. That boundary considered to be the optimum boundary for the case study area is illustrated in Figure 18.5.

Table 18.1: Selected economic activity indicators

	LTVR (escudos)	Mainland (escudos)
Industrial productivity (1994)	19,529,434	11,916,781
Tertiary productivity (1994)	15,663,252	12,567,213
GNP (PIB) (1994)	30,972,511,248	70,169,247,114
GNP(PIB) per capita (1994)	9,408.11	7,487.66

Figure 18.4: Accessibility to Lisbon and highway nodes

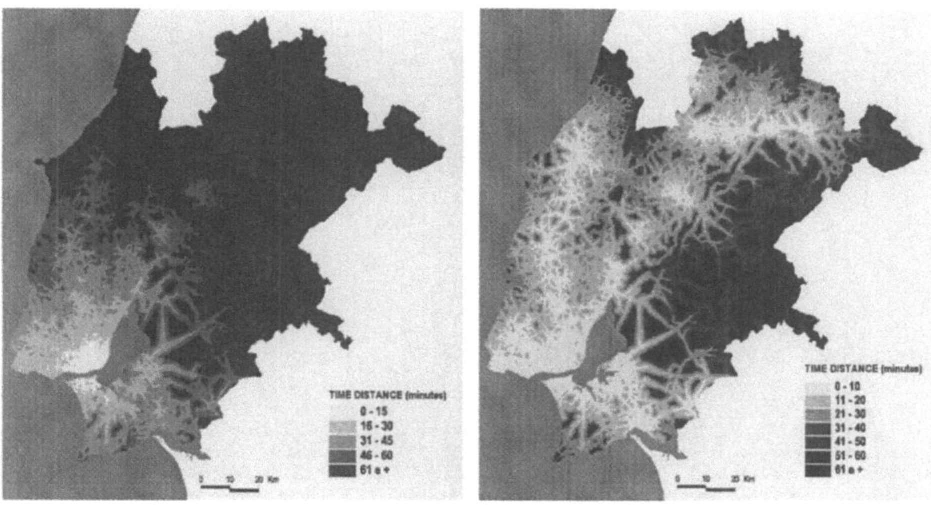

During the last 50 years, Portugal has registered a strong and steady migration flow between rural and metropolitan areas and/or emigration flow to European and North American countries. In the last 100 years, the urbanisation rate has increased from 31.8 per cent in 1890 to 63.5 per cent in 1991 (Soares, 1998). As a result of this process, together with the tertiarisation of the Lisbon core and the return of a large amount of population from former Portuguese colonies (around 2 million persons arrived on the mainland between 1975 and the 1980s), the urban area within the LMA has registered quite a considerable expansion (Figure 18.6). It should be noted, however, that the latest trend in population development is a reverse one. After decades of rapid population growth in the LMA and a stable population around 800,000 for Lisbon itself, the city saw a marked decline of 18 per cent in its population in the 1980s from 807,000 to 663,400 while the metropolitan area grew by no more than 2 per cent (Alden and Pires, 1996). This trend is a by-product of strong rent controls since the 1940s, leading to a large stock of older houses that have fallen into disrepair. The high price for new housing has forced younger people to move out of the centre to the periphery. According to one of the alternative population projections from the Lisbon Municipality for the period 1991-2011, the population of Lisbon could fall to some 400,000 by the year 2011 if the current trends continue. Revitalising the city centre is therefore a major objective of both the PDM and Strategic Plan for Lisbon (Alden and Pires, 1996).

Figure 18.5: Lisbon test site area

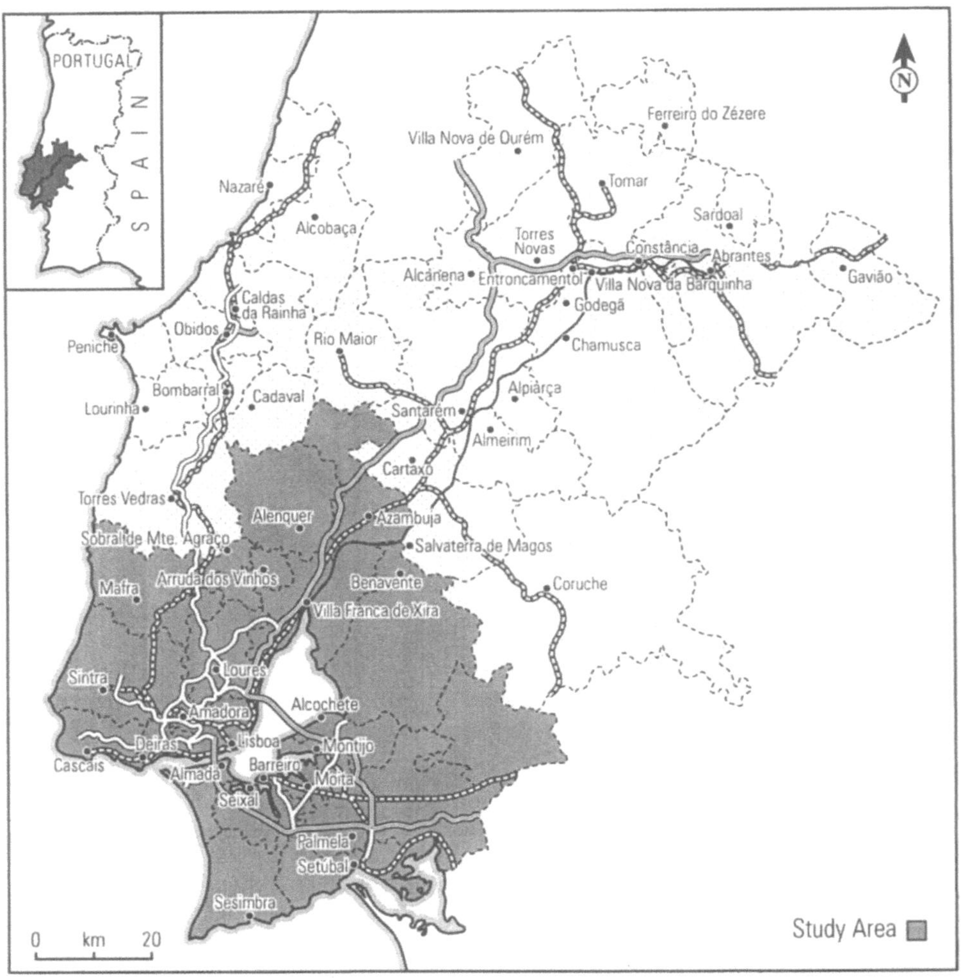

According to several authors, the LMA has a mononuclear structure which means that the centre of Lisbon is the key factor for all its development (Soares, 1998; CCRLVT, 1999). This implies that the accessibility of the centre of Lisbon is the driving force for the urbanisation of the LMA. This accessibility can be divided in to accessibility by road, accessibility by air via the national airport of Lisbon, and accessibility by railway and to railway stations that have good connections with Lisbon. The accessibility characteristics of Lisbon can explain much of the urban growth along the road and railway axis of Cascais, Sintra and Vila Franca de Xira on the north side. On the north side, there are the beginnings of an axis that ends in Loures, supported by the highway A8, but previously structured by the N8. On the south side, the urban growth is clearly connected with the bridge and the highway A2. If one compares the maps in Figure 18.6 with those in Figure 18.4, the relationship between accessibility and urbanisation becomes self-evident. Most of the 'noise' which can be seen in the map image is introduced by the chaotic urbanisation within Amadora and Loures municipalities. If one compares the map of urban development axes (Figure 18.7) with the population density map in Figure 18.3, the urbanisation axes are quite explicit.

Figure 18.6: Urban areas in 1970 and 1990

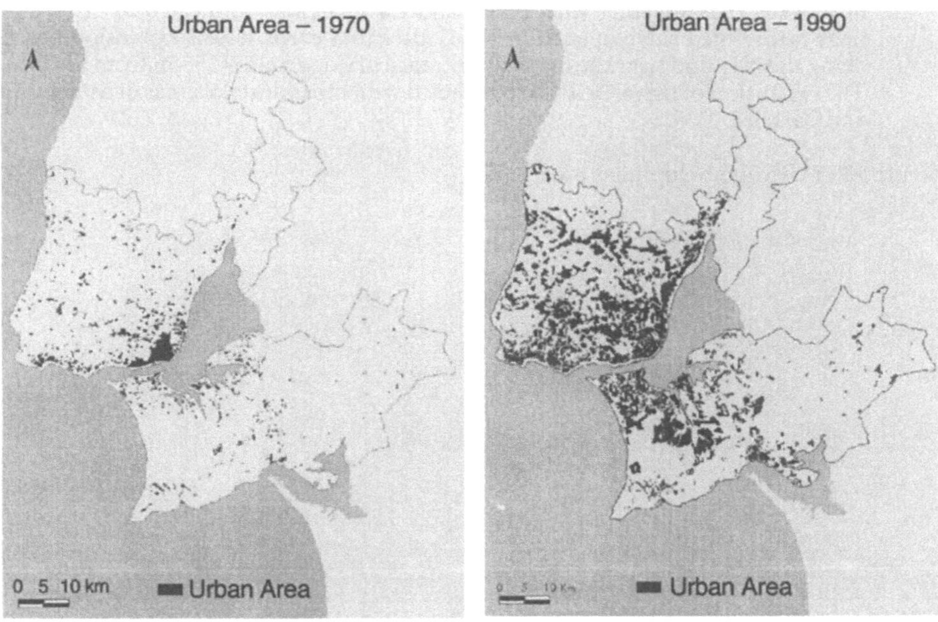

There are five distinguishable axes of expansion in Figure 18.7, each having a different logic:

- *Cascais* is an urbanisation axis structured by the railway line along the coast with the support of two major roads: A5 and N6 (marginal). This was the first area within the LMA to have an urbanisation plan and so its seafront is preserved from chaotic urbanisation. In this axis, a mix can be found between middle and high-income residents. This area is becoming more and more sustainable.

- *Sintra* is the major urbanisation axis within the metropolitan area. It is structured along the railway and the IC19. Highly dense and compact urbanisation or chaotic building accommodates almost half million inhabitants. In this axis, a mix can be found between middle and low-income residents. Some industrial activity and logistics infrastructures can be found in this area.

- *Loures* is a small axis pointing to the north of Lisbon, only supported by road access. Nevertheless, it is densely occupied, especially near Odivelas (the new municipality of the LMA). This axis has a mix between middle and low-income residents. There are some small industrial plots.

- *Vila Franca de Xira* is a heavy industrial axis that connects to the former industrial area of Lisbon (where the EXPO98 was held) supported by railway, N10 and A1. This is a low-income residential area, mainly occupied by industrial infrastructures between the river and the A1 mixed with chaotic residential areas. North of the A1 is one of the many clandestine urbanisation areas of Loures municipality.

- *Almada-Fogueteiro* is the only axis on the south side of Tagus. It is supported by the N10 and the A2 main roads. The oldest part of this urbanisation axis is the city of Almada that, together with Pragal and Cacilhas, was connected to Lisbon by boat before the bridge opened in 1966. After that event, a new expansion began along the N10 and near the highway entrance of Fogueteiro (15 km from Lisbon). The expansion of this area is also connected with the industrial areas of Murgueira and Barreiro.

Figure 18.7: Urban development axes

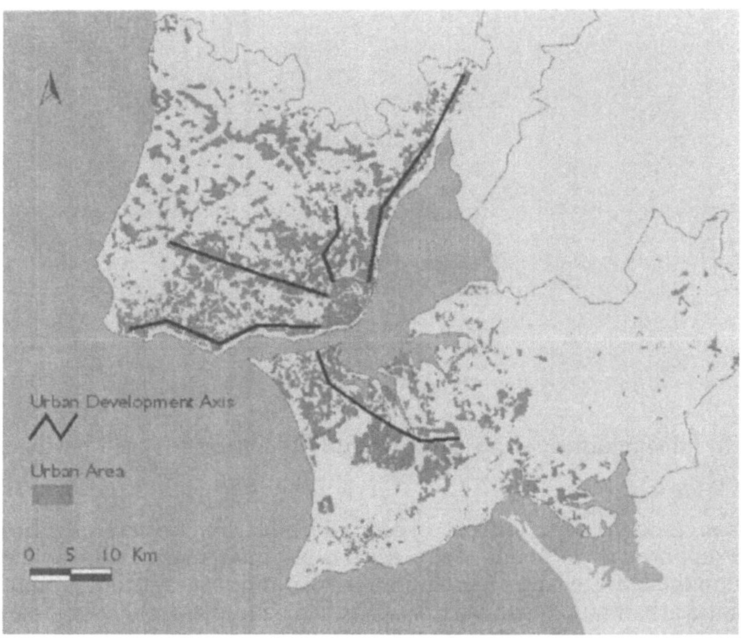

Until recently, the Lisbon central business district (CBD) was considered as the core of all business activity, where residential areas and less profitable activities move out towards the suburbs. These days, thanks to major investments in accessibility infrastructures, the area is getting more attractive again for residential use and other activities. The emergence of new attractive spots, related either to logistics functions or to R&D poles, can be clearly identified.

18.3 Urban and regional plans

Urban and regional planning in Portugal has a very short history, although there are some good examples of plans from the 1930s such as the Master Plan for Costa do Sol (between Lisbon and Cascais). In fact, this was the only existing plan within the LMA. The absence of a planning strategy for the whole metropolitan area, together with socio-political factors in the 1970s, explains the peculiar and chaotic urbanisation of Lisbon's suburbs, especially those on the south side and on the north side along the Vila Franca de Xira axis. The urbanisation of the LMA seems mainly driven by the clandestine and legal plots, not by the existing plans, even if they are mandatory, because they were not approved or executable. For these reasons, the Master Plan for the Lisbon area proposed in 1969 and then again in 1993, was never carried out.

Thus, we can observe that because of the absence of spatial planning the urban-rural border of the LMA is shaped completely by autonomous forces. The consequences of this peculiar history of urban development are even today shown quite clearly by the landscape fragmentation and by the stress over natural areas (Figure 18.8).

Figure 18.8: Clandestine urbanisation in LMA

1.

2.

3.

1. Landscape fragmentation
2. Stress over natural areas: urbanisation in forests
3. Chaotic urbanisation: close to a highway and with incomplete infrastructure (only partly paved).

Source: Photos from CNIG (snig.cnig.pt)

Nowadays, thanks to the mandatory realisation of municipal master plans, the panorama is shifting to a more controlled situation. However, as can be seen in Figure 18.9, these plans have been produced only very recently.

Figure 18.9: Municipal master plans for the LMA

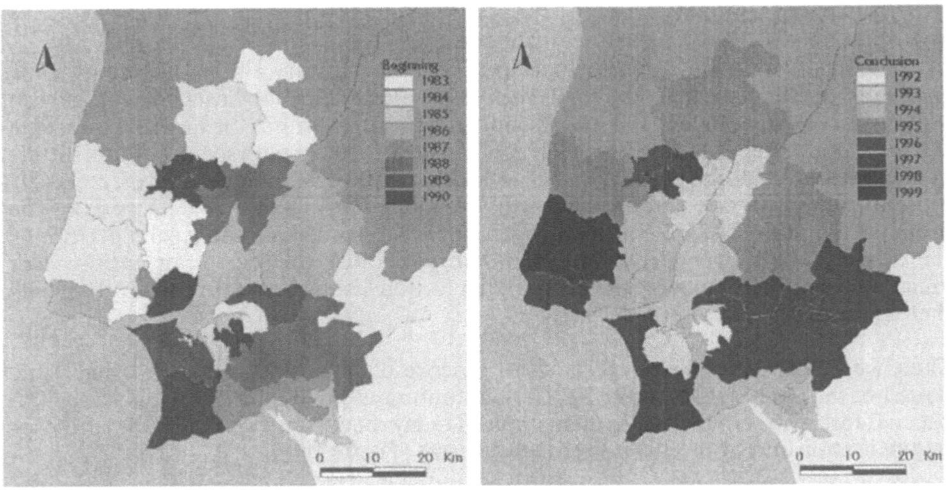

The proposed land use of the municipal master plans shown in Figure 18.10 suggests that the urban area will grow further, especially on the north side of the river. Nevertheless, it is interesting to take a closer look at two cases on the south side. First, there is a clear consolidation of the fifth expansion axis of Lisbon (Almada-Fogueteiro) due to the major infrastructure developments of the railway that now links Lisbon centre to Fogueteiro and the future light railway that will link several urban areas with the Pragal station.

Secondly, there is the proposed tourist area in Benavente (west of Lisbon on the south side of the river), which is in reality nothing but a new urban area proposed as a tourist area to bypass the law. In fact, some of the big farms on the south bank are being converted into residential areas but proposed as areas for tourist development. The degradation of the agricultural income, the accessibility to Lisbon, the open air/nature and the low cost of land are the driving factors for the transformation of these farms into residential areas, which is not allowed, however, according to the law.

To get a better understanding of the role of the driving forces in the territorial evolution of the LMA, a closer look will be taken at the Regional Plan for Lisbon Metropolitan Area or *Plano Regional de Ordenamento do Território da Área Metropolitana de Lisboa* (PROTAML). The PROTAML is being developed by the CCRLVT (LTVR Co-ordination Commission) and its schematic framework is presented in Figure 18.11. It is important to note that this plan will be mandatory for the public administration and, at least in Portugal, the public administration is still an important driving force of development because of all the work done on major infrastructures.

Figure 18.10: Urban, industrial and tourist areas

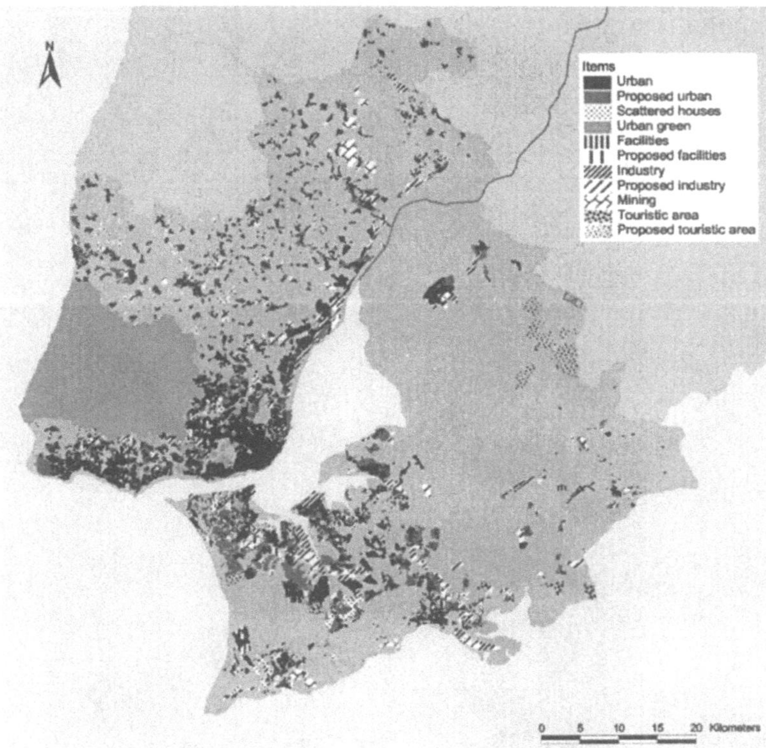

The four major objectives of this plan are:

- *To re-centre the metropolitan area in the Tagus estuary.* In fact the river is quite unused by the inhabitants because it is occupied by heavy industry and the harbour, but recently a rehabilitation of its use has commenced. The major task is to create a city on both banks, integrating the south side with Lisbon and transforming this into the core unit of the LMA.

- *To poli-centre the development and urban areas.* The main idea is to restructure the urban network, reinforcing medium-sized cities and creating new attractive centres within the LMA. To accomplish these developments, the PROTAML identifies the following instruments: supra-municipal business areas (enterprise parks); R&D centres; industrial, logistics and transportation parks; and cultural and economic internationalisation areas.

- *To preserve and promote the territorial diversity and environmental quality.* The LMA has very important natural values (on both national and international levels) that must be protected and promoted as key factors for its valorisation (the idea is to

Figure 18.11: Strategic schema of the Lisbon Metropolitan Area Regional Plan

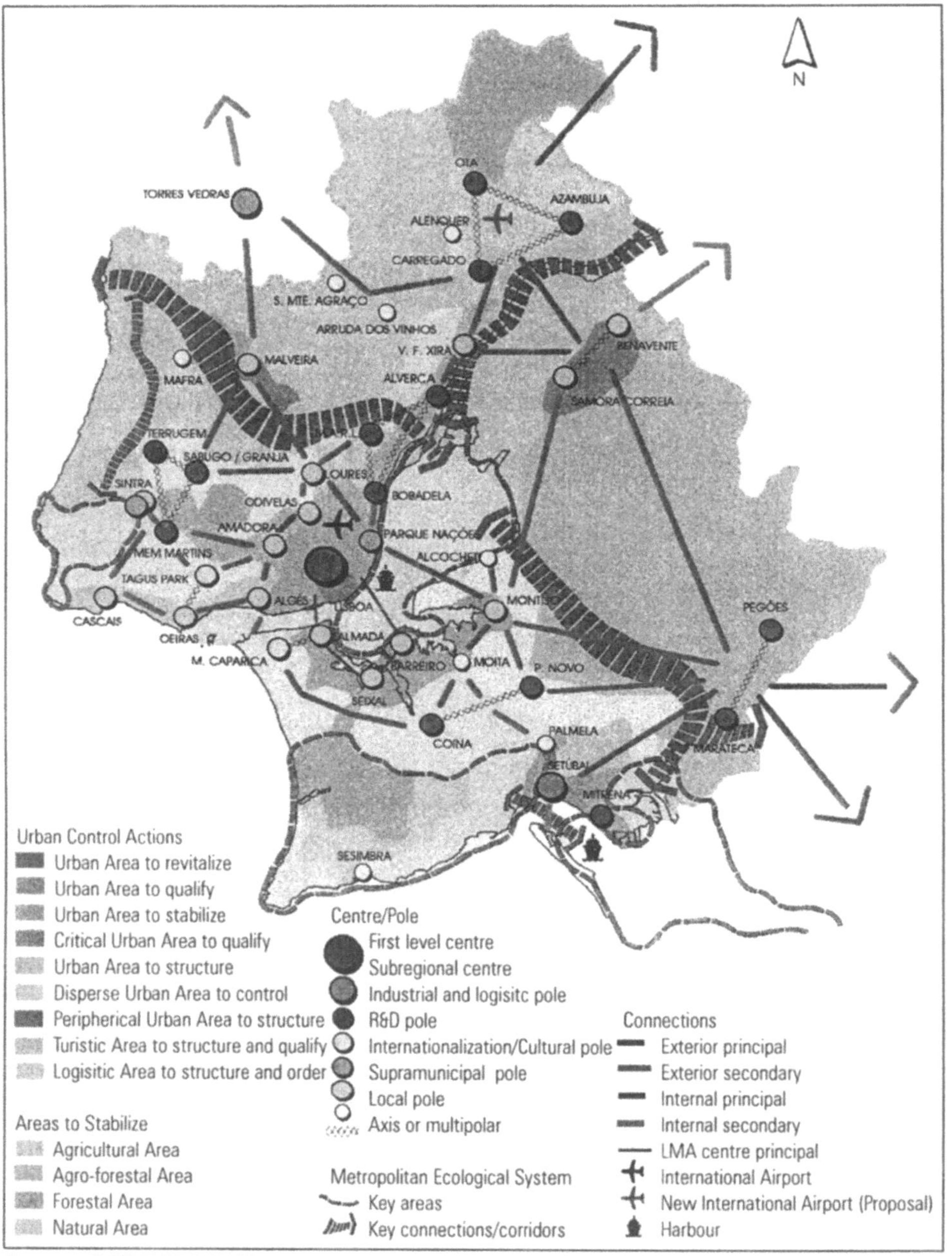

Source: CCRLVT (1999)

use natural heritage as a key resource for regional development) like the waters of the Tagus and Sado estuaries, Sintra and Arrábida.

- *To promote internationalisation infrastructures.* To achieve the LMA international vocation, (Lisbon as the European Atlantic Capital) stressed in the PNDES (Social and Economical Development National Plan), the PROTAML identifies the need for the connection of the LMA with the exterior, by means of high quality transportation services and networks.

18.4 Land use classification of Landsat images between 1984 and 1998

In this section, the results of the land use classification of the Landsat images in 1984 and 1998 are described and discussed. As explained previously, we focus on the process of conversion from rural to urban land use and the resulting spatial patterns of urban 'expansion' and 'densification'. In the urban-rural fringe, different types of transition can be distinguished (Turkstra, 1998). These include conversions from rural/nature to residential; rural/nature to other urban (commercial, recreational, urban green); vacant to residential (vacant areas are without construction but with development indications); vacant to other urban; residential to commercial; commercial to residential; and low or medium density residential to high density residential.

However, the classification method used here does not allow us to make a distinction between urban commercial or urban residential. Only land cover can be identified rather than land uses. The uniform classification method developed for the three test sites allows the following classes to be distinguished:

1 Water
2 Rural bare
3 Rural vegetation low
4 Rural vegetation high
5 Urban bare
6 Urban built-up open
7 Urban built-up
8 Urban vegetation low
9 Urban vegetation high

On the basis of this classification, the following transitions could be measured by overlaying maps from 1984 and 1998 images: (1) no change; (2) expansion; and (3) 'densification'. The following decision rules have been applied:

- *expansion* is where land use in classes 2, 3 and 4 in 1984 is greater than land use in classes 5, 6, 7, 8 and 9 in 1998; and

- *densification* is where land use in class 9 in 1984 is greater than land use in classes 5, 6 and 7 in 1998.

The changes from 1984 (class 8) to 1998 (classes 5, 6 and 7) appeared to be unreliable in the multi-temporal classification and were therefore ignored. Likewise, the transition from urban bare (class 5) to urban built-up (classes 6 and 7) was also ignored. Figure 18.12 shows the results of the Landsat image classification and the change analysis for the Lisbon region (in grid cells of 100 x 100 metres).

The map shows large scattered urban expansion areas to the West and South of Lisbon and along the Tagus. The total urban area of 34,600 hectares in 1984 has expanded by almost 30 per cent (10,350 hectares) during the period 1984-1998, which means an average yearly growth rate of 2.2 per cent. It is particularly interesting that the expansions have not necessarily taken place along the city edges of Lisbon, but have frequently emerged in isolated patches. Urban densification has taken place all around the main city but also along the urbanised coast zones. For 6.5 per cent (2,250 hectares) of the total urban area of Lisbon (34,600 hectares) in 1984, the urban density has increased between 1984 and 1998. A simple indicator for the main growth characteristics of a region is the ratio between urban expansion and densification. In the Lisbon region, the expansion-densification ratio is 4.6 compared with 1.2 in the Paris region and 0.4 in the Randstad region (Schotten *et al.*, 2001), clearly reflecting different growth processes or growth strategies which, in the case of the Randstad area, involve spatial policies aimed at keeping cities compact and the countryside open. However, in the Lisbon region, the opposite is observed; a chaotic expansion has occurred in the open areas west and south of the main city, while a relatively small amount of densification has taken place. This pattern reflects the lack of spatial planning in this area during the last two decades.

Figure 18.12: Land use classes and change between 1984 and 1998 in the Lisbon Metropolitan Area

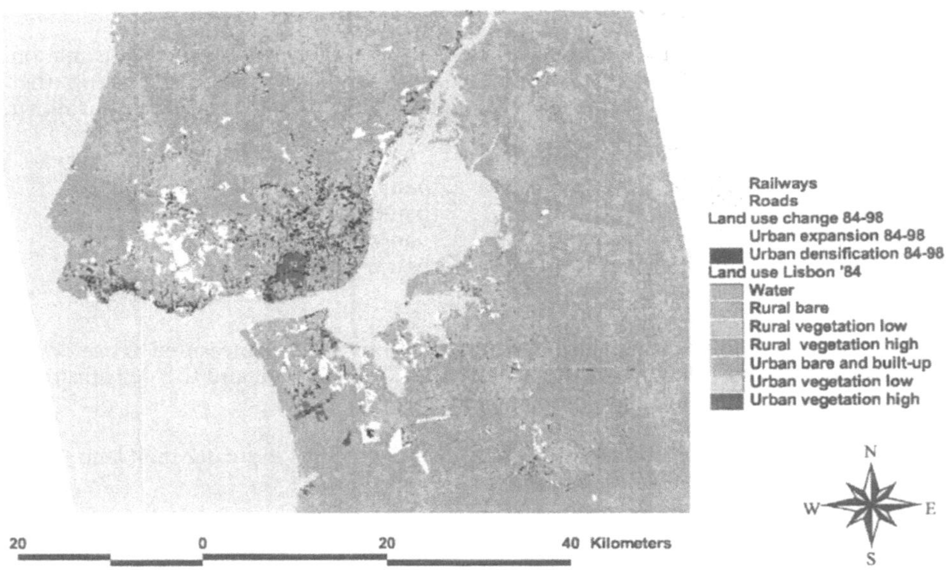

The results of the classification procedure, when compared with the CORINE classification of 1986 (Figure 18.13) show that the Landsat TM '84 classification enlarges the shape of Lisbon, especially to the north and northwest. The shape captured by CORINE is closer to the built-up area of Lisbon, and we suggest that the developed area in the Landsat image is more extensive because the classification used includes all green areas inside the city perimeter.

Figure 18.13: Distribution of land use according to the CORINE 1986 and the Landsat TM 1984 classifications

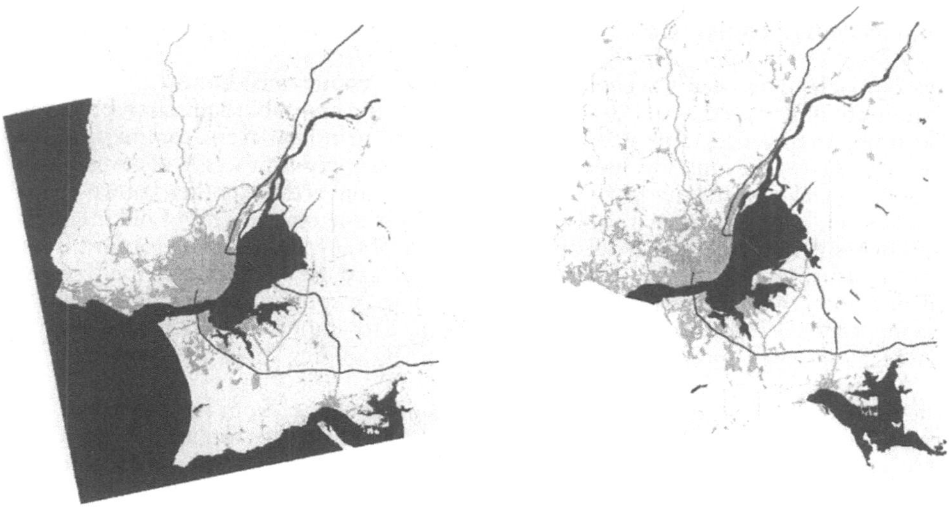

Landsat TM 1984 image classification CORINE 1986 classification

18.5 Identification of driving forces

The next task is to identify the driving forces behind these changes in the rural-urban fringe and to quantify the extent to which these forces explain the changes measured. Subsequently, derived rules can be used for the prediction of land use changes. But first it is necessary to agree what kind of land use changes should be the focus of attention.

The nine land use classes distinguished in the image classification cover all possible land uses. The first four classes (water and rural) also include conservation areas, forests and wastelands. Apart from built-up areas with houses and commercial and industrial units, the urban classes also include parks, sports fields, leisure areas and vacant areas. The definitions of urban and rural uses are implicit in the selected classification and stratification techniques, but we must acknowledge that these definitions are not straightforward. As Tacoli (1998) states, "*... populations and activities described as either 'rural' or 'urban' are more closely linked across space and sectors than is usually thought, and that the categorisations are often misleading*".

It is clear that, in our case, the definition of 'urban' is so general that it will not be possible to find general driving factors that explain the whole spectrum of urban change. Therefore it is necessary to concentrate only on the most dominant classes of urban area, i.e. houses and commercial/industrial units, recognising that an undeveloped urban bare area may be turned into a residential area or a business park with accompanying infrastructure, urban green and sports fields, but never, or seldom, will this occur the other way around. Another possibility, in the case of densification, is the conversion of urban green or sport fields into residential or commercial areas. More recently, even land used for the main infrastructure, like highways and railways inside the city borders, can be extended into commercial and residential areas by layered multiple land use. In the case of expansion outside the city or along the city edge, however, the main infrastructure is often the determining factor in the location and shape of the new commercial/residential areas.

Residential land use seems to be more important than commercial/industrial land use in terms both of its extent and its leading role in expansion and suburbanisation processes. The times when houses were constructed directly around industrial and commercial areas are long past. On the contrary, one can now often see the opposite processes of development of commercial areas taking place near to new residential areas. Another known spatial development that became significant in the 1980s was the replacement of old industrial land by residential or commercial areas, as in some of the old dockland areas in Lisbon. The opposite process, the replacement of residential land use by other land use types is a phenomenon that is seen less often in Europe. A few exceptions can be found including the small parts of residential areas which are sacrificed for important infrastructural works like highways, high-speed railway connections or new landing strips of airports.

The underlying forces that determinine the *amount* of change from rural to urban land use are basically population growth and economic growth. These factors are not considered here because we take the amount of land use change as exogenous and concentrate only on the driving factors behind the *distribution* of this land use conversion.

As Bruckner (2000) has recently argued, the conversion from rural to urban land use depends predominantly on the land's productivity in urban use (which depends on the value of the houses built) relative to the land's productivity in agriculture (as reflected in the value of the farm output). Therefore it seems logical to focus on the social and economic forces behind the conversion to residential land use. Bibby (1997) mentions four reasons to justify the attention on house-building as the driving force behind the overall rate of land conversion. In the first place the residential area usually occupies in general the largest share of the total built-up area. Secondly, as stated previously, other forms of land uses, such as retailing, warehouses, schools and leisure facilities, tend to follow residential developments. Thirdly, it is generally argued that it is the decentralisation of population rather than the decentralisation of employment that is leading the evolution of the settlement pattern. Finally, the spatial planning departments of the different European governments lay particular emphasis on housing allocations from which flow the assessment and allocation of land for other uses.

Not all urban developments start with residential development, however. It therefore makes sense not to focus solely on the driving factors behind residential development but to consider also the driving factors for the other main urban land use types, such as infrastructure and commercial and industrial areas. These areas can usually be easily spatially distinguished from residential areas and have relatively independent driving factors. In the first place we think of spatial policy plans for infrastructure and business parks, but also the proximity of existing commercial and industrial areas because of advantages in the economies of scale (e.g. Mills and Hamilton, 1994).

Finally, we must also consider the factors limiting rural to urban change. These can be such as noise and pollution preventing urban development. Other limiting factors can be policy restrictions for building in or near certain areas, such as conservation areas, or urban growth boundaries that are used to limit urban sprawl.

On the basis of the considerations above and the specific knowledge of the Lisbon region described in Section 18.2, it is possible to define the driving forces for urban expansion and densification in the Lisbon region.

18.5.1 *Identification of driving factors behind residential development*

New residential development is related to available space, spatial policy and personal preferences (of the real estate developer and his target market). Following Wagtendonk and Rietveld (2000), we distinguish the following driving forces:

- Proximity and concentration of existing residential areas measured as the average residential density in the zones within a maximum radius of 500 metres

Although the search radius is quite big (Goetgeluk, 1997), the majority of people are likely to move inside their own neighbourhood or municipality. Accessibility to friends and family and existing services like schools, shops, hospitals, restaurants, but also cultural services like cinemas and theatres, form the basic advantages of living in existing residential areas. Also, the costs of facilities for a new residential area will be lower if the area is connected to an existing residential area. Thus, nearby residential areas should therefore be weighted as an important driving force.

- Accessibility of employment within a maximum radius of 60 km

People attach value to the amount and the accessibility of work in their surrounding area and they also accept, to a certain extent, the disadvantage of commuting because of the better living environment and cheaper housing at the origin end of the commute (Mills and Hamilton, 1994). This implies that the distance to work should be weighted according to a certain distance decay function that depends on the concentration of available work. Different investigations point out that the mean maximum accepted commuting distance in the Netherlands is between 30 and 60 minutes (Gerritsen, 1997). Since no specific figures are known for the Portuguese situation we assume that it will be in a comparable range.

- Accessibility (maximum radius 15 km), quality and size of nature conservation, forest and recreation areas (including surface water) and distance to coast

Leisure, open space, nature and recreational opportunities are relevant determinants. This is shown by the considerable price increases of dwellings near to these kinds of areas. The Lisbon region has a very attractive coastline for beach, water recreation and fishing. The distance to the coast is therefore regarded as an important determinant of both residential as commercial development.

- Distances to railway stations and highway exits

Short travelling distances to work and recreation are important, especially where local forms of transport (by foot, bicycle, tram) are heavily dependent on good connections to the highway and the railway network. Therefore, we consider the distance to the closest connection point as an important factor. Furthermore, it becomes important to distinguish new from old infrastructure because new investments in freeways and other transportation infrastructure encourage the process of suburbanisation because of reduced costs for commuting and transport in areas with low ground prices and cheap housing (Bruckner, 2000). The proximity of railways (except stations), highways (except exits), airports and industries can be a negative factor in the case of noise, air or visual pollution.

- Location of new (planned) towns or expanded towns

In countries or regions with a long planning history, a large part of the spatial distribution of dwellings is explained by the stock created or co-ordinated by past and present governments. In countries or regions that lack a spatial policy, like Portugal until recently, the location of new residential areas is solely determined by the autonomous driving forces mentioned above. In the Portuguese case study, the implication is that that this driving factor is irrelevant in any historical analysis, but considering the recent focus on spatial planning, may be very relevant in the case of future simulation.

- Distance to existing urban area

In many countries, spatial policy since the 1960s has promoted the concept of compact cities and the prevention of uncontrolled urban sprawl. It is therefore expected that the main part of urban expansion takes place along the border or near existing cities. Secondly, the combination of the availability of land with relatively cheap ground prices, and the proximity of the city with its social and cultural facilities and possibilities for work, makes people decide to live as near the city as possible. This trend is visually confirmed by the expansion patterns shown in the Randstad and Paris pilot studies (Schotten *et al.*, 2001). For the Lisbon area, with its less clearly defined urban expansion, only the second motivation is true. Although this location factor overlaps partly with the first factor (proximity and concentration of existing residential areas), it was decided to add this location factor which expresses the distance to the edge of the existing urban area.

- Physical terrain factors

Physical terrain factors like slope, soil stability and zones with natural risks (e.g. risk zones for flooding, landslides, avalanches) can be serious limiting factors for urban development. However, these factors are not considered in the historical analysis because, until recent times, they have not held back many urban developments in the Lisbon region. However, for future simulations, in the situation of a more planned and controlled spatial policy, it makes sense to add these factors as restrictive zones on the map.

18.5.2 *Identification of driving factors behind industrial/commercial development*

The identification of the driving factors behind the development of industrial and commercial areas also begs questions about why the commercial and industrial areas have previously developed where they are. Once again the answer is related to available space, spatial policy and the personal preferences of the real estate developer and his target market. However, unlike residential development, the spatial development of commercial and industrial areas is a more dynamic process in the sense that the locations of these areas are shifting as a result of pull and push factors. At the level of individual cities, for example, in the last 25 years in the Netherlands a process has taken place of suburbanisation of employment from the centre to the edge of the cities (Van Ham, 1999). The push factors include the high ground prices, the congestion and the decreased accessibility, environmental restrictions and the limited possibilities for expansion in the inner city, while the main pull factors of the city edge are exactly the opposite. These deconcentration processes can be found in almost all major European cities including Lisbon (Soares, 1998). Also, at a regional scale, changes are taking place. In Portugal it seems that the core area of Lisbon is starting to become (relatively) less important as an economic centre compared to the adjacent new developing urban areas.

This is, however, a quite recent trend for which less physical evidence is available. From the following factors only the first and the fourth imply the addition of essentially new factors.

- Proximity and concentration of existing commercial/industrial areas (maximum radius 500 metres)

One of the most important determinant factors for the spatial distribution of commercial and industrial development is the proximity and the extension of existing commercial/ industrial areas. For many companies, it is important to be in a concentrated commercial area with only short distances to suppliers and the customers for goods and services. Bigger companies especially tend therefore to concentrate in regional economic centres located in mixed or sector-specific business parks mainly on the edge of the cities. Also new offices tend to concentrate mainly along the city edges as development possibilities in the inner cities are usually quite limited. Industries also tend to cluster, but are more often located further away from the cities and closer to water, road or railway infrastructure for the supply of ground materials. Nearby situated commercial/industrial area should therefore be weighted as an important factor.

- Distance to employees and customers (existing urban area)

For labour-intensive companies and companies delivering goods and services directly to their customers, it is important to be close to densely populated areas where their consumers live and where the commuting distances of their employees are short. The best locations for these types of companies are usually near the city edges of moderate to large cities.

- Distances to railway stations and highway exits and proximity and size of ports and airports

Many companies and industries are heavily dependent on essential and quick transport facilities. Therefore the distances to the access points of a main road, water and air infrastructure are considered very important factors. The distances to highway exits and railway stations are especially important for the accessibility of the commuting employees of companies, offices and industries. Companies and offices tend therefore to be even more concentrated around highway exits and railway stations than residential areas. Another reason that locations along highways are often popular among companies and offices is their great visibility from the highway (Wagtendonk and Schotten, 2000). Ports and airports are also attractive to companies and offices related to different kind of services and transshipments.

- Location in economic development zones and planned commercial areas

The location of economic development zones and business parks is often planned in many countries or regions. Physical planning maps should therefore be included in the data set, where available. For the Lisbon region, once again this refers only to the future simulations.

- Physical terrain factors

As with residential development in some areas, the development of commercial/industrial areas can be hampered by factors like slope, soil stability and zones with natural risks. As mentioned in the historical analysis, these factors are not considered. However, for the Lisbon case study, these factors can easily be integrated into future simulations by adding them as restrictive zones on the map.

18.5.3 *Identification of driving factors behind infrastructure development*
As the development of infrastructure mainly follows residential and commercial development, we do not consider them as separate driving forces in this study. However, infrastructure can also be planned; for example, to develop or access less developed areas. Other factors may also be important; for example, the new bridge in Lisbon (Vasco de Gama) of 1998 was build to create a regional connection within the LMA with the south riverbank of the Tagus river and to divert the north-south traffic from the city centre of Lisbon. Another example is the planned shift in northerly direction of the national airport of Lisbon to make growth of the airport possible. The driving forces behind these particular infrastructure developments are much too specific and unique to integrate in this analysis. If the plans themselves are available, however, they can be integrated in the future simulations.

18.6 Operationalisation of driving forces

In order to explain the historic urban developments in the Lisbon test site, the driving forces or determinant factors discussed in the previous section must be translated into measurable map variables. For this translation, the following three types of maps are used: gravity maps, Euclidean distance maps and Boolean maps.

18.6.1 *Gravity maps*
The effects of distance and area as driving forces are incorporated in the spatial variables by means of distance decay functions. With these functions, we calculate (by summation) a gravity value for each grid cell, which depends on the distance, concentration and area (surface) of the driving force. Only cells around the destination cell within a specified circular radius (minimum and maximum distance) are included in this calculation. In the literature, different types of distance decay functions can be found depending on the type of driving factor. We have used an empirically based log-logistic distance decay function from the research of Hilbers and Verroen (1993) in which the parameters can be adapted according to the type of driving factor:

$$W(x) = 1/(1+e(-a+b*lnx)) \tag{18.1}$$

Where $W(x)$ is the weight to be applied in the accessibility index. The factor x represents distance, so that the weights are proportional to the potential interaction between two grid cells at distance x. To produce the different gravity maps, this function was used in a 'focal-sum' operation in the GIS ArcInfo with the following maximum influence distances:

- gravity maps: work factors (maximum 60 kilometres)

  - industrial/commercial areas, ports, airports;

- gravity maps: recreation factors (maximum 15 kilometres)

  - green urban areas and *high urban vegetation*, areas with sport and leisure facilities, forest areas, natural areas, water and wetland areas; and

- gravity maps: social and commercial concentration factors (maximum 500 metres)

  - urban continuous and discontinuous area, *urban built-up and urban bare/low urban vegetation*, industrial/commercial area.

The gravity maps in the list above written in Italic are maps derived from the Landsat classification, the maps in normal text are derived from the CORINE classification.

18.6.2 *Euclidean distance maps*
Euclidean distance maps were made for driving forces that are only related to the distance to a certain feature, e.g. the nearest railway station. All maps were calculated on the basis of Euclidean distance between the cell centres of rasterised vector maps (500 metre cells). All distances were expressed in kilometres. Separate Euclidean distance maps were calculated for distances to the following: railways and railway stations; highways and highway exits; city edges (based on urban areas Landsat 1984 > 100 hectares); municipality centres; Lisbon centre; and coasts.

If data sets with highway exits were not available, the distances were calculated to the nodes in the highway network instead. This implies that some intersections were assumed to be highway exits while they were not and that some other highway exits not located on intersections were probably missed.

For the accessibility to the infrastructure network it would have been better to use the time distance maps which are available for the Lisbon region. However, because these maps were not available for the Paris and Randstad regions, it was decided not to use them. In the case of the maps with distance to city edges, only connected urban areas larger than 100 hectares were selected, because a significant influence is only expected from reasonable large urban areas.

18.6.3 *Boolean maps*
Boolean maps were composed for driving factors in which information regarding spatial planning or spatial restrictions had to be incorporated. The following Boolean maps were generated:

- the location of a new airport (from PROTAML Lisbon);
- zones with limitations for residential construction (from PROTAML Lisbon); and
- the presence of railways or highways in a grid cell.

18.7 Statistical analysis land use changes between 1984 and 1998

Statistical analysis is carried out to explain the observed spatial pattern of urban expansion and densification in the Lisbon region for the period 1984-1998. The statistical relationships are analysed with the help of a logistic regression model. The application of such a model is possible because it is plausible to consider the observed increase in new urban area in each (500 x 500 metre) map square as the result of a binomial chance process in which the chance of transition from a rural to an urban area or from an urban green to an urban bare/built-up area is estimated, or that no transition occurs at all. Correlation analysis is used to get an initial impression of the statistical relationships between the different variables and to make a first selection of independent variables for the regression analysis in which different sets of variables are tested to find an optimal small set of variables of high explanatory value. The resulting statistical parameters found are subsequently used to make reference or alternative scenarios (in different areas) in future simulations.

18.7.1 *Correlation analysis*
Correlation analysis is used here to find the relationship between the independent variables (the location factors) and the dependent change variables which are urban expansion and densification. Because the definition of urban expansion implies that only cells partly or fully filled with rural land use can transform into urban land use, we select only those grid cells from the data set where change from rural to urban land use is possible, i.e. where the rural category from the Landsat classification is larger than zero. To test the correlation for densification, we take only those grid cells where change from urban vegetation (high) to urban bare/built-up is possible.

Secondly, we use correlation analysis to extract only the real independent variables and to test the comparibility of the CORINE classification and the classification based on our Landsat images. This comparison appeared to be reasonably good for the urban bare/built-up classes and moderately reasonable for the urban vegetation classes. All correlation values of distance variables, like distance to towns and infrastructure, should be interpreted in reverse (i.e. negative correlation should be read as positive correlation) because the shortest distances are represented by the lowest figures (see the last six columns of expansion and densification in Figure 18.14). Only the most significant variables are presented in the figure; Schotten *et al.* (2001) provides the complete correlation table.

The correlation coefficients shown in Figure 18.14 that relate to urban *expansion* are relatively high for the 'proximity and extension of urban bare/built-up area' (r = 0.41) and the 'urban low vegetation area' (r = 0.35). Also, the urban classes of CORINE (not shown) correlate well, and the r value is respectively 0.20 for 'urban continuous' and 0.40 for the 'urban discontinuous area'. The absence of correlation with the 'proximity and extension of industrial and commercial areas' is noteworthy. Only a low positive correlation is found with the '60 km gravity map' of these areas (r = 0.06). This result can be explained by looking at the CORINE map which shows that only very small areas of commercial/industrial areas were classified, which automatically results in low correlations. Strong positive correlations are found, however, for the '(60 km) proximity' and 'extension of ports' (r = 0.32) and 'airports' (r = 0.23).

Both the variables 'accessibility to green urban areas' and 'areas with leisure facilities' correlate strongly (r = 0.31 and 0.32). A less strong but significant positive correlation (r = 0.17) is also found with the '(15 km) accessibility to water areas' variable. Finally, quite strong correlations are also found for the 'travel times' (r = -0.22) (not shown in Figure 18.14) and 'distance to highway exits' (r = -0.20) and 'distance to railway stations' (r = -0.18), 'travel-time' (r = -0.29) and 'distance to Lisbon' (r = -0.25) and the 'travel time' (r = -0.16) and 'distance to the centres of municipalities' (r = -0.13). Also a high correlation is found for the 'distance to town edges' (r = -0.25) and 'to coast recreation' (r = -0.15). Significantly negative correlations were only found for the 'accessibility to forest areas'. On the basis of these correlations, we conclude that a large part of the urban expansion in Lisbon is driven by autonomous forces, where living close to existing towns, green urban areas, the coast, leisure facilities and access to car and rail infrastructure are more important than the proximity and extension of work areas.

Figure 18.14: Correlation of the most significant variables for the Lisbon and Tagus Valley area

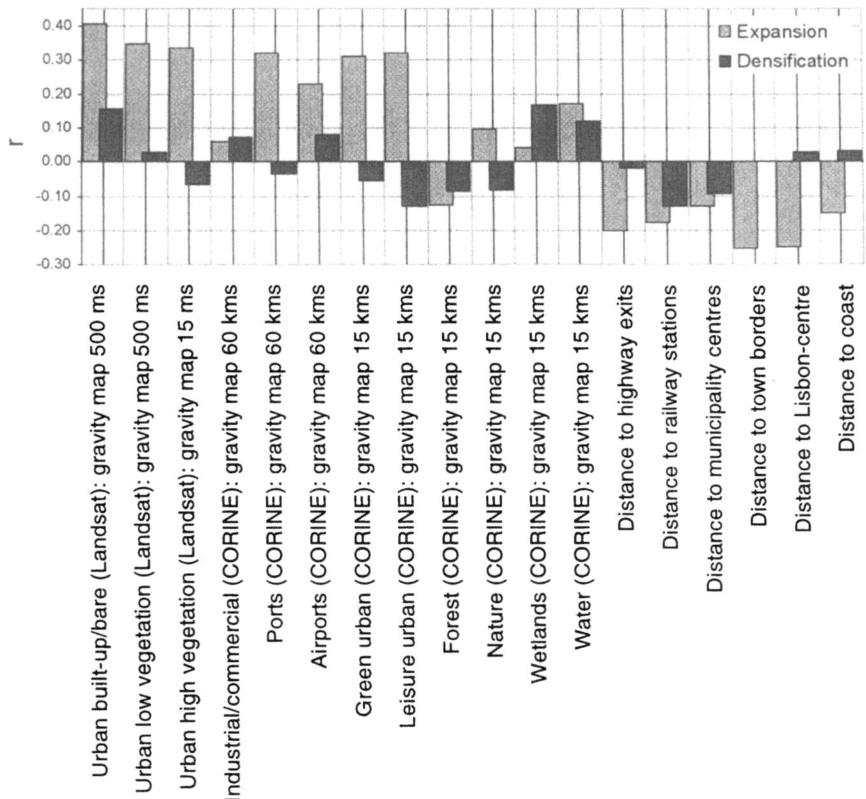

The correlation analysis in relation to urban *densification* shows reasonable correlations with the 'urban bare/built-up area' (r = 0.16) but very low, hardly significant, correlation values for the 'urban low vegetation class of Landsat' (r = 0.03) and even negative correlation values for the 'urban high vegetation class' (r = -0.07). Relatively high correlation values (r = 0.12), however, are found for the 'proximity and extension of urban discontinuous areas' from the CORINE classification (not given in Figure 18.14). Further, densification appears to be only weakly related to 'industrial and commercial areas' (see explanation above) and 'proximity to ports and airports'. The 'accessibility to wetlands' (r = 0.17) and 'water areas' (r = 0.12) correlates reasonably with densification. Also the correlation with 'distance to the nearest railway stations' is quite high (r = -0.13), which is confirmed by physical evidence of large concentrated building (high apartment blocks) directly around railway stations nearby Lisbon, for instance in Barcarena and Cacem. Negative correlations are especially found for the 'accessibility to forest areas' (r = -0.12), green urban areas (r = -0.05) and areas with leisure facilities (r = -0.13). However, these negative correlations seem to be related to the central location of the large forest park 'de Monsanto', where no urban densification is possible. This was already confirmed by the negative correlation of the 'urban high vegetation class of Landsat' which makes up the largest part of this park (Figure 18.12). The correlations found for the 'travel-time' (r = -0.1) and 'distance to highway exits' (r = -0.02) and 'municipality centres' are rather weak and in the case of distance to Lisbon even negative. It makes sense that the relative differences in the correlations of 'travel-times' and 'travel distance' are bigger in the urban area as travel time per distance unit increases towards the centre.

Although the correlation values are low in general, the patterns described seem logical and consistent and therefore the results are considered useful. On the basis of the significance of these correlations and the interpretation of the mutual correlations (Schotten *et al.*, 2001), we decided to select the location factors shown in Table 18.2 as the initial set for the regression analysis.

18.7.2 *Regression analysis*

The second stage in the analysis is a statistical regression in which the strength and the direction of relationships between a set of explanatory variables and the dependent variable are determined. A logistic regression approach is chosen, because this type of regression is relevant when the measurement scales of the independent and the dependent variables are different. In addition, the logistic regression corresponds well with the way the simulation model (*EuroScanner*) allocates land use, in terms of chances between one and zero. For each explanatory variable, the strength and direction of the relation is expressed in the form of a regression coefficient (beta). The quality of the beta coefficient is estimated with the t-test of statistical significance. The resulting beta coefficients are used later on in the extrapolation of the statistical trends in the simulation model.

Prior to the regression analysis, the dependent variables have been transformed to relative values by dividing them by the area available for respectively expansion and densification. Following the applied definitions for expansion and densification, this means that for each cell, the expanded area is divided by the rural area in 1984 and the 'densified' area is divided by the urban vegetation high area in 1984. In this way, values were obtained between zero and one, where one implies maximum expansion.

Table 18.2: Selected location factors

Lisbon and Tagus Valley	
Urban built-up/bare (Landsat): gravity map 500 m	Wetlands (CORINE): gravity map 15 km
Urban vegetation low (Landsat): gravity map 15 km	Water (CORINE): gravity map 15 km
Urban vegetation high (Landsat): gravity map 15 km	Distance to municipalities
Ports (CORINE): gravity map 60 km	Distance to highway exits
Airports (CORINE): gravity map 60 km	Distance to railway stations
Leisure (CORINE): gravity map 15 km	Distance to Lisbon
Forest (CORINE): gravity map 15 km	Distance to towns > 100 hectares*
Nature (CORINE): gravity map 15 km	Distance to coast

* Only expansion

Stepwise regression analysis was used initially where the explanatory value of each variable is tested in a structured way, to get a first indication of the explanatory value of the different variables. The resulting set of variables and parameters of the stepwise regression analysis was subsequently used as a reference set for the regression analyses carried out later on. In the actual regression analysis, variables were removed and added manually from the initial set, on the basis of expected relationships and the actual regression values, until the best explanatory combination was found. The explanatory value of each of the tested sets was evaluated by means of the residual deviance.

Table 18.3 shows the regression results for urban *expansion* in the Lisbon area. The figures show that both the proximity and extension of the existing urban area (concentration factors) as the distance to the town borders, railway stations and road exits are important factors. One important result is the high negative value for the '60 km ports gravity map'. In the correlation analysis, a highly positive value was found for this variable. In the combination of variables, however, the ports have apparently a relative negative influence on the city expansion. The removal of the variable 'distance to central Lisbon', indicates that the accessibility to local economic and social centres and access points to main infrastructure is apparently more important than access to the core of Lisbon. Also the 'accessibility to leisure, forest and water areas' seems to have had a significant influence on the urban expansion in the Lisbon and Tagus Valley area.

Table 18.4 shows the regression results for urban *densification* in the Lisbon area. The relationship between the variables with the growth of densification patterns is less clear. This is probably the result of the few densification processes in Lisbon compared to the large expansion processes which have taken place. Relatively strong negative relationships are found with the 'distance to the coast' and the 'distance to the centre of Lisbon'. A reasonable positive relation is found with the 'distance to highway exits' and to 'railway stations'.

Table 18.3: Regression results for expansion in the Lisbon area

Variable	Beta coefficients	Standard error	t value
Constant	-2.3387	0.28668	-8.1578
Urban built-up/bare (Landsat): gravity map 500m	0.0271	0.00466	5.8182
Urban low vegetation (Landsat): gravity map 500m	0.0429	0.00789	5.4359
Ports (CORINE): gravity map 60km	-11.8735	2.46310	-4.8205
Airports (CORINE): gravity map 60km	3.4233	0.74499	4.5951
Leisure (CORINE): gravity map 15km	1.8375	0.29506	6.2274
Forest (CORINE): gravity map 15km	0.0291	0.00721	4.0379
Wetlands (CORINE): gravity map 15km	-0.7812	0.17450	-4.4767
Water (CORINE): gravity map 15km	0.0835	0.02035	4.1072
Distance to municipalities (in km)	-0.0335	0.0140	-2.3927
Distance to highway exits (in km)	-0.0687	0.0173	-3.9725
Distance to railway stations (in km)	-0.0815	0.0198	-4.1058
Distance to towns > 100 ha (in km)	-0.1016	0.0155	-6.5505

Table 18.4: Regression results for densification in the Lisbon area

Variable	Beta coefficients	Standard error	t value
Constant	-1.3215	0.5406	-2.4442
Urban built-up/bare (Landsat): gravity map 500m	0.0160	0.0040	3.9522
Urban high vegetation (Landsat): gravity map 15km	-0.1936	0.0714	-2.7095
Wetlands (CORINE): gravity map 15km	-0.2898	0.2565	-1.1295
Water (CORINE): gravity map 15km	0.0552	0.0233	2.3633
Distance to highway exits (in km)	-0.0409	0.0265	-1.5400
Distance to railway stations (in km)	-0.0286	0.0298	-0.9606
Distance to Lisbon (in km)	0.0525	0.0211	2.4861
Distance to towns > 100 ha (in km)	-0.0673	0.0337	-1.9965
Distance to the coast (in km)	0.0924	0.0231	3.9923

18.8 Simulations with EuroScanner

One of the main aims of the case study is to show how *EuroScanner* can assist policy-makers in the economic, social and environmental evaluation of different policies and scenarios. The observed land use developments of the past decades should be translated to land use rules to perform future simulations. A second aim is to evaluate the socio-economic and environmental effects of different scenarios and spatial strategies. To address the latter question, decisions have to be made on the type of scenarios and strategies to be investigated.

EuroScanner can be used in different ways. It can be used in *future exploration* involving the simulation of (alternative) spatial perspectives based on expert judgement that is defined here to mean the best possible knowledge about the investigated spatial relations. In future exploration, an imaginable reality is being explored (not the probable or desirable future land use pattern). This method uses the scenario methodology in which a number of possible scenarios are translated into spatial images based on some relatively simple assumptions (where possible based on expert judgement) and a supply-demand model. These assumptions can be based on existing knowledge derived from research literature about well known spatial relations, such as a distance decay function based on travel time to a railway station and the observed population density. However, it can also be newly defined relationships based on the weighting of different factors by an expert panel.

An important argument to use this method is the limited capability of the human mind to transform a written scenario into a complete map image. In the decision-making process, it is important that the different stakeholders have the same image and interpretation of the proposed scenarios. An example of this methodology can be found in Schotten *et al.* (1997).

EuroScanner can be used to *extrapolate historic trends* involving the simulation of a future 'reference-image' or 'base scenario' based on statistical analysis of historic land use data. This empirical way of simulation is undertaken using the statistical methods of correlation and regression analysis to find mathematical rules between the driving forces (independent variables) and the resulting land use changes (the dependent variable). These transition rules form the base for the 'attractiveness' maps that are used for the future simulation of land use (in different scenarios with different land use claims). This methodology leads to a reference image or base scenario, which is based on the extrapolation of the historic trends to the future. The reference image is used for the evaluation of the continuation of current policy and comparison with alternative policies.

A third way of using *EuroScanner* is by simulation of the (short-term) *effects of planned interventions* in landscape or infrastructure, based on expert judgement or statistically based transition rules. In the case of the Lisbon case study, we are dealing with a situation in which a number of more or less fixed strategic and physical plans exist in which interventions in the landscape or the infrastructure are planned to solve spatial problems or to push spatial developments in a desired direction. This can be the reclamation of an entire area or the construction of a new highway or bridge (the new Vasco de Gama bridge, for example), which has implications for future land use (with different time scales). This type of intervention is normally not planned further ahead than approximately 10 years. This means that for the simulation of the resulting land use changes, it could be assumed that changes are influenced by more or less the same driving forces that have been shaping the landscape in recent history. Taking into account all known plans for physical intervention and the expected growth of different types of land use (derived from other models), we can simulate the new land use distribution with the help of existing transition rules and expert judgement. This type of simulation is preferably carried out in different time steps according to the time schedule of the integrated plans.

Fourthly, there is the *comparison of different national or regional spatial strategies.* What happens if the spatial strategy of zone A is carried out in zone B (including transition rules)? This is an experimental method of simulation meant to gain knowledge and understanding about the processes behind land use dynamics. Besides, this method offers the possibility of quick simulation of alternative scenarios giving way to the development of totally new concepts of spatial strategies. An example could be the projection of the transition rules for the allocation of new single and multiple family houses (derived from the Dutch study for the Fifth Memorandum on National Physical Planning), as outlined in the previous chapter, to the Lisbon and Tagus Valley region. What would be the result of applying Dutch planning concepts and Dutch preferences in the choice of living areas? These are interesting questions if one thinks of the long planning history in the Netherlands and the typical living preferences which are influenced by the strong perception of Dutch people of living in a congested country with a minimum amount of (artificial) nature and open space. The enormous urban expansion in the Lisbon and Tagus Valley region, at the cost of natural areas and open space and in parallel with the recent burst of spatial policy plans, might make this comparison not that unrealistic.

From the methods described above, future exploration is clearly distinctive from the others, because transition rules in this approach are based on expectations instead of experiences. A combination of the different methods is possible if we expect that some spatial relations stay the same while others can radically change. It is therefore possible to simulate a new scenario which combines elements from the first three methods. Also the fourth method could be added, for instance just for one type of land use.

If we compare the land use dynamics from the past in the Lisbon region, for instance, with the Rotterdam region in the Netherlands, we see many differences. Both areas have a history of rapid urban expansion, but the way this expansion took place in each area is totally different. In the Netherlands with its long history of spatial planning, this expansion took place in a rather structured way influenced by spatial policies and plans, and bounded by spatial legislation, although the magnitude of this influence seems in reality lower than might be expected (Ransijn and Vreeker, 1998). In Portugal, however, the lack of spatial planning policy in the past and the weak control of spatial legislation caused a relative uncontrolled expansion of the urban area, leading to a more or less random distribution of land use types. This distribution is mainly based on two key factors – the accessibility to Lisbon and the cost of land. These differences in urban development emerge also very clearly in the land use classification maps and the correlation analyses conducted for both test areas (Schotten *et al.*, 2001).

If we look at the current developments in the Lisbon and Tagus Valley region, we see a contrasting situation. Different spatial plans have been developed for the area, which are currently being carried out. We can therefore conclude that a trend break is occurring in the spatial developments of this area. And we can expect other spatial trend breaks to occur in the area, given the new attitude to spatial planning and the changed legislation. These kinds of breaks are the most difficult to predict as they are based on technical, economic and autonomous developments in the way people think and act. What, for instance, will be the effect of a significant increase in teleworking, or radical innovations in ways of transport or alternative energy? Moreover, people's changing desire to improve their quality of life is strongly influencing spatial developments. In the Netherlands, for example, a recent trend is that many companies, after moving out from the congested city centres to the city edges in the 1970s and 1980s, are now moving out of the urban areas in the Randstad altogether to smaller cities outside this area, to the areas where new residential towns were developed in the 1960s and 1970s. This is a totally opposite process to that occurring historically where people moved closer to their work locations. If a process like this were to take place in Portugal, this would make the accessibility maps to the municipality centres much more important (translated in new attractiveness maps by changing the relative weights of the underlying gravity maps) than the accessibility map to the centre of Lisbon. Next to these more urban trends, we can expect major changes in the agricultural development and, in the longer term, spatial consequences from changing climatic factors.

We start the simulations with a validation of the model by simulating the current land use changes in 1998, based on the map variables of 1984. Then, because of the large range of possible scenarios, we decided to start the future simulations in the Lisbon and Tagus Valley region with building a *reference scenario*, by extrapolating the historic trends to the future, to which alternative scenarios can be compared. Secondly, we decided to simulate the *effects of planned interventions* according to the PROTAML of the Lisbon metropolitan area, namely the development of a new airport northeast of Lisbon and the implementation of restriction zones for residential development. Thirdly, we carried out a simulation in which *Dutch policy-based transition rules* for urban expansion and densification were applied in the Lisbon and Tagus Valley region.

The inputs necessary for a simulation are as follows:

- *Land use claims*

Land use claims are ideally derived from demographic, agricultural and economic forecasting models. Because this kind of data was not available for this study, we derived all land use claims from the satellite images. For the calibration and the simulation of the reference scenario, this is done by determining the amount of land that was subject to expansion and densification in the past period and allocating a proportional amount of land depending on the length of the simulation period in the future. For alternative scenarios, we can allocate alternative land use claims.

- *Suitability maps*

Because both expansion and densification will be simulated, two suitability maps have to be derived that show the location probabilities of both urbanisation processes. The suitability maps used in the simulations are derived from the equations found with the statistical analyses. As discussed above, the simulation of the historic expansion and densification is carried out with the same data used to determine the equations. For the simulation of the future urbanisation trend, these variables are updated. For alternative scenarios, additional data are incorporated into the suitability maps.

- *Additional constraints*

Some additional constraints are also used in the simulations. As described in Section 18.7, the definition of expansion and densification implies that expansion only occurs in areas that are classified as rural in the start situation, while densification is limited to vegetated urban areas.

Most of the driving forces change gradually in space and also the map variables derived from these forces do not often have sharp boundaries. This peculiarity implies, together with the properties of the allocation algorithm described in Chapter 16, that the simulated land use changes tend to blur out. In order to limit this, the minimum amount of allocated land is set to 1 hectare or 4 per cent per grid cell. This is in accordance with the minimum area of urban expansion and densification that can be detected with the satellite images. This threshold is operationalised by running the allocation algorithm for each simulation two times: firstly with all land available for expansion and densification, and secondly with the exclusion of grid cells in which less than 4 per cent expansion or densification is allocated. The application of alternative planning scenarios also implies the use of additional data; for example, the new airport in the Lisbon area. The gravity map that is the translation of the attraction of this airport has to be incorporated into the regression equation. But also the boundaries of the airport itself have to be known in order to prevent urban expansion on the airport itself when land use is simulated. In these alternative scenarios, it may even be necessary to exclude certain areas in the proximity of the airport as building ground due to legal constraints concerning noise nuisance or safety.

18.9 Results of simulations

We start the simulations with a validation of the simulation model built for the Lisbon and Tagus valley region, by simulating the current land use changes in 1998 based on the map variables of 1984. In Figures 18.15 and 18.16, the simulated expansion and densification patterns are shown together with the historic expansion and densification for the Lisbon area. The simulated expansion and densification look similar to the observed historic patterns found between 1984 and 1998. Expansion is quite well simulated at the fringe of Lisbon and on the east site of the river Tagus. The concentrations of urban expansion simulated in the surroundings of Lisbon are, however, lower than historically found. This is partly due to the properties of the suitability map as discussed before. High suitability values near Lisbon also cause an underestimation of expansion in villages at a greater distance from the city core.

The simulated densification has greater similarity with the observed patterns and the amount of historic densification per cell. The densification in villages at a reasonable distance from the centre of Lisbon is similar to the densification found historically. The main differences between the simulated and the actual densification patterns are the high densification values simulated near the coast.

In the second phase of simulations, we extrapolated the historic urbanisation trends to the future to create a reference scenario. Figure 18.17 shows the urban expansion and densification patterns from the simulation of the period 1998-2012.

Figure 18.15: Historical (left) and simulated (right) expansion for the period 1984-1998

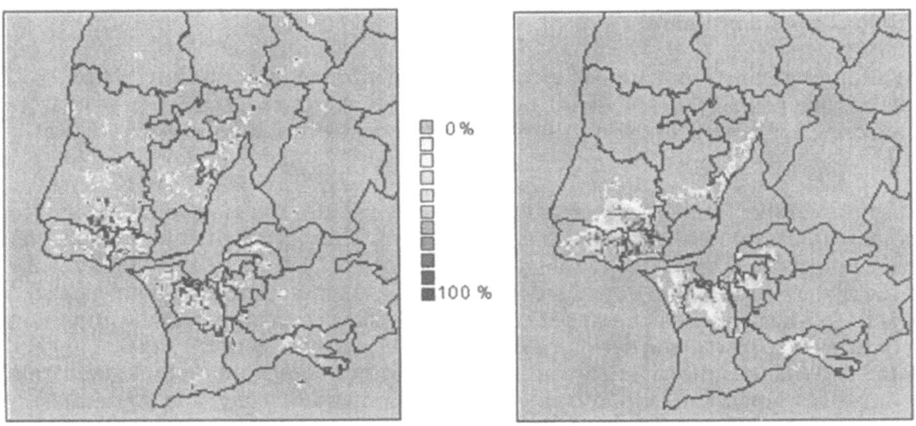

Figure 18.16: Historical (left) and simulated (right) densification for the period 1984-1998

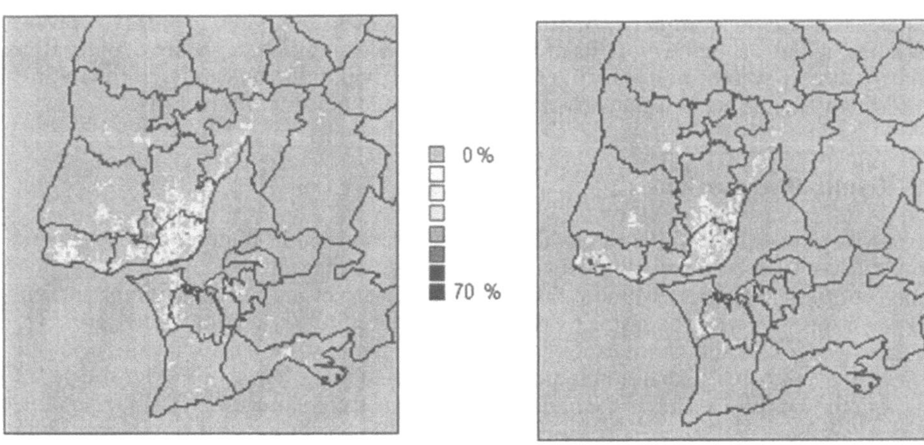

Figure 18.17: Simulated expansion (left) and densification (right) for the period 1998-2012

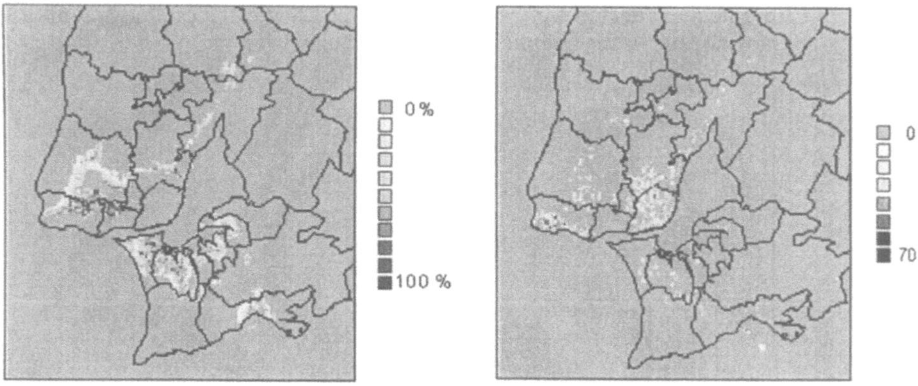

For this simulation, it was assumed that the amount of area subject to rural to urban conversion in the future period is the same as in the period 1984-1998 and that the location factors for expansion and densification have remained the same. Updated are the boundaries of rural and urban areas where expansion and densification respectively, can occur. Furthermore, the variables derived from Landsat TM 1984 images in the regression equations listed in Tables 18.3 and 18.4 are updated with those of 1998. This also includes the distance to towns, where the towns are derived from the satellite images. The variables derived from the CORINE data, however, cannot be updated, as recent CORINE data is not yet available. As can be expected, the simulations show a continuation of the historical trends. Remarkable, however, are the low expansion values at the northern and northwest fringe of Lisbon city, compared to the much higher expansion values in this area found for the simulation over 1984-1998. On the other hand, higher expansion values are simulated over 1998-2012 for the area southeast of the river Tagus.

The third type of simulation carried out was the simulation of the effects of a planned physical intervention. Figure 18.18 shows the effects of one of the elements in the PROTAML of the metropolitan region of Lisbon; the impact (only expansion) of the new airport that is planned north of the Lisbon Metropolitan area. The coefficients used in this simulation are the same as those used for the trend simulation but now also the new airport is added as a variable (with the regression coefficient found for airports in the Lisbon area). Because of the effect of the huge size of the airport in the gravity maps, the suitability for urban expansion will be very high near the airport and expansion elsewhere in the Lisbon area will be diverted away. To temper this effect, we chose to rescale the variable in accordance with the values found for the existing airport. Because the airport will be built in a rather empty area, where few other location factors are effective, few distinct patterns are yet found in the surroundings of the airport. However, this simulation shows that the impact of the new airport will be enormous, this includes a diversion of urban extension from central Lisbon to the area to the north.

In the fourth and last type of simulation carried out, the transition rules from an alternative spatial strategy were projected to the Lisbon and Tagus Valley region. In this case, the transition rules were derived from the Dutch study for the Fifth Memorandum on National Physical Planning, as outlined in the previous chapter. In other words, this simulation studies the spatial effects of applying Dutch planning concepts and Dutch preferences in the choice of living areas in the Lisbon region.

Figure 18.18: Simulated expansion for the period 1998-2012 with the establishment of a new airport in the Lisbon Metropolitan Area

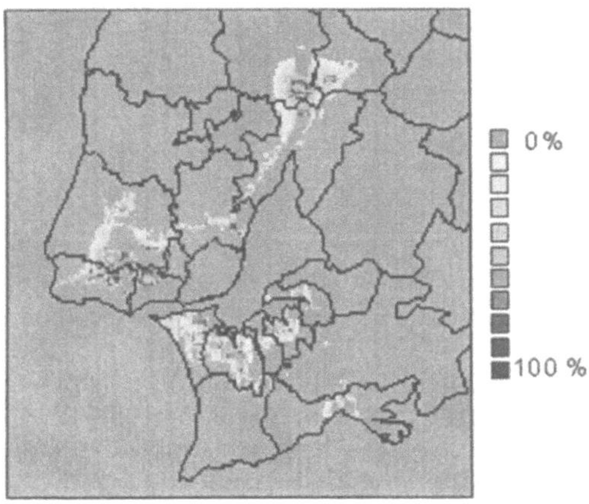

The results of this last simulation in Figure 18.19 show clearly a much more compacted form of urban expansion and densification compared to the historical found expansion and densification. In this alternative scenario, the urban expansion near Lisbon is limited and takes place for a large part at the southeastern bank of the river Tagus, while densification only occurs in Lisbon, with very high densities in the centre of town.

Figure 18.19: Simulated expansion (left) and densification (right) for the period 1984-1998 using the transition rules determined for Randstad Holland

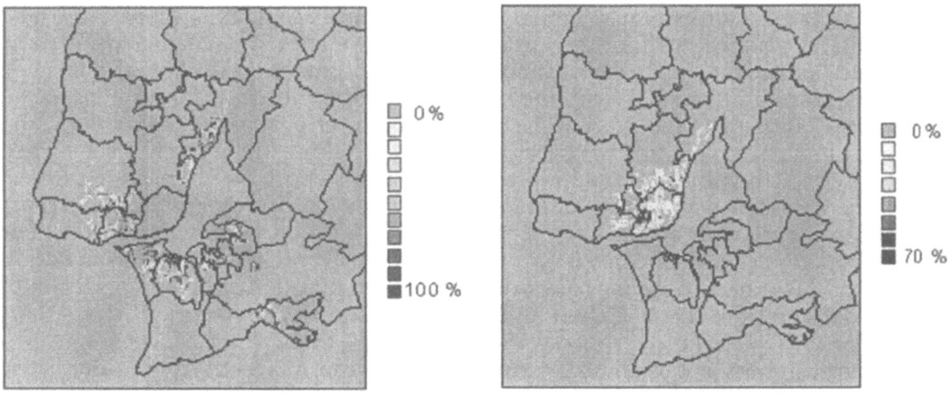

18.10 Conclusions

Chapter 17 has already shown that the linkage of an allocation model with the spatial analytical possibilities of a GIS in *EuroScanner* makes land use modelling a serious instrument for supporting policy-makers in spatial planning, particularly if the same validated data are used and the quality and limitations of intermediate results are known to the researchers. This chapter aimed to show in which ways the *EuroScanner* system can be used for land use simulation on a regional scale, making use of less detailed satellite-derived data and uniform classification methods. The results for the Lisbon and Tagus Valley area test site prove that rural to urban land use dynamics can be simulated in a satisfactory way at the regional scale. Alternative scenarios can be created quickly by using transition rules based on the experience of other areas. The example using the transition rules associated with the 'compact city policy' from the Randstad, shows clearly in which the direction the spatial development of the Lisbon region could turn if new spatial planning policy was introduced along these lines. By implementing quantitative indicators for spatial change, the differences in urban expansion and densification between alternative scenarios can be evaluated easily.

Simulations can be varied between a reference scenario and alternative scenarios by changing parameters based on spatial relations in other areas under different regimes of spatial planning, or changing variables and weights based on expert judgement. At the same time, the case study has shown that it is of paramount importance to have reliable geographic and socio-economic background knowledge on a local level in order to make a sound choice and undertake constructive analysis of the driving forces behind the land use change. Local expert judgement is also indispensable for controlling the classification of satellite images.

Before it becomes possible to generate reliable simulations of land use on a European scale, it is therefore necessary to investigate the land use dynamics of other test sites, of cities and regions under different geographic, economic, social and political circumstances and under different regimes of spatial planning. In this way, European zones and regions can be classified with a blueprint of spatial transition rules according to their characteristic land use dynamics and with local, regional and national plans and expectations realistic simulations can be made drawing new maps and perspectives of European land use. It is also clear that all this work cannot be done without the help of expert knowledge of local land use dynamics and supply of uniform historical and updated geographic data. However, our regional simulation in Portugal suggests that *EuroScanner* has immense potential as a decision support system to assist regional planning throughout Europe.

Acknowledgements

We wish to thank Andy Turner and Joost Buurman for their improvements on the English and the text structure; Kees de Zeeuw and Henk Kramer for the classifications of the satellite images; Piet Rietveld for his comments on the correlation and regression analysis and Camiel Heunks for the help with the data transformations and the simulations with *EuroScanner*. Finally we wish to thank the editors John Stillwell and Henk Scholten for their editorial work and their patience waiting for the results.

References

Alden, J. and Pires, A.D. (1996) Lisbon – strategic planning for a capital city, *Cities*, 13: 25-36.

Baptista, F.O. (1995) Agriculture, rural society and the land question in Portugal, *Sociologia Ruralis*, 35: 309.

Bibby, P. and Shepherd, J. (1997) Projecting rates of urbanisation in England, 1991-2016, *Town Planning Review*, 68(1): 93-124.

Bruckner, J.K. (2000) Urban sprawl: diagnosis and remedies, *International Regional Science Review*, 23(2): 160-171.

CCRLVT (1999) *PROTAML: Ante-Projecto*, CCRLVT, Lisboa.

ESDP (1997) *European Spatial Development Perspective*, First official draft presented at the meeting of EU Ministers responsible for Spatial Planning. Noordwijk, 9 and 10 June 1997, Commission of the European Communities (CEC), Brussels.

Gerritsen, R.A. (1997) *De Toenemende Reistijd Woon-werkverkeer: Er Zijn Grenzen!*, Wereldhave, Den Haag.

Goetgeluk, R. (1997) *Bomen over wonen; Woningmarktonderzoek met beslissingsbomen*, Thesis, Faculteit Ruimtelijke Wetenschappen, Universiteit Utrecht.

Ham, M. van (1999) *Waar Moet dat Heen? Bereikbaar Heid als Indicator*, Urban Research Centre, Utrecht.

Hilbers, H.D. and Verroen, E.J. (1993) *Het Beoordelen van de Bereikbaarheid van Locaties: Definiëring, Maatstaven, Toepassing en Beleidsimplicaties*, TNO Inro, Delft.

Mills, E.S. and Hamilton, B.W. (1994) *Urban Economics*, HarperCollins College Publishers, New York.

Ransijn, M. and Vreeker, R. (1998) *Historische Analyse van Ruimtegebruik en Ruimtelijk Beleid in Nederland*, Vrije Universiteit, Amsterdam.

Schotten, C.G.J., Boersma, W.T., Groen J. and van de Velde, R. (1997) *Simulatie van de Ruimtelijke Perspectieven Nederland 2030*, RIVM-rapport 711901004, RIVM, Bilthoven.

Schotten, C.G.J., Heunks, C., Wagtendonk, A.J., Buurman, J.J.G., de Zeeuw, C.J., Kramer, H. and Boersma, W.T. (2001) *Simulating Europe in the 21st Century*, NRSP Report, BCRS, Delft.

Soares, N.P. (1998) *O Sistema Urbano Português: 1890/1991*, Universidade Nova de Lisboa, Lisboa.

Tacoli, C. (1998) Beyond the rural-urban divide, *Environment and Urbanization*, 10(1): 3-4.

Turkstra, J. (1998) *Urban development and Geographical Information*, Thesis, Utrecht University and ITC, Den Haag.

Wagtendonk, A.J. and Rietveld, P. (2000) *Ruimtelijke Ontwikkelingen Woningbouw Nederland, 1980-1995*, Een historisch-kwantitatieve analyse van de ruimtelijke ontwikkelingen in de woningbouw in de periode 1980-1995, ter ondersteuning van de Omgevingseffectrapportage Vijfde Nota Ruimtelijke Ordening, Vrije Universiteit Amsterdam.

Wagtendonk, A.J. and Schotten, C.G.J. (2000) *Bedrijfsterreinen weg van de Snelweg?*, Een historische analyse van de ruimtelijke veranderingen van bedrijfsterreinen in de periode 1981-1993, op het ruimtelijk schaalniveau van 500 meter gridcellen, RIVM rapport nr. 711901028, Bilthoven.

Alfred J. Wagtendonk
SPINLab: Spatial Information Laboratory
Vrije Universiteit Amsterdam
De Boelelaan 1115
1081 HV Amsterdam
The Netherlands

Rui Pedro Julião
Universidade Nova de Lisboa
Faculdade de Ciencias Sociais e Humanas
Departamento de Geografia e Planeamento Regional
Avenida De Berna 26 C
1069-061 Lisboa
Portugal

Kees G.J. Schotten
National Institute of Public Health and the Environment (RIVM)
Postbus1,
3720 BA Bilthoven
The Netherlands

SUBJECT INDEX

The GeoJournal Library

1. B. Currey and G. Hugo (eds.): *Famine as Geographical Phenomenon.* 1984
ISBN 90-277-1762-1
2. S.H.U. Bowie, F.R.S. and I. Thornton (eds.): *Environmental Geochemistry and Health.* Report of the Royal Society's British National Committee for Problems of the Environment. 1985 ISBN 90-277-1879-2
3. L.A. Kosiński and K.M. Elahi (eds.): *Population Redistribution and Development in South Asia.* 1985 ISBN 90-277-1938-1
4. Y. Gradus (ed.): *Desert Development.* Man and Technology in Sparselands. 1985
ISBN 90-277-2043-6
5. F.J. Calzonetti and B.D. Solomon (eds.): *Geographical Dimensions of Energy.* 1985
ISBN 90-277-2061-4
6. J. Lundqvist, U. Lohm and M. Falkenmark (eds.): *Strategies for River Basin Management.* Environmental Integration of Land and Water in River Basin. 1985
ISBN 90-277-2111-4
7. A. Rogers and F.J. Willekens (eds.): *Migration and Settlement.* A Multiregional Comparative Study. 1986 ISBN 90-277-2119-X
8. R. Laulajainen: *Spatial Strategies in Retailing.* 1987 ISBN 90-277-2595-0
9. T.H. Lee, H.R. Linden, D.A. Dreyfus and T. Vasko (eds.): *The Methane Age.* 1988
ISBN 90-277-2745-7
10. H.J. Walker (ed.): *Artificial Structures and Shorelines.* 1988 ISBN 90-277-2746-5
11. A. Kellerman: *Time, Space, and Society.* Geographical Societal Perspectives. 1989
ISBN 0-7923-0123-4
12. P. Fabbri (ed.): *Recreational Uses of Coastal Areas.* A Research Project of the Commission on the Coastal Environment, International Geographical Union. 1990
ISBN 0-7923-0279-6
13. L.M. Brush, M.G. Wolman and Huang Bing-Wei (eds.): *Taming the Yellow River: Silt and Floods.* Proceedings of a Bilateral Seminar on Problems in the Lower Reaches of the Yellow River, China. 1989 ISBN 0-7923-0416-0
14. J. Stillwell and H.J. Scholten (eds.): *Contemporary Research in Population Geography.* A Comparison of the United Kingdom and the Netherlands. 1990
ISBN 0-7923-0431-4
15. M.S. Kenzer (ed.): *Applied Geography.* Issues, Questions, and Concerns. 1989
ISBN 0-7923-0438-1
16. D. Nir: *Region as a Socio-environmental System.* An Introduction to a Systemic Regional Geography. 1990 ISBN 0-7923-0516-7
17. H.J. Scholten and J.C.H. Stillwell (eds.): *Geographical Information Systems for Urban and Regional Planning.* 1990 ISBN 0-7923-0793-3
18. F.M. Brouwer, A.J. Thomas and M.J. Chadwick (eds.): *Land Use Changes in Europe.* Processes of Change, Environmental Transformations and Future Patterns. 1991
ISBN 0-7923-1099-3

The GeoJournal Library

19. C.J. Campbell: *The Golden Century of Oil 1950–2050.* The Depletion of a Resource. 1991 ISBN 0-7923-1442-5
20. F.M. Dieleman and S. Musterd (eds.): *The Randstad: A Research and Policy Laboratory.* 1992 ISBN 0-7923-1649-5
21. V.I. Ilyichev and V.V. Anikiev (eds.): *Oceanic and Anthropogenic Controls of Life in the Pacific Ocean.* 1992 ISBN 0-7923-1854-4
22. A.K. Dutt and F.J. Costa (eds.): *Perspectives on Planning and Urban Development in Belgium.* 1992 ISBN 0-7923-1885-4
23. J. Portugali: *Implicate Relations.* Society and Space in the Israeli-Palestinian Conflict. 1993 ISBN 0-7923-1886-2
24. M.J.C. de Lepper, H.J. Scholten and R.M. Stern (eds.): *The Added Value of Geographical Information Systems in Public and Environmental Health.* 1995 ISBN 0-7923-1887-0
25. J.P. Dorian, P.A. Minakir and V.T. Borisovich (eds.): *CIS Energy and Minerals Development.* Prospects, Problems and Opportunities for International Cooperation. 1993 ISBN 0-7923-2323-8
26. P.P. Wong (ed.): *Tourism vs Environment: The Case for Coastal Areas.* 1993 ISBN 0-7923-2404-8
27. G.B. Benko and U. Strohmayer (eds.): *Geography, History and Social Sciences.* 1995 ISBN 0-7923-2543-5
28. A. Faludi and A. der Valk: *Rule and Order. Dutch Planning Doctrine in the Twentieth Century.* 1994 ISBN 0-7923-2619-9
29. B.C. Hewitson and R.G. Crane (eds.): *Neural Nets: Applications in Geography.* 1994 ISBN 0-7923-2746-2
30. A.K. Dutt, F.J. Costa, S. Aggarwal and A.G. Noble (eds.): *The Asian City: Processes of Development, Characteristics and Planning.* 1994 ISBN 0-7923-3135-4
31. R. Laulajainen and H.A. Stafford: *Corporate Geography.* Business Location Principles and Cases. 1995 ISBN 0-7923-3326-8
32. J. Portugali (ed.): *The Construction of Cognitive Maps.* 1996 ISBN 0-7923-3949-5
33. E. Biagini: *Northern Ireland and Beyond.* Social and Geographical Issues. 1996 ISBN 0-7923-4046-9
34. A.K. Dutt (ed.): *Southeast Asia: A Ten Nation Region.* 1996 ISBN 0-7923-4171-6
35. J. Settele, C. Margules, P. Poschlod and K. Henle (eds.): *Species Survival in Fragmented Landscapes.* 1996 ISBN 0-7923-4239-9
36. M. Yoshino, M. Domrös, A. Douguédroit, J. Paszynski and L.D. Nkemdirim (eds.): *Climates and Societies – A Climatological Perspective.* A Contribution on Global Change and Related Problems Prepared by the Commission on Climatology of the International Geographical Union. 1997 ISBN 0-7923-4324-7
37. D. Borri, A. Khakee and C. Lacirignola (eds.): *Evaluating Theory-Practice and Urban-Rural Interplay in Planning.* 1997 ISBN 0-7923-4326-3

The GeoJournal Library

---

38. J.A.A. Jones, C. Liu, M-K. Woo and H-T. Kung (eds.): *Regional Hydrological Response to Climate Change.* 1996 ISBN 0-7923-4329-8
39. R. Lloyd: *Spatial Cognition.* Geographic Environments. 1997 ISBN 0-7923-4375-1
40. I. Lyons Murphy: *The Danube: A River Basin in Transition.* 1997 ISBN 0-7923-4558-4
41. H.J. Bruins and H. Lithwick (eds.): *The Arid Frontier.* Interactive Management of Environment and Development. 1998 ISBN 0-7923-4227-5
42. G. Lipshitz: *Country on the Move: Migration to and within Israel, 1948–1995.* 1998 ISBN 0-7923-4850-8
43. S. Musterd, W. Ostendorf and M. Breebaart: *Multi-Ethnic Metropolis: Patterns and Policies.* 1998 ISBN 0-7923-4854-0
44. B.K. Maloney (ed.): *Human Activities and the Tropical Rainforest.* Past, Present and Possible Future. 1998 ISBN 0-7923-4858-3
45. H. van der Wusten (ed.): *The Urban University and its Identity.* Roots, Location, Roles. 1998 ISBN 0-7923-4870-2
46. J. Kalvoda and C.L. Rosenfeld (eds.): *Geomorphological Hazards in High Mountain Areas.* 1998 ISBN 0-7923-4961-X
47. N. Lichfield, A. Barbanente, D. Borri, A. Khakee and A. Prat (eds.): *Evaluation in Planning.* Facing the Challenge of Complexity. 1998 ISBN 0-7923-4870-2
48. A. Buttimer and L. Wallin (eds.): *Nature and Identity in Cross-Cultural Perspective.* 1999 ISBN 0-7923-5651-9
49. A. Vallega: *Fundamentals of Integrated Coastal Management.* 1999 ISBN 0-7923-5875-9
50. D. Rumley: *The Geopolitics of Australia's Regional Relations.* 1999 ISBN 0-7923-5916-X
51. H. Stevens: *The Institutional Position of Seaports.* An International Comparison. 1999 ISBN 0-7923-5979-8
52. H. Lithwick and Y. Gradus (eds.): *Developing Frontier Cities.* Global Perspectives – Regional Contexts. 2000 ISBN 0-7923-6061-3
53. H. Knippenberg and J. Markusse (eds.): *Nationalising and Denationalising European Border Regions, 1800–2000.* Views from Geography and History. 2000 ISBN 0-7923-6066-4
54. R. Gerber and G.K. Chuan (eds.): *Fieldwork in Geography: Reflections, Perspectives and Actions.* 2000 ISBN 0-7923-6329-9
55. M. Dobry (ed.): *Democratic and Capitalist Transitions in Eastern Europe.* Lessons for the Social Sciences. 2000 ISBN 0-7923-6331-0
56. Y. Murayama: *Japanese Urban System.* 2000 ISBN 0-7923-6600-X
57. D. Zheng, Q. Zhang and S. Wu (eds.): *Mountain Geoecology and Sustainable Development of the Tibetan Plateau.* 2000 ISBN 0-7923-6688-3

The GeoJournal Library

58. A.J. Conacher (ed.): *Land Degradation.* Papers selected from Contributions to the Sixth Meeting of the International Geographical Union's Commission on Land Degradation and Desertification, Perth, Western Australia, 20–28 September 1999. 2001 ISBN 0-7923-6770-7
59. S. Conti and P. Giaccaria: *Local Development and Competitiveness.* 2001 ISBN 0-7923-6829-0
60. P. Miao (ed.): *Public Places in Asia Pacific Cities.* Current Issues and Strategies. 2001 ISBN 0-7923-7083-X
61. N. Maiellaro (ed.): *Towards Sustainable Buiding.* 2001 ISBN 1-4020-0012-X
62. G.S. Dunbar (ed.): *Geography: Discipline, Profession and Subject since 1870.* An International Survey. 2001 ISBN 1-4020-0019-7

KLUWER ACADEMIC PUBLISHERS – DORDRECHT / BOSTON / LONDON

MIX
Papier aus verantwortungsvollen Quellen
Paper from responsible sources
FSC® C105338

If you have any concerns about our products,
you can contact us on
ProductSafety@springernature.com

In case Publisher is established outside the EU,
the EU authorized representative is:
Springer Nature Customer Service Center GmbH
Europaplatz 3, 69115 Heidelberg, Germany

Printed by Libri Plureos GmbH
in Hamburg, Germany